44.55

The Facts On File

DICTIONARY
of
ATOMIC
and
NUCLEAR PHYSICS

The Facts On File

DICTIONARY
of
ATOMIC
and
NUCLEAR PHYSICS

Edited by
Richard Rennie

☑®
Facts On File, Inc.

The Facts On File Dictionary of Atomic and Nuclear Physics

Copyright © 2003 by Market House Books Ltd

Facts On File, Inc.
132 West 31st Street
New York NY 10001

Library of Congress Cataloging-in-Publication Data

The Facts on File dictionary of atomic and nuclear physics / edited by Richard Rennie.
    p. cm.
  Includes bibliographical references.
  ISBN 0-8160-4916-5
      1. Nuclear physics—Dictionaries.   2. Atoms—Dictionaries.   I. Title: Dictionary of atomic and nuclear physics.   II. Rennie, Richard.

  QC772.F33 2002
  539.7'03—dc2l   2002032545

Compiled and typeset by Market House Books Ltd, Aylesbury, UK

Printed in the United States of America

  MP   10 9 8 7 6 5 4 3 2 1

This book is printed on acid-free paper

PREFACE

This dictionary is one of a series covering the terminology and concepts used in important branches of science. *The Facts On File Dictionary of Atomic and Nuclear Physics* is planned as an additional source of information for students taking Advanced Placement (AP) Science courses in high schools, and will also be helpful to older students taking introductory college courses.

This volume covers the whole area of modern atomic and nuclear physics, including atomic theory, the structure of matter, states of matter, spectroscopy, quantum theory, nuclear physics, and the new developments in particle physics and cosmology. It also includes the basic concepts necessary for an understanding of mechanics, electricity, magnetism, and wave motion. The definitions are intended to be clear and informative and, where possible, we have provided helpful diagrams and examples. The book also has a selection of short biographical entries for people who have made important contributions to the field. There are a number of useful appendices including fundamental constants, elementary particles, and chemical elements. We have also added lists of webpages and an informative bibliography.

The book will be a helpful additional source of information for anyone studying either of the AP Physics courses or the AP Chemistry course. However, we have not restricted the content to these syllabuses. Modern atomic and nuclear physics is a fast-moving exciting subject covering some of the fundamental questions about the nature of the Universe and of reality itself. We have attempted to give accounts of these, with entries on such topics as Bell's paradox, quantum entanglement, string theory, and the Higgs particle. We hope that the reader will find these informative as well as useful.

ACKNOWLEDGMENTS

Contributors

John Daintith B.Sc., Ph.D.
Eric Deeson M.Sc., F.C.P., F.R.A.S.

CONTENTS

A

AAS *See* atomic absorption spectroscopy.

Abel, Niels Henrik (1802–29) Norwegian mathematician who was one of the early workers on the theory of GROUPS. *Abelian groups* are named for him.

Abelian gauge theory *See* gauge theory.

Abelian group *See* group.

ab initio calculation Any calculation of the properties of atoms or molecules from the basic principles of QUANTUM MECHANICS, without using experimental parameters such as those found from spectroscopy. The fundamental constants of Nature are used in *ab initio* calculations. A large amount of computation is needed for such calculations, with the computing time required increasing rapidly as the size of the atom or molecule increases. As the power of computers has increased, the number of atoms and molecules that can be dealt with accurately has also increased. Consequently *ab initio* calculations have replaced SEMI-EMPIRICAL CALCULATIONS for many purposes. It is possible to use *ab initio* calculations to determine bond angles and bond lengths in molecules by calculating the total energy of the molecule for many different molecular geometries and finding which particular configuration has the lowest energy.

absolute space A fundamental concept underlying NEWTONIAN MECHANICS is that there is a preferred FRAME OF REFERENCE that all observations should be referred to, with this space existing independently of any bodies in the Universe. This preferred frame is called *absolute space*. The concept of absolute space has been criticized by many people concerned with the foundations of mechanics, including LEIBNIZ and MACH. In GENERAL RELATIVITY THEORY the assumption that absolute space exists is replaced by the EQUIVALENCE PRINCIPLE.

absolute temperature A temperature that is based on an *absolute scale*, i.e. a scale based on ABSOLUTE ZERO. The usual absolute temperature scale used is known as the *thermodynamic temperature* scale. It is also called the *Kelvin temperature* scale, after Lord KELVIN. The unit of absolute temperature is the *kelvin*, which used to be called the *degree absolute*. The size of the kelvin is the same as the degree Celsius. There is also a less commonly used absolute temperature scale known as the *Rankine* scale, named after the nineteenth-century engineer W. J. M. Rankine, in which the unit of temperature has the same size as in the Fahrenheit scale.

absolute time The analog for time of ABSOLUTE SPACE for space. Like absolute space, absolute time is a fundamental concept that underlies NEWTONIAN MECHANICS. The concept was criticized by LEIBNIZ and others. As with absolute space, the assumption that absolute time exists is replaced in GENERAL RELATIVITY THEORY by the EQUIVALENCE PRINCIPLE.

absolute zero The zero temperature on an ABSOLUTE TEMPERATURE scale. Absolute zero is the lowest temperature that is theoretically attainable. At this temperature the energy of atoms and molecules is the minimum possible (*see* zero-point energy). It is equivalent to –273.15°C or –459.67°F. In many systems the effects of quantum mechanics become very pronounced at temperatures close to absolute zero.

absorption The process by which energy of light or other types of electromagnetic radiation is transferred to the material in a medium as the radiation passes through the medium. Absorption occurs because the number of photons of radiation passing through the medium is reduced as a result of some of the photons causing transitions to EXCITED STATES of the atoms or molecules of the medium.

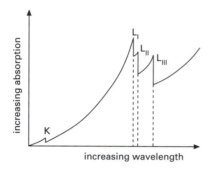

Absorption edge

absorption edge A sudden decrease in value of the amount of absorption of X-rays in a specimen, occurring at certain wavelengths characteristic of the material. As the wavelength increases the quantum energy of the radiation becomes less. An absorption edge occurs at the point at which the quantum energy falls below the energy required to eject an electron from a particular level. For example, if the wavelength is greater than the K edge then the photons have insufficient quantum energy to eject electrons from the K shell.

absorption spectrum *See* spectrum.

abundance 1. The relative amount of an ELEMENT among others, usually either in the crust of the Earth or the Universe as a whole. It is usual to express abundances as percentages based either on mass or on numbers of atoms. In the context of the Universe, the abundance is called the *cosmic abundance*.

The measure of the cosmic abundance is based on studies of the Sun and other bodies in the Solar System. It has been found that, in terms of the number of atoms, the *solar abundances*, i.e. the abundances of elements in the Solar System, are 90.8% hydrogen, 9.1% helium, and 0.1% for all other elements put together. It has also been found, using spectroscopy, that these abundances are very similar to those of other stars, with old stars that were formed when the Universe was young having even lower abundances of elements heavier than helium.

The cosmic abundances are a consequence of NUCLEOSYNTHESIS. Elements that are heavier than iron have very low cosmic abundances because they are only formed in SUPERNOVA explosions. For example, there are three atoms of gold in the Universe for every ten million atoms of sulfur.

The abundances found in the crust of the Earth have been determined by the processes of the Earth's evolution. The abundances for the Earth as a whole are different from those of the crust. (There is a large amount of iron in the Earth's core.) 2. The ratio of the number of atoms of a particular ISOTOPE of an element to the total number of atoms of that element in a given sample. This ratio is frequently expressed as a percentage. The *natural abundance* of an isotope of an element is the abundance of that isotope that occurs naturally, i.e. in the absence of ENRICHMENT. For example, chlorine has two stable isotopes, their mass numbers being 35 and 37. The natural abundance of ^{35}Cl is 75.5%, with that of ^{37}Cl being 24.5%. The natural abundances of isotopes accounts for the RELATIVE ATOMIC MASS of the element. In certain cases the abundance of a particular nuclide depends on the source.

accelerator A device that is used to increase the kinetic energy of electrically charged PARTICLES. Accelerators are used in nuclear physics and elementary particle physics. In an accelerator, beams of high-energy particles either smash into a fixed target or into each other. In a *cyclic accelerator* magnetic fields are used to keep the particles moving in circular or spiral paths. The CYCLOTRON, the SYNCHROCY-CLOTRON, and the SYNCHROTRON are examples of cyclic accelerators. In a LINEAR ACCELERATOR the particles are accelerated in a straight line. The earliest linear accelerator was the COCKCROFT–WALTON MACHINE.

acceptor *See* semiconductor.

a.c. Josephson effect *See* Josephson effect.

actinic radiation Radiation that can cause a chemical reaction; for example, ultraviolet radiation is actinic.

actinide contraction The smooth decrease in the atomic or ionic radius that occurs as the proton number increases in the ACTINIDES. The actinide contraction occurs because the increase in the number of protons with increasing proton number attracts the inner electrons towards the nucleus, with this reduction considerably outweighing the increase in atomic or ionic radius caused by adding *f*-electrons. A similar effect (the lanthanide contraction) occurs in the LANTHANIDE elements.

actinides (actinoids) A series of chemical elements in the PERIODIC TABLE, usually taken to range in proton number from 90 (thorium) to 103 (lawrencium). Sometimes, actinium itself (proton number 89) is also included in the series.

actinium A soft silvery-white radioactive metallic element that immediately precedes the ACTINIDES in the periodic table. Its most stable isotope ^{227}Ac has a half-life of 21.77 years. The element, which glows in the dark, occurs in very small quantities in uranium ores. It can be produced by bombarding radium with neutrons and is used as a source of alpha particles.

Symbol: Ac. Melting pt.: 1050±50°C. Boiling pt.: 3200±300°C. Relative density: 10.06 (20°C). Proton number: 89. Most stable isotope: ^{227}Ac (half-life 21.77 y). Electronic configuration: [Rn]6d^{1}7s^2.

actinium series *See* radioactive series.

actinoids *See* actinides.

action The product of momentum and displacement (or the product of kinetic energy and time). The action is the integral over time of the LAGRANGIAN of a system. The PRINCIPLE OF LEAST ACTION means that action is a quantity of fundamental importance in classical mechanics and classical field theory. Action is also an important quantity in quantum theory, particularly in the PATH INTEGRAL formulation of quantum mechanics.

action at a distance An effect in which one body has an influence on another separate body, with no apparent mechanism for transmitting a force between the two bodies. An action-at-a-distance theory is one in which only the effect is considered, not the underlying mechanism. COULOMB'S LAW of electrostatics and NEWTON'S LAW OF GRAVITY are the main examples of action-at-a-distance theories. Classical electrodynamics, as described by Maxwell's equations, and general relativity theory are examples of theories that do *not* involve action at a distance.

adiabatic change In thermodynamics, a change in which no HEAT is gained or lost by the system undergoing the change. The concept of an adiabatic change is an idealization that cannot be realized in practice. However, it is possible to obtain a good approximation to an adiabatic process by having good thermal insulation and by carrying out the change rapidly so that there is little time for heat to flow into or out of the system. In an adiabatic change, any work done on or by the system changes its internal energy, and hence its temperature. *See also* isothermal.

adiabatic approximation An approximation that is used in quantum mechanics when a parameter of the system is changing slowly with time. An example is an electronic transition in a molecule. The distance between the nuclei of atoms in the molecule is changing slowly in comparison with the speed of the electronic transition, and the nuclei can be regarded as stationary during the transition (*see* Born–Oppenheimer approximation). In the adiabatic approximation the solution of SCHRÖDINGER'S EQUATION at one time goes over continuously to the solution at a later time. The adiabatic approximation was formulated by Max Born and Vladimir Fock in 1928 an example of its use is the BORN–OPPENHEIMER APPROXIMATION.

adiabatic demagnetization A technique used to produce temperatures very close to ABSOLUTE ZERO. In adiabatic demagnetization a paramagnetic salt is first cooled in liquid helium in a strong magnetic field to about 1 degree KELVIN. The heat produced in the process of magnetization is absorbed by the helium. The salt is then thermally insulated and the magnetic field is removed. The resulting demagnetization of the salt causes a fall in its temperature. It is possible to obtain temperatures of a few millikelvin in this way. Even lower temperatures (of about 10^{-6} kelvin) may be achieved by using the magnetic properties of the nucleus.

The possibility of obtaining very low temperatures using adiabatic demagnetization was suggested independently in 1926 by Peter Debye and William Francis Giaque and first demonstrated by Giaque in 1933.

advanced gas-cooled reactor *See* nuclear reactor.

AES *See* atomic emission spectroscopy.

aether *See* ether.

AFM *See* atomic force microscope.

age of the Earth Estimates using RADIOACTIVE DATING suggest that the age of the Earth, and the rest of the Solar System, is about 4.6 billion years. To be more precise, the oldest rocks that have been found in the crust of the Earth have been estimated, using radioactive dating, to be slightly less than four billion years old. Rock samples from the Moon have been dated similarly, with the oldest sample being about 4.5 billion years old. The Earth must be older than the age of the oldest rocks found because of the time during which it cooled down. The figure of 4.6 billion years for the age of the Earth includes an estimate of the cooling time. As well as providing a method for dating rocks, radioactivity also provides a source of heat which must be taken into account when estimating the rate at which the Earth cooled.

age of the Universe A time that is estimated from HUBBLE'S CONSTANT, which gives the rate at which the Universe is expanding. The age of the Universe would be given simply by the inverse of Hubble's constant if the rate of expansion of the Universe had been constant since the BIG BANG. However, the attractive gravitational interaction between galaxies means that the rate of expansion has slowed down with time. This means that the age of the Universe calculated from the present value of Hubble's constant gives an overestimate of the age of the Universe. Taking this difficulty into account, observations of the Hubble constant lead to estimates of the age of the Universe lying between 12 and 15 billion years, with 18 billion years being an upper limit to its age.

If the MULTIVERSE idea is correct then the time that has elapsed since the big bang is not the true age of the Universe as a whole but merely the age of our particular Universe.

AGR *See* nuclear reactor.

Aharonov–Bohm effect An effect that occurs in a variant of the DOUBLE SLIT EXPERIMENT in which a long thin solenoid is inserted between the slits. In this experiment, the magnetic field denoted by B is almost entirely contained within the solenoid. Nevertheless, in 1959 Yakir Aharonov and David Bohm predicted that, although the beams pass through regions where $B = 0$, there should be a change in the INTERFERENCE pattern as B is changed. The Aharonov–Bohm effect, i.e. the change in the pattern as B varies, was first observed experimentally in 1960. The effect

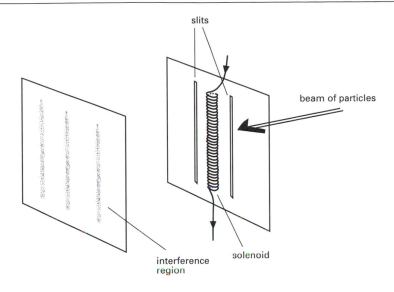

slits

beam of particles

interference region

solenoid

Aharonov–Bohm effect

occurs because the vector potential A does not completely vanish in the regions through which the beams pass. An important aspect of the effect is that it indicates that the vector potential has real physical significance.

algebra At the most elementary level algebra is the branch of mathematics in which symbols such as x and y are used for numbers. In this sense, it is a generalization of arithmetic. In more advanced mathematics algebra means an *algebraic structure*, i.e. a formal mathematical system consisting of certain elements with operations that satisfy given axioms. This advanced form of algebra is sometimes called *abstract algebra*. In abstract algebra the elements need not be numbers; they can be matrices, symmetry operations, operators, etc. A GROUP is an example of an algebraic structure.

algebraic field theory A formulation of QUANTUM FIELD THEORY in terms of its ALGEBRAIC STRUCTURE. Algebraic field theory has been developed since the 1960s and has given many useful insights into quantum field theory. It has also enabled results such

as the SPIN–STATISTICS THEOREM to be derived in a very general way.

algebraic structure *See* algebra.

alkali metals The elements of group 1 of the PERIODIC TABLE. All the alkali metals have an electron configuration equivalent to that of a noble gas together with a single outer s-electron. This electron can easily be lost to form a stable cation with a noble gas electron configuration.

alkaline earth metals The elements of group 2 of the PERIODIC TABLE. All the alkaline earth metals have an electron configuration equivalent to that of a noble gas with two outer s-electrons. These electrons can easily be lost to form a stable cation with a noble gas electron configuration.

allotropy The property of certain ELEMENTS of existing in two or more different physical forms (*allotropes*). For example, diamond and graphite are allotropes of carbon.

allowed transition A transition between two QUANTUM STATES that is allowed by the SELECTION RULES associated with group theory. *See also* forbidden transition.

alpha decay A type of RADIOACTIVE DECAY in which an alpha particle (helium nucleus) is emitted from a nucleus. The nuclide that results has a mass number that is decreased by 4 and an proton number decreased by 2 with respect to the original nuclide. The helium nuclei emitted in alpha decay are known as *alpha particles* and streams of such particles constitute *alpha radiation*. Alpha radiation is strongly ionizing and not highly penetrating (a few centimeters of air at STP).

Alpha decay occurs in heavy nuclei, which may spontaneously disintegrate, for example:
$$^{238}_{92}U \rightarrow \,^{234}_{90}Th + \,^{4}_{2}He$$
In many cases of alpha decay the nuclide is produced in an excited state, from which it decays by emission of gamma ray photons. Also, the electrons orbiting the nucleus will have different energy levels in the product nucleus than in the parent. This may result in emission of x-ray photons.

The process of alpha decay was one of the earliest discoveries to be made in nuclear physics. It is the process that produces helium on Earth (e.g. in certain oil wells) and it is also one of the heat sources in the Earth's core. Elements that emit alpha particles are known as *alpha emitters* and are used in certain instruments (e.g. smoke detectors).

The mechanism of alpha decay was given by GAMOW in an early application of quantum mechanics. The alpha particle escapes through an energy barrier by quantum-mechanical TUNNELING. The theory accounts for the wide variation of lifetimes found in alpha emitters (from nanoseconds to gigayears).

alpha particle *See* alpha decay.

Alpher, Ralph Asher (1921–) American physicist who worked with George GAMOW and Robert HERMAN on the early evolution of the Universe in terms of the BIG-BANG THEORY. In particular, Alpher *et al.* calculated how elements could have been formed soon after the big bang (*see* nucleosynthesis) and predicted the existence of residual background radiation left over from the big bang.

aluminum A silvery white metallic element that belongs to group 3 of the PERIODIC TABLE. It is the most abundant metallic element in the Earth's crust (8.1% by weight). ^{27}Al is the stable isotope. A number of radioactive isotopes can be produced. The most stable of these is ^{26}Al, with a half-life of 7.2×10^5 y. The spelling *aluminum* is the US spelling. The official scientific spelling was set as *aluminium* in 1990 (to bring it into line with the spellings of other chemical elements ending in 'ium'), and this is also the usual British spelling of the name.

Symbol: Al. Melting pt.: 660.37°C. Boiling pt.: 2470°C. Relative density: 2.698 (20°C). Proton number: 13. Relative atomic mass: 26.981539. Electronic configuration: [Ne]3s²3p¹.

Alvarez, Luis Walter (1911–88) American physicist who won the 1968 Nobel Prize for physics for his work on elementary particle physics, particularly the development of the BUBBLE CHAMBER. He became well known in the 1980s for his suggestion that the extinction of the dinosaurs was caused by a large meteorite hitting the Earth.

americium A radioactive metallic element belonging to the ACTINIDES. It is a TRANSURANIC ELEMENT and does not occur naturally on Earth, but can be made from plutonium. Its most stable isotope ^{243}Am has a half-life of 7.37×10^3 years and is produced by bombarding plutonium-239 with neutrons. Another isotope, ^{241}Am, can be obtained by prolonged bombardment of plutonium-241 with neutrons. Americium-241 has a half-life of 432.2 years. It is used in medical diagnosis (e.g. gamma-ray radiography) and radiotherapy.

Symbol: Am. Melting pt.: 1172°C. Boiling pt.: 2607°C. Relative density: 13.67 (20°C). Proton number: 95. Most stable isotope: ^{243}Am (half-life 7370 y). Electronic configuration: [Rn]5f⁷7s².

Umbrella inversion of ammonia

ammonia clock A type of ATOMIC CLOCK using the vibrational motion of ammonia molecules. A molecule of ammonia (NH_3) consists of a pyramid with one hydrogen atom at each corner of the base of the pyramid and the nitrogen atom at the apex. The nitrogen atom can pass backward and forward through the base of the pyramid. (This is known as the *umbrella inversion*). This vibrational motion has a frequency of 23 870 hertz. Energy is fed to the gas at this frequency using a circuit containing a crystal oscillator. Using a suitable feedback mechanism it is possible for the oscillator to be locked to this frequency exactly.

amount of substance Symbol: n A measure of the number of entities present in a substance. *See* mole.

Ampère, André-Marie (1775–1836) French mathematician, chemist and physicist who is best known for his pioneering contributions to ELECTRODYNAMICS. In particular, Ampère made the relation between an electric current and the magnetic field associated with that current quantitative, (*see* Ampère's law). The SI unit of electric current, the *ampere*, is named for him.

Ampère's law **1.** (Ampère–Laplace law) The elemental force, dF, between two current elements, I_1dl_1 and I_2dl_2, parallel to each other at a distance r apart in free space is given by:
$$d = \mu_0 I_1 d_1 l_1 I_2 dl_2 \sin\theta / 4\pi r^2$$
Here μ_0 is the permeability of free space and θ is the angle between either element and the line joining them.

2. The principle that the sum or integral of the magnetic flux density B times the path length along a closed path round a current-carrying conductor is proportional to the current I. For a circular path of radius r round a long straight wire in a vacuum, $B.2\pi r = \mu_0 I$. (μ_0 is the magnetic permeability of free space.) Ampère's law enables the value of B inside a solenoid to be calculated using the equation $B = n\mu_0 I$, where n is the number of turns per unit length.

amplitude *See* wave.

a.m.u. *See* atomic mass unit.

Anderson, Carl David (1905–91) American physicist noted for his pioneering contributions to particle physics, particularly the discovery of the POSITRON and the MUON. Anderson shared the 1936 Nobel Prize for physics with Victor HESS.

Anderson, Philip Warren (1923–) American physicist who has made many contributions to theoretical physics, particularly the theory of solids. Areas in which he has worked include SUPERCONDUCTIVITY, SUPERFLUIDITY, magnetic properties, and disordered systems. He also helped to clarify the concept of BROKEN SYMMETRY. He shared the 1977 Nobel Prize for physics with Nevill MOTT and John Van Vleck.

Anderson localization *See* electronic structure of solids.

angstrom Symbol: Å A unit of length defined as 10^{-10} meter. The angstrom is sometimes used for expressing wavelengths of light or ultraviolet radiation or for interatomic distances and the sizes of molecules.

angular momentum The analog of MOMENTUM for rotational motion. The angular momentum, denoted L, of a particle of mass m that is rotating about some axis O is the VECTOR PRODUCT of the instantaneous values of the momentum of the particle and its position vector r, where the position vector is the perpendicular distance from O to the direction of the velocity v. Thus $L = m\,v \times r$. In the case of a rigid body that is rotating about an axis O the angular momentum for the whole body is given by the product of the MOMENT OF INERTIA of the body, denoted I, and its *angular velocity*, denoted ω, i.e. the rate of change of angular displacement: $\omega = d\theta/dt$. Thus, $L = I\omega$.

In quantum mechanics angular momentum is quantized and has a discrete set of values for a given system. The quantum theory of angular momentum is closely asociated with the ROTATION GROUP and is important in the theory of atoms, molecules, and the nucleus. *See also* conservation of angular momentum.

anharmonic oscillator An oscillating system that is not a HARMONIC OSCILLATOR. The problem of an anharmonic oscillator cannot, in general, be solved exactly either in classical mechanics or in quantum mechanics. However, many systems are nearly harmonic oscillators. For such systems the *anharmonicity*, i.e. the deviation of the system from being a perfect harmonic oscillator, is small and can be calculated using PERTURBATION THEORY. If the anharmonicity is large then other approximation techniques have to be used.

anion A negatively charged ion. In electrolysis anions are attracted to the anode. *Compare* cation.

anisotropic Describing a medium in which certain physical quantities are different in different directions. Wood, for example, is an anisotropic medium: its strength along the grain is different from its strength perpendicular to the grain. Many crystals are anisotropic with respect to their electrical and POLARIZATION properties.

anisotropic superconductor A type of superconductor. An anisotropic superconductor has a system of FERMIONS in which COOPER PAIRS form in a state in which there is relative orbital motion and possibly nonzero total SPIN. There is evidence that the materials that exhibit HIGH-TEMPERATURE SUPERCONDUCTIVITY are anisotropic superconductors.

anisotropic superfluid A system that exhibits a type of SUPERFLUIDITY resulting from a BOSE–EINSTEIN CONDENSATE of Cooper pairs in which there is relative orbital motion and possibly nonzero total spin. SUPERFLUID Helium-3 is an example of an anisotropic superfluid.

annihilation The destruction of a particle and its antiparticle when they collide. The energy released in the process of annihilation is equal to the sum of the rest energies of the particle and antiparticle and their kinetic energies at the point of collision. For example, the collision of an electron and a positron creates two photons, which move away in opposite directions. Momentum is conserved in the process. The radiation produced in this way is called *annihilation radiation* and is in the gamma-ray region of the electromagnetic spectrum; each of the photons produced in electron–positron annihilation has an energy of 0.511 MeV. The energy associated with the annihilation of a NUCLEON and its antiparticle is much larger than for electron–positron annihilation and results in the production of MESONS.

annihilation operator An OPERATOR that is used to describe the annihilation of particles and antiparticles in the formalism known as SECOND QUANTIZATION.

anode The electrode that has a relative positive potential. In an electrical system the anode is the place at which electrons flow out of the system. In electrolysis the anode attracts negative ions. *Compare* cathode.

anomalous dispersion   *See* dispersion.

anomalous magnetic moment   A MAGNETIC MOMENT of a particle that differs from the predictions of the DIRAC EQUATION.

In the case of a charged LEPTON, such as an electron or a muon, deviations (0.2%) from the predictions of the Dirac equation occur because of processes involving virtual particles, such as VACUUM POLARIZATION, which affect the ratio of electric charge to mass of the particle. The anomalous magnetic moments of charged leptons can be measured extremely precisely. They can also be calculated to a high degree of accuracy using perturbation theory in QUANTUM ELECTRODYNAMICS (QED). The remarkable agreement (to many decimal places) between the very accurate measurements and the very accurate calculations is not exceeded in any other branch of physical science and is a major triumph for QED.

PROTONS and NEUTRONS have anomalous magnetic moments in the sense that their magnetic moments are different from those expected for pointlike particles obeying the Dirac equation. In the case of the proton there is a large deviation from this expected value. In the case of the neutron the expectation from the Dirac equation is that the value of the magnetic moment should be 0 since the neutron is electrically neutral whereas, in reality, it has a substantial nonzero value. These results for the proton and the neutron occur because they are not pointlike particles but composite particles made of QUARKS.

anomalous Zeeman effect   *See* Zeeman effect.

anomaly   In QUANTUM FIELD THEORY there is said to be an *anomaly* if a SYMMETRY that is present in a classical field theory is not present in the corresponding quantum field theory. There are many important consequences of anomalies in quantum field theory, including an expla-

nation of the decay of neutral PIONS and the requirement that the number of quarks and leptons in each family of elementary particles must be the same. Anomalies and their consequences can be calculated using PERTURBATION THEORY; they are associated with nontrivial geometry and topology, particularly INDEX THEOREMS.

anthropic principle   The principle that the Universe has to be the way that it is since otherwise we would not be here to observe it. For example, if any of the fundamental constants of Nature, such as the GRAVITATIONAL CONSTANT, had different values then the Universe would not have evolved in a way for conscious life to form. There are several different versions of the anthropic principle. The validity of the principle is a matter of great controversy among physicists, with many regarding it with great scepticism. Both the hope of finding a unique THEORY OF EVERYTHING and the concept of the MULTIVERSE are opposite in spirit to the anthropic principle.

antibonding orbital   *See* orbital.

antiferromagnetism   A state of magnetic LONG-RANGE ORDER in which neighboring atoms have ANTIPARALLEL SPINS. The LATTICE in an antiferromagnetic material can be regarded as consisting of two equivalent sublattices, with equal numbers of entities having equal magnetic moments in opposite directions, resulting in zero net magnetization at low temperatures. As the temperature is raised the magnetic susceptibility rises. The critical temperature for a given material at which antiferromagnetism breaks down is called the *Néel temperature*. Above this point the material exhibits PARAMAGNETISM.

antimatter   Matter that is made up of ANTIPARTICLES. For example, an *antihydrogen atom* consists of an antiproton with a positron orbiting it. When matter collides with antimatter there is mutual ANNIHILATION. *See also* antiparticle; matter–antimatter asymmetry.

antimony An element that belongs to group 15 of the periodic table. It exists in three allotropic forms, of which the most stable form is a brittle metal. Antimony is used in alloys and in certain semiconductor devices. There are two natural isotopes: ^{121}Sb with a natural abundance of 57.3% and ^{123}Sb with a natural abundance of 42.7%. A large number of radioactive isotopes can be formed: ^{124}Sb has a half-life of 60.3 days and is used in research.

Symbol: Sb. Melting pt.: 630.74°C. Boiling pt.: 1635°C. Relative density: 6.691. Proton number: 51. Relative atomic mass: 112.74. Electronic configuration: [Kr]$4d^{10}5s^25p^3$.

antineutrino The ANTIPARTICLE of a NEUTRINO. It is customary to call the neutrino that accompanies the emission of an electron in BETA DECAY an 'antineutrino' and the neutrino that accompanies the emission of a positron in beta decay a 'neutrino' because of the conservation of LEPTON NUMBER. Each of the three types of neutrino has its own corresponding antineutrino.

antineutron The ANTIPARTICLE of the NEUTRON. An antineutron consists of two anti-down quarks and one anti-up quark.

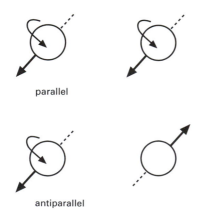

parallel

antiparallel

Parallel and antiparallel spins

antiparallel spins Particle spins that have different senses, so that the magnetic moments are parallel but act in opposite directions. The state of ANTIFERROMAGNETISM is an example of the occurrence of antiparallel spins. *See also* parallel spins; exchange.

antiparticle A PARTICLE with the same mass and spin as a given particle but with equal and opposite electric charge (and other properties). For example, the antiparticle of the ELECTRON is the POSITRON, which is a spin-½ particle with the same mass as an electron but with a positive charge that is equal in magnitude to the negative charge of the electron. Similarly, the ANTIPROTON is a spin-½ particle with the same mass as a PROTON but with a negative charge equal in magnitude to the positive charge of the proton. In the case of unstable particles, such as a NEUTRON in isolation, the half-life of both the particle and the antiparticle is the same. When particles and antiparticles collide ANNIHILATION occurs. For charged particles the distinction between particles and their antiparticles may be indicated by a superscript giving the sign of the electric charge. For example, a positron is denoted by e$^+$ (the electron is e$^-$). Alternatively, a bar is used over the symbol. The bar can also be used for charged particles. Thus it is customary to denote antiprotons by $\bar{p}$ (and refer to them as 'p-bars'). The existence of antiparticles is predicted by RELATIVISTIC QUANTUM MECHANICS.

antiproton The ANTIPARTICLE of the PROTON. The antiproton was discovered in 1955 by Owen CHAMBERLAIN and Emilio SEGRÈ at the University of California, Berkeley. The discovery of the antiproton led to the discovery of the ANTINEUTRON in 1956 by the reaction:

$$p + \bar{p} \rightarrow n + \bar{n}.$$

antiscreening *See* asymptotic freedom.

antisymmetric wave function A WAVE FUNCTION ψ for a system in quantum mechanics with n identical particles in which interchanging any two of these particles changes the sign of the wave function: $\psi \rightarrow -\psi$. Particles that have antisymmetric wave functions are called FERMIONS. If the wave function for a system with n identical particles is not an antisymmetric wave function then it must be a symmetric wave function and the particles must be BOSONS. The PAULI EXCLUSION PRINCIPLE is a consequence of the existence of antisymmetric wave functions. *See also* boson; fermion; Pauli exclusion principle; spin–statistics theorem; symmetric wave function.

anyon   *See* quantum statistics.

approximation technique   Any of various mathematical techniques for obtaining approximate solutions to problems that cannot be solved exactly. PERTURBATION THEORY is the most-used type of approximation technique in physics.

argon   An element that is one of the NOBLE GASES. It is a colorless odorless chemically inert monatomic gas present in the air (0.93% by volume). The main isotope is ^{40}Ar (natural abundance 99.6%). There are two other natural isotopes: ^{36}Ar (0.337%) and ^{38}Ar (0.063%). A number of radioactive isotopes exist: ^{37}Ar (half-life 35 days) is used in research.

Symbol: Ar. Melting pt.: $-189.37°$C. Boiling pt.: $-185.86°$C. Density: 1.784 kg m^{-3} (0°C). Proton number: 18. Relative atomic mass: 39.95. Electronic configuration: [Ne]3s^{2}3p^6.

Arrhenius, Svante August (1859– 1927) Swedish chemist who was the first person to realize that an ELECTROLYTE conducts electricity because it exists as or is dissociated into IONS. Arrhenius also made other contributions to physical chemistry, including the theory of the rate of chemical reactions. He was one of the first people to predict the possibility of the 'greenhouse effect', i.e. the change in the climate of the Earth caused by carbon dioxide in the atmosphere. Arrhenius was awarded the 1903 Nobel Prize for chemistry.

arrow of time   A phrase coined by Sir Arthur EDDINGTON to express the asymmetry between past and future. At the MICROSCOPIC level the laws of physics, except those governing the WEAK INTERACTIONS (*see* time reversal), do not distinguish between past and future. Nevertheless, at the MACROSCOPIC level the distinction between past and future is very clear. For example, if a glass falls from the edge of a table and is smashed on the floor we know that it will not come together again spontaneously.

The origin of the arrow of time, i.e. how IRREVERSIBILITY arises, is understood in terms of thermodynamics and statistical mechanics. In particular, the second law of thermodynamics can be stated in the form that the ENTROPY in an isolated system increases in an irreversible process. It became clear from the work of James Clerk MAXWELL and Ludwig BOLTZMANN in the second half of the nineteenth century that the second law of thermodynamics, and hence the arrow of time, is statistical in origin, with entropy being a measure of the disorder of a system.

arsenic   An element belonging to group 15 of the periodic table. It exists in three allotropic forms, of which the most stable is a gray metal. The stable isotope is ^{75}As. A number of radioactive isotopes are known: ^{73}As with a half-life of 80.3 days is used in research.

Symbol: As. Melting pt.: 817°C (gray at 3 MPa). Sublimation pt.: 616°C (gray). Relative density: 5.78 (gray at 20°C). Proton number: 33. Relative atomic mass: 74.92159. Electronic configuration: [Ar]3d^{10}4s^{2}4p^3.

artificial radioactivity (induced radioactivity)   Radioactivity produced by nuclear reactions. Various methods are available.

Aspect experiment   *See* Bell's inequality.

associated production   A process first found in bubble chambers and cloud chambers in the 1950s, which led to the concept of STRANGENESS. It was found that negtive PIONS and protons reacted to give two types of neutral particles, denoted x^0 and X^0:

$$\pi^- + p \rightarrow x^0 + X^0,$$

with these neutral particles decaying:

$$x^0 \rightarrow \pi^+ + \pi^-,$$
$$X^0 \rightarrow \pi^- + p.$$

It was found that the masses of the x^0 and X^0 particles are about 500MeV and 1100 MeV respectively. An important aspect of the production of these particles is that they are always formed in association with each other rather than singly in processes such as

$$\pi^- + p \rightarrow x^0 + n,$$
$$\pi^- + p \rightarrow \pi^0 + X^0$$

which are *not* observed. The x° is a MESON and the X° is a BARYON known as the LAMBDA PARTICLE, denoted Λ°. The hypothesis of associated production, rather than single production, of such particles was put forward by Abraham Pais in 1952. In terms of the QUARK model of elementary particles it is thought that associated production occurs because quarks have labels known as FLAVORS, with one of these labels being called STRANGENESS. Quantities such as strangeness are conserved in STRONG INTERACTIONS but violated in WEAK INTERACTIONS. The x° is a K-MESON with the composition $(d\tilde{O})$ where d denotes a down quark and $\tilde{O}$ denotes an anti-strange quark while the Λ° particle has the composition (uds), where u denotes an up quark and s denotes a strange quark. Thus, associated production occurs because the conservation of strangeness in the strong interactions means that the production of a particle with a strange quark must be associated with the production of a particle with an anti-strange quark, with the conservation of strangeness being violated in the decays of K° and Λ° via weak interactions.

astatine   An element belonging to group 7 of the periodic table. It is radioactive and has many short-lifetime isotopes.
Symbol: At. Melting pt.: 302°C (est.). Boiling pt.: 337°C (est.). Proton number: 85. Most stable isotope ^{210}At (half-life 8.1 h). Electronic configuration: $[Xe}4f^{14}5d^{10}6s^26p^5$.

Aston, Francis William (1877–1945) English physicist who invented MASS SPECTROSCOPY, which he used to show that many elements come in several forms called ISOTOPES having different atomic masses but the same proton number. Aston was awarded the 1922 Nobel Prize for chemistry.

astrophysics   The branch of physics that is concerned with physical and chemical processes in astronomy, such as the formulation and evolution of stars and galaxies.

asymptotic freedom   The result in certain GAUGE THEORIES, notably QUANTUM CHROMODYNAMICS (QCD), that the forces between particles such as quarks decrease the closer the quarks are. This is the opposite to what occurs in electricity and gravity, in which the forces increase the closer the bodies are. It is reminiscent of the stretching of an elastic band. In QUANTUM ELECTRODYNAMICS the interaction between particles means that there is SCREENING between particles. Asymptotic freedom means that there is *antiscreening* between quarks as a result of the nature of the VACUUM STATE in a NON-ABELIAN GAUGE THEORY.

The successes of the PARTON MODEL of pointlike objects inside HADRONS is explained by asymptotic freedom, with systematic corrections to the parton model being calculated using PERTURBATION THEORY in QCD. The result of asymptotic freedom that the interaction between quarks increases as the distance between them increases has led to the hypothesis of QUARK CONFINEMENT.

An important result concerning asymptotic freedom is the *Coleman–Gross theorem*, which states that non-Abelian gauge theories in which the gauge symmetry is unbroken are the only RENORMALIZABLE quantum field theories that can have asymptotic freedom. Thus, the WEINBERG–SALAM MODEL does not have asymptotic freedom. The theorem was stated by Sidney Coleman and David Gross in 1973. Asymptotic freedom was discovered in 1973 by Gross and Frank Wilczek and independently by David Politzer.

asymptotic series   A series of the type
$$A_0 + A_1/x + A_2/x^2 + \ldots + A_n/x^n$$
is an asymptotic series for a function $f(x)$ if, for every value of n the limit as $x \rightarrow \infty$ of $x^n[f(x) - S_n(x)] = 0$, where S_n is the sum of the series up to the term A_n/x^n. An asymptotic series is not necessarily a CONVERGENT SERIES. Many of the series that arise in physical applications of PERTURBATION THEORY are asymptotic series.

Atiyah, Sir Michael Francis (1929–). English mathematician who has made major contributions to GEOMETRY and TOPOLOGY, particularly INDEX THEOREMS, and their applications to GAUGE THEORIES. He was one of the leading figures in the revival of mutual interest between geometry and physics, which started in the mid 1970s.

Atkinson, Robert D'escourt (1898–1982).   Welsh physicist who showed with Frity HOUTERMANS in the late 1920s that NUCLEAR FUSION could explain the production of energy inside stars. An essential feature of their calculations was that TUNNELING, which had been used by George GAMOW and others to explain alpha decay, could also explain how nuclei could come sufficiently close to combine in spite of the coulomb barrier between them.

atom   The smallest part of a chemical ELEMENT that can exist, and hence take part in a chemical reaction. An atom consists of a small dense nucleus of protons and neutrons surrounded by moving electrons. The number of electrons in an atom is equal to the number of protons, i.e. an atom is electrically neutral overall. *See also* atomic structure; electronic structure.

atomic absorption spectroscopy (**AAS**)   A technique in SPECTROSCOPY in which a sample of material is vaporized, with the unexcited ATOMS absorbing electromagnetic radiation at certain specific characteristic wavelengths.
   A typical technique is to spray a solution of the sample into a hot flame through which is passed a monochromatic beam of light or ultraviolet radiation at a specific wavelength, corresponding to a characteristic absorption energy for the element in question. Absorption of the radiation allows identification of the element, and the amount of absorption can be used to determine the amount of element present (with appropriate calibration of the instrument using standard samples).

atomic beam   A beam of atoms. *See* molecular beam.

atomic bomb   *See* nuclear weapons.

atomic clock   A device for measuring time based on a periodic phenomenon in atoms or molecules. *See* ammonia clock; cesium clock.

atomic emission spectroscopy (**AES**)   A technique in SPECTROSCOPY in which a sample of material is vaporized, with the atoms of the material being detected by their emission of electromagnetic radiation at certain characteristic wavelengths.

atomic force microscope (**AFM**)   A type of microscope that has a small probe, made of a very small chip of diamond, held on a spring-loaded cantilever very close to the surface of a sample. The force between the probe and the surface is monitored as the probe is moved slowly across the surface and the distance of the probe from the surface is adjusted so that the force is constant. This enables a 'contour' map of the surface to be generated and it is possible to resolve individual molecules in this way. The device is similar to a SCANNING TUNNELING MICROSCOPE except that it uses mechanical forces rather than electrical signals. This means that AFM can be used to investigate nonconducting materials, such as biological samples.

atomicity   The number of atoms in a molecule of a compound. For example, ammonia (NH_3) has an atomicity of 4, methane (CH_4) has an atomicity of 5.

atomic lithography   Modification of the surface of a solid at the atomic level.

There are various ways in which this can be done, including electron beams, x-rays, the ATOMIC FORCE MICROSCOPE, and the SCANNING TUNNELING MICROSCOPE.

atomic mass unit (a.m.u.) A unit of MASS used to express the relative masses of ATOMS. It is equal to 1/12 of the mass of an atom of the isotope carbon-12, which is equal to 1.66033×10^{-27} kg. This unit is sometimes called the *Dalton* after John DALTON.

atomic number *See* proton number.

atomic orbital *See* orbital.

atomic pile An early type of NUCLEAR REACTOR that used uranium rods as the fuel and a pile of graphite blocks as the moderator.

atomic radius A somewhat imprecise concept, that is usually taken to be half the distance between atoms of the same element in a molecule or crystal. There are several variants of the concept of atomic radius depending on the nature of the chemical bond between the atoms: *covalent radius*, *ionic radius*, or *metallic radius*. Within any period of the periodic table the atomic radius decreases on going from left to right, in spite of the number of electrons in an atom increasing, because of the increase in the number of protons exerting a strong attraction to the inner electrons of the atom.

atomic spectrum A spectrum produced by emission from atoms or absorption by atoms. Atomic spectra consist of sets of sharp lines. The analysis of atomic spectra in terms of quantum mechanics enables a great deal of information to be obtained about the ENERGY LEVELS of atoms. *See also* electronic structure of atoms; fine structure; hydrogen spectrum; hyperfine structure.

atomic structure The internal structure of an ATOM. This consists of a small, dense NUCLEUS at the center made up of protons and neutrons surrounded by a cloud of moving electrons. The number of electrons in a neutral atom is equal to the number of protons. This model of atomic structure emerged from the experiments of RUTHERFORD, followed by the discoveries of the proton and the neutron in the first third of the twentieth century. The expression 'atomic structure' is sometimes taken to mean the ELECTRONIC STRUCTURE OF ATOMS, which is the arrangement of electrons in orbits around the nucleus.

atomic time A measure of time as determined by an ATOMIC CLOCK.

atomic trap A device for trapping isolated ATOMS. Such a device was first constructed in the 1970s using an oscillating ELECTRIC FIELD. *See* laser cooling.

atomic units A set of units, introduced by Douglas Hartree in 1928, which is convenient for calculations involving atoms. These units are built up from combinations of the charge and mass of the electron and the Planck constant h, which is sometimes replaced by the DIRAC CONSTANT $h/2\pi$. The unit of mass is m, the mass of the electron and the unit of charge is e, the charge of the electron. The unit of length is a, the BOHR RADIUS. The unit of energy is $e^2/a = 4\pi^2 me^4/h^2$, where the mass of the nucleus is taken to be infinite. This unit of energy is equal to twice the IONIZATION POTENTIAL of the hydrogen atom. A unit of energy that is half that of the Hartree atomic unit for energy is frequently used, called the *Rydberg unit*.

atomic volume The RELATIVE ATOMIC MASS of a chemical element divided by its density.

atomic weight *See* relative atomic mass.

attractor The point, or set of points, in PHASE SPACE to which a point in a dynamical system tends as the system evolves with time. It is possible for the attractor to be a single point, in which case the system reaches a fixed state, a closed curve known as a *limit cycle*, in which case the system has periodic behavior, or a FRACTAL, known as a *strange attractor*, in which case the system has chaotic behavior. *See* chaos theory.

Aufbau principle The 'building up' principle that determines the order in which ORBITALS are filled with electrons in an atom. This order of filling is 1s, 2s, 2p, 3s, 3p, 4s, 3d, 4p, 5s, 4d, 5p, 6s, 4f, 5d, 6p, 7s, 5f, 6d, 7p. Each s level can contain up to 2 electrons, each p-level can contain up to 6, and each d-level can contain up to 14.

The Aufbau principle, combined with the PAULI EXCLUSION PRINCIPLE, determines the ELECTRON CONFIGURATIONS of elements. There are several exceptions to the principle involving *d*- and *f*-electrons.

Auger effect The ejection of an electron from an atom or ion that occurs because of the de-excitation of an excited state. For example, if an electron is removed from an inner energy level of an atom or molecule, e.g. by bombardment with photons or high-energy electrons, then the atom is in an excited state. It may revert to the ground state by emission of a photon but, in the Auger effect, it actually emits a second electron (*Auger electron*).

An Auger electron is emitted at a specific energy for a particular type of atom. This is used in a type of electron spectroscopy called *Auger spectroscopy* to investigate the energy levels of atoms and ions. Auger spectroscopy is also used to detect the presence of elements in surface layers of solids.

The effect is named for the French physicist Pierre Auger (1899–1994) who discovered it in 1925.

Avogadro, Count Amedeo (1776–1856) Italian physicist and chemist who discovered AVOGADRO'S LAW and after whom the AVOGADRO CONSTANT is named.

Avogadro constant (Avogadro's number) Symbol: N_A The number of atoms or molecules in one mole of a substance, i.e. the amount of that substance with a mass in grams that is equal to the ATOMIC WEIGHT or MOLECULAR WEIGHT of that substance. The value of this number is 6.02252×10^{23}. For example, 12.0 grams of carbon contains this number of carbon atoms. Since hydrogen gas exists as a diatomic molecule, 2.0 grams of hydrogen gas contains this number of hydrogen molecules. The Avogadro constant is sometimes denoted *L*, in honor of Joseph Loschmidt who first determined its value.

Avogadro's law (Avogadro's hypothesis) The principle that, at the same pressure and temperature, equal volumes of all gases contain equal numbers of molecules. It was first stated by Avogadro in 1811 and he emphasized at the time that the individual particles of the gas might be molecules rather than single atoms. His work did not make much impact in his lifetime but was revived by his compatriot Stanislao CANNIZZARO in 1858. The law is strictly true only for IDEAL GASES, but is a very good approximation for real gases.

axial vector (pseudo-vector) A VECTOR that does not change its sign when the COORDINATE SYSTEM is changed to a new coordinate system by a reflection in the origin. An example of an axial vector is the ANGULAR MOMENTUM vector *L* formed by the VECTOR PRODUCT $L = r \times p$, where *r* is the position vector of the particle and *p* is its MOMENTUM vector of the particle, with both *r* and *p* being POLAR VECTORS. *See also* pseudo-scalar.

axiomatic field theory A somewhat misleading name given to the formulation and derivation of results in RELATIVISTIC QUANTUM FIELD THEORY in a general way using principles such as CAUSALITY without using any APPROXIMATION TECHNIQUE. The SPIN–STATISTICS THEOREM and the CPT THEOREM have been derived using axiomatic field theory.

axion A type of hypothetical elementary particle that arises in a solution which has been put forward to solve the STRONG CP PROBLEM. Axions have not been found experimentally in spite of extensive searches.

They have potentially important consequences in astrophysics and cosmology, such as having a possible effect on the cooling of stars and making up part or all of the DARK MATTER in the Universe. Some of these consequences enable limits to be calculated for the values of the mass and other properties of axions. *See also* strong CP problem.

azimuthal quantum number   *See* electronic structure.

baby universe A region of SPACE–TIME that is connected to another region of space–time by a WORMHOLE. It has been postulated that when a massive object in the Universe collapses to a black hole it could go through the SINGULARITY at the center of the black hole and create an expanding baby universe in another region of space–time. INFLATION could allow such a baby universe to become large very rapidly. It may be the case that the Universe we live in originated in this way. Although the idea of baby universes is highly speculative and requires a theory of QUANTUM GRAVITY before it can be formulated definitively, it has been investigated by a number of people including Stephen HAWKING and Andrei LINDE. *See also* multiverse.

background radiation Low-intensity IONIZING RADIATION that is always present in the atmosphere and at the surface of the Earth as a result of COSMIC RAYS and the presence of RADIOISOTOPES. In astronomy, the term background radiation is taken to mean the COSMIC MICROWAVE BACKGROUND.

Balmer, Johann Jakob (1825–98) Swiss mathematics teacher who was the first person to give a mathematical expression for the frequencies of lines in atomic spectra. In 1885 Balmer published a paper in which he put forward a simple empirical formula for the frequencies of the visible spectral lines of hydrogen (*see* hydrogen spectrum). Attempts to understand Balmer's formula stimulated the development of the application of quantum mechanics to atomic structure.

Balmer series *See* hydrogen spectrum.

band spectrum A spectrum that appears as a number of bands of emitted or absorbed radiation. Band spectra are characteristic of molecules. Often each band can be resolved into a number of closely spaced lines. The bands correspond to changes of electron orbit in the molecules. The close lines seen under higher resolution are the result of different vibrational states of the molecule. *See also* spectrum.

band structure *See* electronic structure of solids.

Bardeen, John (1908–91) American physicist who was one of the inventors of the transistor along with Walter BRATTAIN and William SHOCKLEY. He also formulated the theory of SUPERCONDUCTIVITY along with Leon COOPER and Robert SCHRIEFFER. Bardeen, Brattain, and Shockley won the 1956 Nobel Prize for physics for their work on the transistor. Bardeen, Cooper, and Schrieffer won the 1972 Nobel Prize for physics for their theory of superconductivity. Bardeen is the only person so far to be awarded two Nobel Prizes for physics.

Bardeen–Cooper–Schrieffer theory (BCS theory) A theory of SUPERCONDUCTIVITY put forward by John BARDEEN, Leon COOPER, and Robert SCHRIEFFER in 1957. When an electron moves through a crystal lattice there is a small distortion of the lattice caused by coulomb interaction between the negatively charged electron and the positively charged ions in the lattice. It is possible that the distortion of the lattice caused by one electron can attract a second electron. In 1956 Cooper showed that this can effectively result in a net attraction between the two electrons and that the current in superconductors is not carried by single electrons but by bound pairs of electrons formed in this way (*Cooper pairs*). The pairs condense at a certain CRITICAL TEMPERATURE, which is usually a few degrees above absolute zero. In the BCS theory this condensation of the Cooper pairs is represented by a wave function in which all the moving electrons are paired. The theory is very successful in describing the properties of superconductors. It was also influential in the development of the theory of elementary particles, particularly the HIGGS MECHANISM. Although the BCS theory is successful in describing the superconductivity at very low temperatures of metals and alloys, it is generally thought that another mechanism is required to describe HIGH-TEMPERATURE SUPERCONDUCTIVITY.

bare charge *See* renormalization

bare mass *See* renormalization.

barium A dense low-melting reactive metal; the fifth member of group 2 of the periodic table and a typical alkaline-earth element.
Symbol: Ba. Melting pt.: 729°C. Boiling pt.: 1640°C. Relative density: 3.594 (20°C). Proton number: 56. Relative atomic mass: 137.327. Electronic configuration: [Xe]$6s^2$.

Barkla, Charles Glover (1877–1944) British physicist who made important contributions to our understanding of X-RAYS, for which he was awarded the 1917 Nobel Prize for physics. For example, in 1906 Barkla showed that polarization occurs for x-rays, in accord with the suggestion made by George Johnstone STONEY that x-rays are ELECTROMAGNETIC WAVES with shorter wavelengths than visible light.

barn Symbol: b A unit of area sometimes used to express the CROSS-SECTIONS in nuclear reactions. The value of 1 barn is 10^{-28} square meter. The word barn comes from the expression 'side of a barn' (i.e. easy to hit).

baryogenesis *See* matter–antimatter asymmetry.

baryon Any member of the subclass of HADRONS that consists of heavy, strongly interacting spin-½ particles. Baryons are subdivided into NUCLEONS and HYPERONS. Baryons are not genuinely elementary particles but are made up of QUARKS and GLUONS. These constituents and their interactions that give rise to baryons are described by QUANTUM CHROMODYNAMICS. Baryons have antiparticles, called antibaryons, which are made up of three antiquarks.

baryon number A QUANTUM NUMBER designating BARYONS. The baryon number is +1 for baryons and -1 for antibaryons. Particles such as GAUGE BOSONS, MESONS, and LEPTONS, which are not baryons, have baryon number 0. Baryon number is observed to be a conserved quantity in all known particle interactions. However, GRAND UNIFIED THEORIES allow processes such as PROTON DECAY in which baryon number would not be conserved. In addition, explanations of MATTER–ANTIMATTER ASYMMETRY necessitate the violation of baryon-number conservation.

Basov, Nikolai Gennadiyevich (1922–2001) Russian physicist who developed LASERS and MASERS, for which he shared the 1964 Nobel Prize for physics. In 1958 he suggested the use of semiconductors to make lasers.

BCS theory *See* Bardeen–Cooper–Schrieffer theory.

beam A group of particles or rays of light, or other types of radiation, moving in the same direction.

beam hole A hole in the shielding of a NUCLEAR REACTOR that allows a NEUTRON BEAM, or beam of other particles, to escape. Beams of neutrons obtained in this way can be used for experiments.

beam-splitter A device for splitting

beams. Beam-splitters are used extensively in optics.

beauty *See* bottom quark.

becquerel Symbol: Bq   The SI unit of RA-DIOACTIVITY. One becquerel is one disintegration per second.

Becquerel, Antoine Henri (1852–1908)   French physicist who discovered RADIOACTIVITY in 1896. Following the discovery of X-RAYS in 1895 Becquerel was interested in whether crystals that exhibit fluorescence could produce x-rays. One of the substances he used to investigate this problem was a salt of uranium and he found that RADIATION was emitted by the salt due to the presence of uranium, even in the absence of any light to stimulate fluorescence. Becquerel's discovery of radioactivity stimulated other scientists, notably Pierre and Marie CURIE, to investigate radioactivity. In 1900 Becquerel showed that the radioactivity of radium consists of a stream of electrons. He also found evidence that radioactivity transforms one chemical element into another. Becquerel shared the 1903 Nobel Prize for physics.

Bednorz, George (1950–)   German-born physicist who discovered HIGH-TEMPERATURE SUPERCONDUCTIVITY with Alex MÜLLER in the mid-1980s. Bednorz and Müller shared the 1987 Nobel Prize for physics for this discovery.

Bekenstein, Jacob (1947–)   Israeli physicist who put forward the view in the early 1970s that the relation between the area of the EVENT HORIZON of a black hole and ENTROPY is not merely an analogy but an identity, with this area being a measure of the entropy of a black hole. This suggestion was initially greeted with a great deal of scepticism but was vindicated by the theoretical discovery of HAWKING RADIATION. Bekenstein has also used information theory to analyze entropy in black holes.

Bekenstein bound   A result that relates the area of a surface and the maximum amount of information about the UNIVERSE on one side of the surface that can pass through to an observer on the other side. It states that the number of bits of information an observer can gain must be less than or equal to one quarter of the surface area in PLANCK UNITS. This result is related to the ENTROPY of black holes since entropy is related to both information theory and the area of the EVENT HORIZON of a black hole. The Bekenstein bound was derived and discussed by Jacob BEKENSTEIN in papers he wrote in the mid 1970s but its importance was not appreciated until 20 years later, when it was one of the inspirations for the HOLOGRAPHIC HYPOTHESIS.

Bell, John Stewart (1928–1990)   Irish physicist who is best known for his work on the foundations of quantum mechanics, particularly in devising a test for whether a HIDDEN VARIABLES formulation underlying quantum mechanics based on locality is possible. This test was stated in the form of an inequality, now known as BELL'S INEQUALITY. Bell's work stimulated a great deal of subsequent theoretical and experiment work on the foundations of quantum mechanics. He also made important contributions to RELATIVISTIC QUANTUM FIELD THEORY and produced a proof of the CPT THEOREM in 1955.

Bell's inequality   An inequality concerning the probabilities of two events both occurring in two parts of a system that are well separated. In the mid 1960s John BELL gave a theoretical analysis of a modified version of the EPR EXPERIMENT in which he showed that a certain mathematical inequality is obeyed in a HIDDEN VARIABLES theory based on locality but violated in quantum mechanics. Bell's inequality is in accord with 'common sense' since it suggests that measurements of two well separated parts of a system are not correlated, whereas Bohr's interpretation of quantum mechanics implies that measuring one part of the system automatically fixes the value of the other part of the system. In the early 1980s Alain Aspect and his colleagues performed experiments on the polarization of well-separated photons that clearly showed that Bell's inequality is violated, and hence that hidden variables theories based on locality cannot be correct. This set of experiments, known collectively as the *Aspect experiment*, stimulated a great deal of interest in the foundations of quantum mechanics, since the violation of Bell's inequality showed that quantum mechanical ENTANGLEMENT is a real physical phenomenon.

Bell's theorem A theorem which states that there is no HIDDEN VARIABLES theory based on locality that can make the same predictions in complete agreement with QUANTUM MECHANICS. In particular, quantum mechanics predicts the violation of BELL'S INEQUALITY. John BELL proved this theorem in the mid 1990s along with his inequality.

berkelium A silvery radioactive transuranic element of the actinoid series of metals, not found naturally on Earth. Several radioisotopes have been synthesized. The metal reacts with oxygen, steam, and acids.
  Symbol: Bk. Melting pt.: 1050°C. Proton number: 97. Relative density: 14.79 (20°C). Most stable isotope: ^{247}Bk (half-life 1400 years). Electronic configuration: [Rn]$5f^9 7s^2$.

Bernoulli, Daniel (1700–1782) Dutch physicist and mathematician who made important contributions to kinetic theory, probability theory and the theory of differential equations. Daniel Bernoulli was one of the greatest of a family of renowned mathematicians. His great work *Hydrodynamica* was published in 1738. In this book he pioneered kinetic theory by showing that it is possible to derive BOYLE'S LAW from the assumption that a gas consists of a very large number of very small particles. He also stated a result, now known as *Bernoulli's theorem* concerning energy in a fluid which was an early example of the principle of the conservation of energy.

beryllium A light metallic element, similar to aluminum but somewhat harder; the first element in group 2 of the periodic table.

Symbol: Be. Melting pt.: 1278±5°C. Boiling pt.: 2970°C (under pressure). Relative density: 1.85 (20°C). Proton number: 4. Relative atomic mass: 9.012182. Electronic configuration: [He]$2s^2$.

Berzelius, Jöns Jakob (1779–1848) Swedish chemist who made many important contributions to the development of chemistry. He devised the system of symbols for chemical ELEMENTS that is still in use. Berzelius also determined many ATOMIC WEIGHTS and was involved in the discovery of several chemical elements.

beta decay A type of radioactive decay caused by the WEAK INTERACTIONS in which an unstable atomic NUCLEUS turns into a nucleus with the same mass number but a different proton number. Beta decay involves either the conversion of a neutron into a proton, with the emission of an electron and an electron antineutrino (n → p + e⁻ + v_e) or the conversion of a proton into a neutron, with the emission of a positron and an electron neutrino (p → n + e⁺ + v_e). The electrons and positrons emitted in beta decay are called *beta particles*. Streams of beta particles are called *beta rays* or *beta radiation*.

beta particle *See* beta decay.

beta transformation The transformation of a nucleus by beta decay. Also the decay of a neutron to a proton, an electron, and an antineutrino:
$$n \rightarrow p + e^- + \bar{v}$$

betatron A device for producing high-energy electrons (greater than 300 MeV). Electrons from some source are injected into a toroidal (doughnut-shaped) ring from which the air has been removed. The ring is held between the poles of an electromagnet. The electrons are accelerated by an increasing magnetic field (i.e. by electromagnetic induction). This type of accelerator was developed by D. W. Kerst in 1939, with the biggest such accelerator being completed in 1950 at the University of Illinois.

Bethe, Hans Albrecht (1906–) German-born American theoretical physicist who has made important contributions to many branches of physics, particularly nuclear physics and its application to understanding the generation of energy in stars. In 1929 he initiated CRYSTAL FIELD THEORY. In the 1930s and 1940s he made important contributions to the theory of phase transitions and other parts of the theory of atoms and solids. His main interest since the mid 1930s has been nuclear physics. Bethe won the 1967 Nobel Prize for physics for his work in the late 1930s on the production of energy in stars.

Bevatron A large particle ACCELERATOR (proton synchrotron) at the Lawrence Berkeley Laboratory, California. It began operating in 1954 and accelerated protons to an energy of 6 GeV.

Bhabha scattering The scattering of electrons by positions ($e^+ e^- \rightarrow e^+ e^-$). The cross-section for this process in quantum electrodynamics (QED) was first calculated by Homi Jehangir Bhabha in 1935. The calculations of QED for this process are in very good agreement with experimental results.

biaxial crystal A birefringent crystal in which there are two optic axes. Examples of biaxial crystals are mica and selenite. Biaxial crystals have three refractive indices. The optical distinctions between optically isotropic crystals can be understood in terms of the structure of crystals and the interaction between electromagnetic waves and the electrons of the atoms in the crystal.

big-bang theory The cosmological theory that asserts that the Universe, and everything in it, originated from a state of extremely high temperature and density a finite time ago. This birth of the Universe is known as the *big bang*. It is thought that the big bang occurred roughly 15 billion years ago and that the Universe has been expanding ever since. Impressive evidence in favor of the big-bang theory comes from: (1) the observed expansion of the Universe; (2) the existence of the COSMIC MICROWAVE BACKGROUND; (3) the observed abundances of light elements such as helium, which were formed soon after the big bang (*see* nucleosynthesis).

The big-bang theory was first put forward by Georges Éduard Lemaître in 1927 and revived by George Gamow and his colleagues in the 1940s. The expansion of the Universe in the theory is described by general relativity theory.

In spite of its successes the big-bang theory is not a complete theory of cosmology, not least because it does not provide an explanation of why the event occurred. *See also* early Universe; inflation.

big crunch The time reversal of the big bang in which space–time collapses to a singularity. As the big crunch approaches galaxies would start to merge and the temperature associated with the background radiation would increase. The big crunch will occur at the end of time if there is enough mass in the Universe to make it closed. It is currently thought that this is not the case.

binary pulsar A binary star system in which one of the stars is a pulsar. Although more than 20 binary pulsars are known astronomers use the expression 'the binary pulsar' to refer to the first such system to be discovered, known as PSR 1913+16. Its discovery in 1974 by Russell Hulse and Joseph Taylor has enabled very accurate tests of GENERAL RELATIVITY THEORY to be performed. In particular, the binary pulsar has provided indirect evidence for the existence of gravitational radiation by the decrease in the orbital period associated with the binary pulsar losing energy in the form of gravitational radiation.

binary stars A pair of stars revolving around their common center of mass.

binding energy The energy required to disperse a system of particles into its constituent particles; i.e. a measure of how strongly a system of particles is held together. The binding energy is the energy equivalent of the MASS DEFECT. The concept

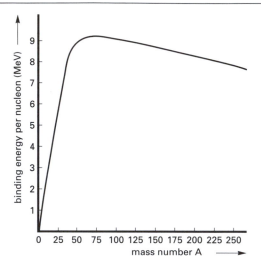

The variation of binding energy with mass number

of binding energy is used extensively in dealing with atomic nuclei. For example, the mass of a nucleus of helium-4 is less than the sum of the masses of two protons and two neutrons.

A convenient quantity for describing nuclei is the *binding energy per nucleon*.

Binnig, Gerd Karl (1947–) German physicist who shared the 1986 Nobel Prize for physics with his colleague Heinrich ROHRER for developing the SCANNING TUN-NELING MICROSCOPE.

birefringence *See* double refraction.

bismuth A brittle pinkish metallic element belonging to group 15 of the periodic table. It occurs native and in the ores Bi_2S_3 and Bi_2O_3. The element has the property of expanding when it solidifies. Compounds of bismuth are used in cosmetics and medicines.

Symbol: Bi. Melting pt.: 271.35°C. Boiling pt.: 1560±5°C. Relative density: 9.747 (20°C). Proton number: 83. Relative atomic mass: 208.98037. Electronic configuration: $[Xe]4f^{14}5d^{10}6s^26p^3$.

Bjorken, James David (1934–) American physicist who was one of the main figures in the development of the PARTON MODEL, which led on to QUANTUM CHRO-MODYNAMICS.

black body A hypothetical, ideal body that absorbs all the radiation falling on it.

black-body radiation The electromagnetic radiation emitted by a hot black body. The problem of determining the distribution of energy as a function of wavelength at a given temperature for black-body radiation was historically very important since it led to quantum mechanics (*see* Planck's radiation law). This energy distribution has a maximum at a

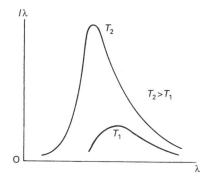

Black-body radiation

certain wavelength for a given temperature. The position of the maximum goes to higher frequencies for hotter bodies.

Blackett, Patrick Maynard Stuart (Lord Blackett) (1897–1974) British physicist who developed the CLOUD CHAMBER, which he used to study COSMIC RAYS. He was awarded the Nobel Prize for physics in 1948 for this work.

black hole   A region of space–time from which nothing, including light, can escape because of a strong gravitational field. The boundary of a black hole is called an *event horizon*. This name arises because events that occur within an event horizon, i.e. in the interior of a black hole, cannot be observed from outside the black hole. It is thought that a black hole could form from the gravitational collapse of a massive star has exhausted its nuclear 'fuel' provided that the mass of the star is greater than the OPPENHEIMER–VOLKOFF LIMIT for a neutron star or the CHANDRASEKHAR LIMIT for a white dwarf.

Although the existence of black holes was predicted in the eighteenth century by John MICHELL and Pierre Simon LAPLACE on the basis of Newtonian gravity, the theoretical study of black holes requires general relativity theory. An important result is the *no-hair theorem*, which states that a black hole is characterized completely by three qualities: its mass, electrical charge, and angular momentum. Black holes are described mathematically by solutions of EINSTEIN'S FIELD EQUATIONS. The *Schwarzschild solution* describes spherical, nonrotating electrically neutral black holes. The *Reissner–Nordstrom solution* describes nonrotating electrically charged black holes. Rotating black holes are nonspherical. The *Kerr solution* describes rotating electrically neutral black holes. The *Kerr–Newman solution* describes rotating electrically charged black holes.

Results have been established for black holes that are analogous to the laws of thermodynamics. In particular, the area of the event horizon of a black hole is analogous to ENTROPY in thermodynamics. This branch of black-hole physics is called *black-hole thermodynamics*.

General relativity theory predicts that there is a SINGULARITY at the center of a black hole. Since a singularity indicates the breakdown of general relativity theory it is necessary to have a theory of QUANTUM GRAVITY to understand black holes fully. Although such a theory does not exist at the present time, there are some important indications of how quantum theory modifies the results of general relativity theory for black holes. A notable example of this is the idea of HAWKING RADIATION. The discovery of this process in 1974 vindicated the earlier suggestion of Jacob BEKENSTEIN that there is an *identity*, rather than merely an analogy, between entropy and the area of a black hole. In the second half of the 1990s important progress was made in understanding the entropy of black holes in terms of SUPERSTRING THEORY and LOOP QUANTUM GRAVITY.

Observational evidence of a black hole comes from its effect on the matter surrounding it. There is now substantial evidence for the existence of black holes. It is thought that very large (*supermassive*) black holes are at the centers of many galaxies and that such black holes power QUASARS.

The possible existence of *mini black holes*, with masses of about 10^{12} kilogram and radii of about 10^{-15} meter, has been postulated. Such objects might have been formed in the EARLY UNIVERSE soon after the big bang. Quantum effects would be important for mini black holes. However, there is no direct evidence for the existence of mini black holes. This result, in itself, gives information about the early Universe since the formation of mini black holes would require the early Universe to be rougher and more irregular.

black-hole thermodynamics   *See* black hole.

blanket   A layer of fertile material surrounding the core in a breeder reactor.

Bloch, Felix   (1905–83) Swiss-born American physicist who made fundamental contributions to the theory of the ELECTRONIC STRUCTURE OF SOLIDS, electrical conductivity of metals, and magnetism. In 1946 Bloch, and independently Edward Mills PURCELL, discovered NUCLEAR MAGNETIC RESONANCE. Bloch and Purcell shared the 1952 Nobel Prize for physics for this discovery.

Bloch's theorem *See* electronic structure of solids.

Bloembergen, Nicolaas (1920–) Dutch-born American physicist who pioneered LASER SPECTROSCOPY independently of Arthur SCHAWLOW in the 1970s. Bloembergen and Schawlow shared the 1981 Nobel Prize for physics for this work.

blue shift *See* red shift.

B-meson A meson in which there is a bottom QUARK (or anti-bottom quark). A B^0 meson consists of a down quark and an anti-bottom quark. Its antiparticle, denoted $\overline{B}^0$ consists of a bottom quark and an anti-down quark. A B^+ consists of an UP QUARK and an anti-bottom quark. Its antiparticle, denoted $\overline{B}^-$, consists of an anti-up quark and a bottom quark. B-mesons have masses of about 5.27 GeV and lifetimes of about 10^{-13} s. They are used to study CP VIOLATION.

Bohm, David Joseph (1917–92) American-born theoretical physicist who made important contributions to the foundations of quantum mechanics. He revived the idea of a HIDDEN VARIABLES theory underlying quantum mechanics. He also contributed to the theory of PLASMAS and electron CORRELATION in solids and was one of the discoverers of the AHARONOV–BOHM EFFECT.

Bohr, Aage Niels (1922–) Danish theoretical physicist who made major contributions to our understanding NUCLEAR STRUCTURE. A son of Niels Bohr, Aage Bohr developed a model of nuclear structure with Ben MOTTELSON which combined the single-particle aspect of nuclear structure with the collective LIQUID DROP MODEL of

the nucleus, in which the shape of the whole nucleus changes. Related work was done by Leo RAINWATER, with Bohr, Mottelson, and Rainwater sharing the 1975 Nobel Prize for physics for this work.

Bohr, Niels Hendrik David (1885–1962) Danish theoretical physicist who was one of the founding fathers of quantum mechanics. In 1913 Bohr put forward a model of the atom which provided a solution for the difficulties of the RUTHERFORD MODEL (*see* Bohr model).

Bohr's work was a major influence on the emergence of quantum mechanics in the mid 1920s. He was also one of the leading figures in developing the interpretation of quantum mechanics known as the COPENHAGEN INTERPRETATION, with his concept of COMPLEMENTARITY being one of the main ingredients of this interpretation.

In the second half of the 1930s Bohr developed the LIQUID DROP MODEL of the nucleus. In 1939 he analyzed NUCLEAR FISSION in terms of this model in a classic paper with John Archibald WHEELER.

Bohr won the 1922 Nobel Prize for physics for his pioneering work on atomic structure.

bohrium A synthetic radioactive element first detected by bombarding a bismuth target with chromium nuclei. Only a small number of atoms have ever been produced.

Symbol: Bh. Proton number: 107. Most stable isotope: ^{262}Bh (half-life 0.1 s). Electronic configuration: [Rn]$5f^{14}6d^57s^2$.

Bohr magneton Symbol: μ A unit of MAGNETIC MOMENT that is used in atomic physics. It is defined by $\mu_B = eh/4\pi mc$, where e is the charge of an electron, h is the Planck constant, m is the rest mass of the electron and c is the speed of light in a vacuum.

Bohr model A pioneering attempt to apply quantum theory to the study of atomic structure, by Niels Bohr (1913). He assumed the newly introduced theory that an atom consisted of a massive positively charged nucleus with orbital electrons, and first considered in detail the simplest atom, that of hydrogen, with only one electron. The electron was supposed to move in circular orbits of radius r at speed v. By simple mechanics the total energy of such an orbit (kinetic plus potential) is shown to be $-e^2/8\pi\varepsilon_0 r$, where e is the electron charge. Bohr assumed that only certain orbits were possible, and tried to find a quantum rule to determine which ones by making assumptions (soon seen to be false) concerning the frequencies of radiation emitted and absorbed by the atoms. His calculations led however to the correct formula for the energies of the allowed states of the atom, and showed that the angular momenta were quantized in units of $h/2\pi$ where h is the Planck constant. For the nth quantum state the angular momentum is $nh/2\pi$ and the energy E_n is given by

$$E_n = -me^4/8\varepsilon^2_0 h^2 n^2$$

Assuming that the frequency of emitted radiation was given by $h\nu = E_{n1} - E_{n2}$, Bohr found that transitions to the orbit with $n = 2$ from higher values corresponded to the lines in the visible spectrum of hydrogen, while transitions to $n = 3$ gave lines in the near infrared. He predicted that transitions to $n = 1$ would give lines in the ultraviolet, which were soon to be discovered by spectroscopists.

The theory was soon successfully extended to the spectrum of singly-ionized helium and to the characteristic lines in the x-ray spectrum (see Moseley's law). Over the next twelve years the theory was developed far enough to suggest that very many facts in physics and chemistry might be explained in such terms, but innumerable difficulties were encountered with the detailed calculations. Various attempts to overcome these problems led to the modern theory of quantum mechanics pioneered by Heisenberg (1926), Schrödinger (1926), and Dirac (1928). Although Bohr's theory has been superseded it is of outstanding historical and philosophical interest.

Bohr radius Symbol: a The radius of the circular orbit of an electron nearest the nucleus in the BOHR MODEL of the hydrogen atom; i.e. the radius for the lowest energy level. It is given by $a = h^2/4\pi^2 me^2 = 5.29 \times 10^{-11}$ m, where h is the Planck constant, m is the rest mass of the electron and e is the charge of the electron. Although the concept of an electron having a precise orbit characterized by the size of its radius was subsequently found to be invalid the Bohr radius is still a useful approximate measure of the size of an atom.

Bohr–Sommerfeld model The model of atomic structure introduced by Arnold SOMMERFELD in papers published in 1915–16 as a modification of the BOHR MODEL. The main modification was to allow orbits to be ellipses rather than restricting them to being circles. This had the consequence that energy levels were characterized not just by one quantum number but by three quantum numbers: the principal quantum number, the azimuthal quantum number (orbital quantum number), and the magnetic quantum number. Although the concept of precise elliptical orbits was subsequently found to be invalid in quantum mechanics the result that energy levels of atoms are associated with these three quantum numbers (and also the spin quantum number) has remained valid.

Bohr–van Leeuwen theorem The result that the application of MAXWELL–BOLTZMANN STATISTICS to any dynamical system leads to the value of the MAGNETIC SUSCEPTIBILITY being zero. This result means that DIAMAGNETISM and PARAMAGNETISM require quantum mechanics to be understood properly. The theorem was proved by Niels BOHR in 1911 and independently by Hendrika Johanna van Leeuwen in 1919.

boiling-water reactor See nuclear reactor.

Boltzmann, Ludwig (1844–1906) Austrian theoretical physicist who made major contributions to thermodynamics, statistical mechanics, and the kinetic theory of gases. In particular, Boltzmann developed the view that thermodynamic properties such as entropy can be derived by applying statistics and probability theory to a very large number of atoms.

Boltzmann constant Symbol: k, sometimes written k_B The ratio of the universal gas constant (R) to the Avogadro constant (N_A). Thus $k = R/N_A = 1.381 \times 10^{-23}$ JK^{-1}.

Boltzmann equation A fundamental equation of NONEQUILIBRIUM STATISTICAL MECHANICS used to calculate TRANSPORT COEFFICIENTS such as conductivity. It is expressed in terms of the distribution function f of a system. The equation was first put forward by Ludwig BOLTZMANN in 1872.

bond *See* chemical bond.

bond energy The amount of energy associated with a chemical bond in a compound. It can be obtained from the HEAT OF ATOMIZATION. For example, the bond energy of the C–H bond in methane is a quarter of the enthalpy of the process:

$$CH_4(g) \rightarrow C(g) + 4H(g).$$

The bond energy depends to a certain extent on the molecule, with the C–H bond energy in methane being slightly different from that in ethane.

The *bond dissociation energy* is a different quantity, being the energy needed to break a particular bond. An example is the energy for the process:

$$CH_4(g) \rightarrow CH_3(g) + H(g).$$

Bondi, Sir Hermann (1919–) Austrian-born, British mathematician and physicist who put forward the STEADY-STATE THEORY of the Universe with Thomas GOLD and Fred HOYLE in the 1940s. He also made contributions to the theory of GRAVITATIONAL WAVES in the 1950s and 1960s.

bonding orbital *See* orbital.

Born, Max (1882–1970) German theoretical physicist who was one of the main pioneers of quantum mechanics. In particular, he was one of the leading figures in the development of the version of quantum mechanics known as MATRIX MECHANICS.

Born also established the link between probability and the wave function ψ in quantum mechanics (*see* Copenhagen interpretation). He shared the 1954 Nobel Prize for physics for his interpretation of quantum mechanics.

Born–Oppenheimer approximation An approximation very extensively used in molecular and solid state theory that the motion of atomic nuclei is so much slower than the motion of electrons that the nuclei can be regarded as being in fixed positions. This approximation was justified using PERTURBATION THEORY by Max BORN and Robert OPPENHEIMER in 1927.

Born's interpretation *See* Copenhagen interpretation.

boron A hard rather brittle metalloid element of group 3 of the periodic table. Boron is of low abundance (0.0003%). High-purity boron for semiconductor applications is obtained by conversion to boron trichloride, which can be purified by distillation, then reduction using hydrogen. Only small quantities of elemental boron are needed commercially; the vast majority of boron supplied by the industry is in the form of borax or boric acid. Natural boron consists of two isotopes, ^{10}B (18.83%) and ^{11}B (81.17%). These percentages are sufficiently high for their detection by splitting of infrared absorption or by NMR spectroscopy.

Symbol: B. Melting pt.: 2300°C. Boiling pt.: 2658°C. Relative density: 2.34 (20°C). Proton number: 5. Relative atomic mass: 10.811. Electronic configuration: [He]$2s^2 2p^1$.

boron chamber A device for detecting low-energy neutrons, in which a compound of boron (usually the gas BF$_3$) fills an ionization chamber. The ^{10}B nuclei, which constitute 18% of natural boron, absorb neutrons and emit alpha particles, which are detected by the ionization they cause.

Bose, Satyendra Nath (1894– 1974) Indian theoretical physicist who discovered what came to be known as BOSE–EINSTEIN STATISTICS in 1924. He made this discovery while deriving PLANCK'S RADIATION LAW in a consistent way. His results were put forward in a paper that Albert EINSTEIN translated into German and arranged to be published.

Bose–Einstein condensation A phenomenon which can occur in a system of BOSONS, the number of which cannot change, at a sufficiently low temperature in which a substantial fraction of the particles occupy a single quantum state. Bose–Einstein condensation has been observed at very low temperatures (of about 2×10^{-7} K) when several thousand atoms behave like a single entity. The phenomenon has been observed experimentally for rubidium and for lithium atoms. It was named after Satyendra Nath BOSE and Albert EINSTEIN and predicted by Einstein in 1925. It is of fundamental importance in explaining SUPERFLUIDITY.

Bose–Einstein distribution The DISTRIBUTION FUNCTION for a system of BOSONS.

Bose–Einstein statistics *See* quantum statistics.

boson A particle that obeys Bose–Einstein statistics. By the SPIN–STATISTICS THEOREM this means that a boson is an integer spin particle, which can either be an elementary particle or an even number of fermions (*see* Wigner's rule). A system of identical bosons is associated with a SYMMETRIC WAVE FUNCTION.

Bothe, Walther Wilhelm Georg Franz (1891–1957) German physicist who developed a particle detector called the coincidence counter, which was derived from the Geiger counter. He used this device for detecting cosmic rays. He shared the Nobel Prize for physics in 1954 for this work.

bottom quark *See* quark.

boundary condition A condition that the solution of an equation describing that system must satisfy at the boundary of that system.

bound state A system consisting of two or more parts that are bound together. In a bound state, energy is required to split the parts. A molecule is an example of a bound state since it consists of two or more atoms bound together.

Boyle, Robert (1627–1691) Influential pioneering English chemist. His book *The Sceptical Chymist*, published in 1661, was important in reviving the idea that matter is made up of atoms. He rejected the ancient Greek idea of the four elements and clarified the distinction between elements and compounds. He is best remembered for BOYLE'S LAW for the behavior of gases.

Boyle's law At a constant temperature, the pressure (p) of a given mass of gas is inversely proportional to its volume (V), i.e. $pV = K$, where K is a constant. Boyle's law is strictly true only for an IDEAL GAS but is a good approximation for real gases at low pressures and high temperatures. It was stated by Robert BOYLE in 1662. It can be derived using the KINETIC THEORY of gases.

Brackett series *See* hydrogen spectrum.

Bragg, Sir William Henry (1862–1942) and **Bragg, Sir William Lawrence** (1890–1971) Father and son English physicists who initiated X-RAY CRYSTALLOGRAPHY after the discovery of X-RAY DIFFRACTION by Max VON LAUE in 1912. This work led to their award of the 1915 Nobel Prize for physics, the only father–son team to have won this prize, with Lawrence Bragg (at age 25) the youngest person to receive the prize. The pioneering work of the Braggs in determining the structure of crystals using x-rays had an enormous impact in subsequent work on chemistry, mineralogy, and molecular biology, with the elucidation of the structure of DNA being an important example of this.

Bragg peak A peak in the scattering intensity of x-rays from a crystal. The intensity of Bragg peaks is proportional to the square of the number of scatterers. The existence of these peaks in the x-ray scattering from a solid indicates that the solid has long-range periodic order. Bragg peaks are named after Sir Lawrence BRAGG, who discovered them in 1912. The scattering that produces Bragg peaks is known as *Bragg scattering*.

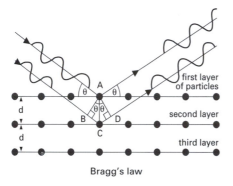

Bragg's law

Bragg's law The law relating the scattering angle θ, and the wavelength λ of x-rays being scattered from a crystal in which the crystal planes are a distance d apart. Constructive interference occurs at angles θ where $n\lambda = 2d \sin \theta$, where n is an integer. Bragg's law was derived in 1912 by Sir Lawrence BRAGG. It is of fundamental importance in X-RAY CRYSTALLOGRAPHY.

brane world A theory of elementary particles in which it is postulated that the nongravitational forces and matter exist in a four-dimensional space–time, which is a surface of a higher-dimensional space–time, with gravity existing in the bulk of the higher-dimensional space–time. This theory, which is associated with the HOLOGRAPHIC HYPOTHESIS in the context of M-THEORY, has attracted a great deal of attention since it was put forward in various forms by a number of physicists in the late 1990s. At present it is a speculative theory, but evidence for it may emerge in the first few decades of the twenty-first century.

Brattain, Walter Houser (1902– 1987) American physicist who developed the transistor with John BARDEEN and William SHOCKLEY in the 1940s. Brattain shared the 1956 Nobel Prize for physics with Bardeen and Shockley.

breeder reactor *See* nuclear reactor.

bremsstrahlung German for 'braking radiation'. The emission of electromagnetic waves, usually x-rays, when electrically charged particles, usually electrons, are accelerated or decelerated. An example of bremsstrahlung is the emission of x-rays when rapidly moving electrons hit the target in an x-ray tube. The x-rays emitted have a continuous range of wavelengths that extend to a minimum value, which depends on the kinetic energy of the electrons. Bremsstrahlung occurs when electrons are deflected by another charged particle such as an atomic nucleus. *See also* synchrotron radiation.

Brillouin, Léon (1889–1969) French theoretical physicist who made several important contributions to physics. In 1926 he was one of the inventors of the SEMI-CLASSICAL APPROXIMATION. In the early 1930s he carried out important investigations in the quantum theory of electrons in crystals, which emphasized the relation between energy bands in crystals (*see* electronic structure of solids) and x-ray diffraction. In work performed after World War II he investigated the relation between entropy and information theory.

Brockhouse, Bertram Neville (1918–) Canadian physicist who worked on neutron scattering. In the 1950s and early 1960s Brockhouse investigated the structure of solids and liquids using the inelastic scattering of slow neutrons. He shared the 1994 Nobel Prize for physics with Clifford SHULL who was (independently) a pioneer in the field.

broken symmetry A situation in which there is less symmetry in the ground state of a many-body system or the vacuum state of a relativistic quantum field theory than the Hamiltonian or Lagrangian defining the system. Because of this, broken symmetry is sometimes known as *hidden symmetry*. There are many examples of broken symmetry in the theory of condensed matter such as ferromagnetism, superconductivity, and superfluidity, with there being a phase transition at some critical temperature. Below this temperature the system exists in a state of broken symmetry, while above this temperature the system exists in a state of UNBROKEN SYMMETRY. The concepts of symmetry and broken symmetry are of fundamental importance in the theory of elementary particles, particularly in theories that attempt to unify different types of forces and particles. There are several close analogies between broken symmetry in certain systems in condensed matter theory and in elementary particle theory. *See also* Goldstone boson; Goldstone's theorem; Higgs boson; Higgs mechanism; spontaneous symmetry breaking; unbroken symmetry.

bromine   A deep red, moderately reactive element belonging to the halogens; i.e. group 17 of the periodic table. Bromine is a liquid at room temperature (mercury is the only other element with this property). It occurs in small amounts in seawater, salt lakes, and salt deposits but is much less abundant than chlorine.

Symbol: Br. Melting pt.: $-7.25°C$. Boiling pt.: $58.78°C$. Relative density: 3.12 (20°C). Proton number: 35. Relative atomic mass: 79.904. Electronic configuration: $[Ar]3d^{10}4s^24p^5$.

Brookhaven National Laboratory   A laboratory sited on Long Island, New York. An early accelerator built there in 1953 was one of the first synchrocyclotrons and was able to accelerate protons to energies greater than 1 GeV.

brown dwarf   An astronomical object that has a mass intermediate between a large planet such as Jupiter and a star. A brown dwarf can shine faintly due to the conversion of gravitational energy into heat as it slowly contracts under its own weight (*see* Kelvin–Helmholtz time-scale) but is not sufficiently large to start nuclear fusion reactions. It is thought that their masses are greater than a few times the mass of Jupiter but less than 80 times the mass of Jupiter. It has been speculated that brown dwarfs contribute to the DARK MATTER in the Universe. There is now observational evidence for the existence of brown dwarfs.

Brownian motion   The random motion of small particles suspended in a fluid. An example of Brownian motion is given by smoke particles in air. Brownian motion was first observed in 1827 when the Scottish botanist Robert Brown noticed that when pollen grains suspended in water are examined in a microscope there is a continuous random motion of the grains. It was suggested by various physicists in the 1860s that this motion is a result of bombardment of the particles by molecules of the fluid they are suspended in. Albert EINSTEIN gave a mathematical description of Brownian motion in 1905.

bubble chamber   A device that detects high-energy particles moving through a fluid by the trail of bubbles they leave as they move. A liquid, which is frequently liquid hydrogen, is kept in a chamber slightly above its boiling point by using sufficiently high pressure. If the pressure is reduced suddenly the liquid remains liquid, but at this reduced pressure such a liquid is a superheated liquid, and hence unstable. If an electrically charged particle moves through the liquid it creates a trail of bubbles along its path. The pressure is then increased again so that all the liquid does not boil and the bubble trail lasts long enough to be photographed. Magnetic fields can be used to alter the paths of the charged particles. The bubble chamber was invented in 1952 by Donald GLASER. *See also* cloud chamber.

Bunsen, Robert Wilhelm   (1811–1899) German chemist and physicist who was a pioneer of spectroscopy.

butterfly effect   *See* chaos theory.

BWR   *See* nuclear reactor.

C

cadmium A transition metal obtained as a by-product during the extraction of zinc. It is used to protect other metals from corrosion, as a neutron absorber in nuclear reactors, in alkali batteries, and in certain pigments. It is highly toxic. There are eight stable isotopes and a large number of radioactive isotopes are also known.

Symbol: Cd. Melting pt.: 320.95°C. Boiling pt.: 765°C. Relative density: 8.65 (20°C). Proton number: 48. Relative atomic mass: 112.411. Electronic configuration: $[Kr]4d^{10}5s^2$.

Calabi–Yau manifold A type of shape into which it has been suggested that the higher dimensions of space required by SUPERSTRING THEORY curl. These manifolds are named after Eugenio Calabi and Shing-Tung Yau who investigated them in a purely mathematical context before superstring theory was introduced. The hypothesis that the higher dimensions in superstring theory curl up into a Calabi–Yau manifold was put forward by Edward WITTEN and three colleagues in 1984. It is a plausible and attractive hypothesis, but one for which there is no experimental evidence at the present time. A particularly attractive feature for the hypothesis is that it provides an explanation for the low number of GENERATIONS of elementary particles in terms of the TOPOLOGY of Calabi–Yau manifolds.

calcium A moderately soft, low-melting reactive metal; the third element in group 2 of the periodic table. Calcium is widely distributed in the Earth's crust and is the third most abundant element. There are six stable isotopes and a number of radioactive ones, the most stable of which is ^{41}Ca (half-life 1.03×10^5 y).

Symbol: Ca. Melting pt.: 839°C. Boiling pt.: 1484°C. Relative density: 1.55 (20°C). Proton number: 20. Relative atomic mass: 40.0878. Electronic configuration: $[Ar]4s^2$.

calculus A set of mathematical techniques invented independently by Isaac NEWTON and Gottfried LEIBNIZ to deal with infinitesimal quantities and the limits associated with such quantities. The use of calculus is ubiquitous in physical science. *Differential calculus* is concerned with the rate of change of one quantity with respect to another quantity. For example, the acceleration of a body is the rate of change of the velocity of the body with respect to time. The process of finding the rate of change of one quantity with respect to another quantity is called *differentiation*.

Integral calculus deals with the inverse process, and *integration* is the inverse of differentiation. For example, the integral of the velocity of a body between two instants of time t_1 and t_2 gives the distance traveled by the body between t_1 and t_2. Integral calculus is useful in finding the areas and volumes of figures.

californium A silvery radioactive transuranic element of the actinoid series of metals, not found naturally on Earth. Several radioisotopes have been synthesized, including californium-252, which is used as an intense source of neutrons in certain types of portable detector and in the treatment of cancer.

Symbol: Cf. Melting pt.: 900°C. Proton number: 98. Most stable isotope: ^{251}Cf (half-life 900 years). Electronic configuration: $[Rn]5f^{10}7s^2$.

Cameron, Alastair (1925–) Canadian-born American physicist. In the 1950s Cameron gave a theoretical description of the way elements are produced in stars independently of Fred HOYLE and his colleagues. *See* nucleosynthesis.

canal rays Streams of positive ions obtained from a discharge tube by boring small holes in the cathode. The ions being attracted to the cathode can thus pass through it forming positive rays.

candela Symbol: cd   The SI base unit of luminous intensity; the luminous intensity in a given direction of a source that emits monochromatic radiation of frequency 540×10^{12} hertz and that has radiant intensity in that direction of 1/683 watt per steradian. Formerly, the unit was defined as the intensity (in the perpendicular direction) of the black-body radiation from a surface of 1/600 000 square meter at the temperature of freezing platinum and at a pressure of 101 325 pascals.

Cannizzaro, Stanislao (1826–1910) Italian chemist who was influential in the development of chemistry and atomic theory. He is particularly remembered as the first to appreciate the earlier work of Amedeo AVOGADRO. He showed that common gases such as hydrogen exist as molecules rather than isolated atoms and was able to resolve the confusion about atomic weights.

capture Any of various processes in which one entity absorbs a particle to form another entity. For example, a neutral atom or molecule may capture an electron to form a negative ion:
$$M + e^- \rightarrow M^-$$
Similarly a positive ion may capture an electron to give a neutral atom or molecule:
$$M^+ + e^- \rightarrow M$$
   Capture also occurs for nuclei. A particular nucleus may capture a neutron under certain conditions to produce a new nucleus, which may be unstable. If the new nucleus is in an excited state it can decay by emission of a gamma-ray photon. this is an example of *radiative capture*. the process known as *K-capture* occurs when a nucleus of an atom captures an electron from the innermost shell (K-shell) of the atom. This leaves the atom in an excited state and decay occurs with emission of an x-ray photon.

carbon The first element of group 14 of the periodic table. Carbon is a universal constituent of living matter and the principal deposits of carbon compounds are derived from living sources; i.e., carbonates (chalk and limestone) and fossil fuels (coal, oil, and gas). It also occurs in the mineral dolomite. The element forms only 0.032% by mass of the Earth's crust. Minute quantities of elemental carbon also occur as the allotropes graphite and diamond. A third allotrope, buckminsterfullerene (C_{60}), also exists. Naturally occurring carbon has the isotopic composition ^{12}C (98.89%), ^{13}C (1.11%), and ^{14}C (minute traces in the upper atmosphere produced by slow neutron capture by ^{14}N atoms). ^{14}C is used for radiocarbon dating because of its long half-life of 5730 years.

   Symbol: C. Melting pt.: 3550°C. Boiling pt.: 4830°C (sublimes); C_{60} sublimes at 530°C. Relative density: 3.51 (diamond), 2.26 (graphite), 1.65 (C_{60}) (all at 20°C). Proton number: 6. Relative atomic mass: 12.011. Electronic configuration: $[He]2s^2 2p^2$.

carbon cycle A series of six nuclear reactions that is thought to be responsible for the production of energy in many stars. The net effect of the carbon cycle is that four hydrogen nuclei combine to form helium nuclei, with the release of energy. The series of reactions is called the carbon cycle because carbon-12 acts like a catalyst, which is reformed at the final reaction of the series:
$$^{12}_{6}C + ^{1}_{1}H \rightarrow ^{13}_{7}N + \gamma$$
$$^{13}_{7}N \rightarrow ^{13}_{6}C + e^+ + \mu_e$$
$$^{13}_{6}C + ^{1}_{1}H \rightarrow ^{14}_{7}N + \gamma$$
$$^{14}_{7}N + ^{1}_{1}H \rightarrow ^{15}_{8}O + \gamma$$
$$^{15}_{8}O \rightarrow ^{15}_{7}N + e^+ + \mu_e$$
$$^{15}_{7}N + ^{1}_{1}H \rightarrow ^{12}_{6}C + ^{4}_{2}He$$
See also proton–proton reaction.

carbon dating (radiocarbon dating)   A method of estimating the age of material that contains matter of biological origin.

Carbon dating uses the fact that radioactive carbon-14 is continuously formed in the atmosphere by neutron bombardment of nitrogen nuclei as a result of cosmic rays:

$$^{14}_{7}N + n \rightarrow {}^{14}_{6}C + p$$

Some of the ^{14}C is taken up by living organisms (e.g. by photosynthesis). When the organism dies the amount of radioactive carbon decreases exponentially with time because of its radioactivity. The ratio of ^{12}C to ^{14}C present in the material enables an estimate to be made of its age. Carbon dating has been used to estimate the ages of specimens up to about 40 000 years old. This technique, which depends on assumptions about past levels of cosmic-ray activity, was developed by Willard LIBBY and his colleagues just after World War II.

Cartan, Élie (1869–1951) French mathematician who made important contributions to geometry, topology, and the theory of LIE GROUPS. Much of his mathematical work, such as his classification of Lie groups and the discovery of SPINORS, subsequently turned out to be relevant to the theory of elementary particles and/or general relativity theory. He also made direct contributions to general relativity theory.

Cartesian coordinates A technique for locating a point P in analytic geometry. The position of the point is specified with respect to a set of perpendicular straight lines called *axes*. In two dimensions there is a horizontal axis called the x-axis and a vertical axis called the y-axis, with the point where they intersect being called the origin 0. Values of y on the y-axis that are lower than 0 are negative. Values of x on the x-axis that are to the left of 0 are negative. A point P is specified by its perpendicular distances from the axes. In three dimensions there is a third perpendicular axis through 0 called the z-axis. Cartesian coordinates are named for René DESCARTES.

Casimir, Hendrik Brugt Gerhardt (1909–) Dutch physicist who made wide-ranging contributions to physics, notably the CASIMIR EFFECT.

Casimir effect An effect in which a force occurs between two parallel closely spaced metal plates. In the 1940s Hendrik CASIMIR predicted that the virtual photons in the quantum-mechanical vacuum state could have an observable consequence. Having the metal plates in position imposes boundary conditions such that only certain wavelengths of photons can fit the gap between the plates, whereas there is no such restriction for wavelengths outside the gap. This results in a very small attractive force between the plates. The Casimir effect was detected by M. J. Sparnaay in 1958, thus demonstrating the physical reality of the VACUUM STATE in quantum field theory.

cathode An electrode with a negative potential. The cathode is the electrode from which electrons emerge. Cations are attracted to the cathode in electrolysis. *See also* anode.

cathode rays Streams of electrons given off at the cathode in a tube at low pressure. Cathode rays were originally observed in gas-discharge tubes at low pressure. If a cathode in a vacuum tube is heated then cathode rays are given off by THERMIONIC EMISSION.

cathode-ray tube (CRT) An electron tube that converts electrical signals into a pattern on a screen. The cathode-ray tube forms the basis of the cathode-ray oscilloscope and the television receiver. It consists of an electron gun, which produces an electron beam. The electrons are focused onto and moved across a luminescent screen by magnetic and/or electric fields, to give a small moving spot of light.

cation A positively charged ion. In electrolysis a cation is attracted to the cathode. *See also* anion.

causality The principle that effect cannot come before cause. Important applications of causality occur in the analysis of the results of scattering experiments and in optics.

Cavendish, Henry (1731–1810) English scientist who made a number of important early discoveries in chemistry and in physics, including a determination of the density of the Earth and the strength of the GRAVITATIONAL CONSTANT. Cavendish was a wealthy recluse whose unpublished research anticipated the work of Coulomb and Faraday.

Cayley, Arthur (1821–95) English mathematician generally considered to be the inventor of matrices. Cayley was a wide-ranging and prolific mathematician whose work included developing the geometry of n dimensions, a subject which, like matrix algebra, has many applications in modern physics.

celestial mechanics The study of the motion of celestial bodies, such as planets in the Solar System. It is largely based on NEWTONIAN MECHANICS and NEWTONIAN GRAVITY, with small corrections made for the effects of special and general relativity theory. Analogies with celestial mechanics were important in the development of the OLD QUANTUM THEORY.

cell A system that has two electrodes in contact with an ELECTROLYTE. In an *electrolytic cell* a chemical reaction is produced by passing a current through the electrolyte. In a *voltaic cell* (or *galvanic cell*) an e.m.f. is produced by chemical reactions at the electrodes.

Cerenkov, Pavel Alekseyevic (1904–90) Russian physicist who discovered what is now known as CERENKOV RADIATION in 1934. Cerenkov shared the 1958 Nobel Prize for physics with Ilya FRANK and Igor TAMM for this work.

Cerenkov counter (**Cerenkov detector**) A particle detector that detects and counts rapidly moving electrically charged particles. The particles pass through a liquid (frequently water) and emit CERENKOV RADIATION, which is detected by a photomultiplier.

Cerenkov radiation Electromagnetic radiation, usually in the form of bluish light, caused by electrically charged particles moving through a transparent medium at a speed that is greater than the speed of light in that medium. Cerenkov radiation is the optical analog of a sonic boom. It is named after Pavel CERENKOV who discovered it in 1934.

cerium A ductile malleable gray element of the lanthanoid series of metals. It occurs in association with other lanthanoids in many minerals, including monazite and bastnasite. The metal is reactive; it reacts with water and tarnishes in air. It is used in several alloys (especially for lighter flints), as a catalyst, and in compound form in carbon-arc searchlights, etc., and in the glass industry. There are four stable isotopes and a large number of radioactive ones, of which the most stable is ^{144}Ce (half-life 284.6 d).

Symbol: Ce. Melting pt.: 799°C. Boiling pt.: 3426°C. Relative density: 6.7 (hexagonal structure, 25°C). Proton number: 58. Relative atomic mass: 140.15. Electronic configuration: [Xe]4f5d6s^2.

CERN (Counseil Européen pour la Recherche Nucléaire) The center for experimental research on elementary particles in Europe. It is based near Geneva and is supported by many European nations. A number of important discoveries have been made with the accelerators at CERN, including the discoveries of W and Z bosons. *See also* LEP.

cesium A soft golden highly reactive low-melting element of group 1 of the periodic table. It is found in several silicate minerals. The metal oxidizes in air and reacts violently with water. Cesium is used in photocells, as a catalyst, and in the cesium atomic clock. The main isotope occurring is ^{133}Cs. The radioactive isotopes ^{134}Cs (half-life 2.065 years) and ^{137}Cs (half-life 30.3 years) are produced in nuclear reactors and are potentially dangerous atmospheric pollutants. The British spelling is *caesium*, although 'cesium' is the official scientific spelling.

Symbol: Cs. Melting pt.: 28.4°C. Boiling pt.: 678.4°C. Relative density: 1.873 (20°C). Proton number: 55. Relative atomic mass: 132.91. Electronic configuration: [Xe]6s^1.

cesium clock An ATOMIC CLOCK that makes use of the fact that two of the energy levels of cesium atoms have transitions between them in which the frequency is 9 192 631 770 hertz. The unit of time, the second, is now defined in terms of this atomic frequency.

CESR (Cornell Electron Synchrotron Ring) A particle accelerator at Cornell University.

Chadwick, Sir James (1894–1974) British physicist who discovered the neutron in 1932. Subsequently he was able to establish that the mass of a neutron is slightly greater than that of a proton. He won the Nobel Prize for physics in 1935.

chain reaction A reaction that is self-sustaining because some of the products of one step initiate a subsequent step. Chain reactions can occur both in chemical reactions and in nuclear reactions.

In nuclear fission, if neutrons produced by the fission of one nucleus cause fission in nearby nuclei a chain reaction may result. To take the most notable example, the disintegration of one uranium-235 nucleus results in the production of neutrons, which cause fission to occur in other such nuclei nearby, resulting in the production of more neutrons, etc.

If each fission causes exactly one other fission the chain reaction is said to be *critical*. A chain reaction in which one fission causes less than one other fission, on average, is said to be *subcritical*. Chain reactions in which one fission causes more than one fission are referred to as *supercritical*. If each fission causes the fission of at least two other nuclei then there is a *runaway chain reaction*, resulting in a nuclear explosion. If a MODERATOR is added to FISSILE

MATERIAL then some of the neutrons are absorbed. If the correct amount of moderator is added the chain reaction can be made exactly critical. A chain reaction that is controlled in this way is called a *controlled chain reaction*.

chalcogens *See* group 16 elements.

Chamberlain, Owen (1920–) American physicist who, together with Emilio SEGRÈ, discovered the ANTIPROTON in 1955. Chamberlain and Segré shared the 1959 Nobel Prize for physics for this discovery.

Chandrasekhar, Subrahmanyan (1910–95) Indian-born American physicist who made important contributions to the theory of stars and black holes. In particular, he is remembered for the concept of the CHANDRASEKHAR LIMIT. He won the 1983 Nobel Prize for physics for his work in astrophysics.

Chandrasekhar limit The maximum possible mass that a WHITE DWARF star can have for DEGENERACY PRESSURE to prevent gravitational collapse. This mass is about 1.4 times the mass of the Sun. For masses higher than the Chandrasekhar limit a star can collapse to either a NEUTRON STAR or a BLACK HOLE. The concept of the Chandrasekhar limit was put forward by Subrahmanyan CHANDRASEKHAR in 1931. *See also* Oppenheimer–Volkoff limit.

change of state (**change of phase**) A change from one phase (solid, liquid, or gas) of matter into another.

chaos theory The theory that describes the unpredictable, apparently random, behavior of a system governed by deterministic laws. Systems that exhibit chaotic behavior are governed by nonlinear equations. This means that such systems are characterized by *sensitivity to initial conditions*, i.e. a very small change in the initial conditions can have a very large effect on the future state of the system. This feature of chaos theory is called the *butterfly effect*. This name arises because it has been suggested that the equations that describe the weather are so sensitive to initial conditions that a butterfly flapping its wings in one part of the world can determine whether a tornado occurs in another part of the world. Chaos theory has many applications in physical science.

charge A basic property of some elementary particles that gives rise to an interaction between them called an *electrical interaction*. There are two types of charge. It is conventional to call these types *positive* and *negative*. Two particles with the same type of charge (i.e. both negative or both positive) repel each other whereas two particles with the opposite type of charge (i.e. one is positive and the other is negative) attract. The interaction between charges is expressed quantitatively by COULOMB'S LAW.

charge conjugation Symbol: C   The operation of exchanging particles and their antiparticles in quantum field theory. This operation can be applied to electrically neutral particles, such as neutrons, as well as to charged particles. Strong and electromagnetic interactions are always symmetrical under charge conjugation but weak interactions are not. This means that the decay rate of a weak interaction process is not exactly equal to the rate of the analogous process when the particles are replaced by their antiparticles. *See also* CPT theorem; CP violation.

charge independence The feature of strong nuclear interactions that they are independent of the electric charges of the particles. This means that the strong interaction between two protons, two neutrons, or a proton and a neutron are all the same. Strong evidence for charge independence comes from the binding energies of MIRROR NUCLEI.

charged current An interaction in which the boson mediating the interaction is electrically charged. For example, weak interactions such as beta decay, which are mediated by W^+ or W^- bosons, involve charged currents. *See also* neutral current.

Charles' law The original statement of this law was that at constant pressure, the volume of a fixed mass of gas expands by a constant fraction of its volume at 0°C for each Celsius degree rise in temperature. For an ideal gas this fraction is approximately 1/273. This law was discovered by Jacques Alexandre César Charles from experiments in the late eighteenth century and established more firmly several years later by Joseph GAY-LUSSAC and independently by John DALTON. Charles' law is sometimes known as *Gay-Lussac's law*. It is simpler to state this law in the form that at constant pressure, the volume of a fixed mass of gas is proportional to the absolute temperature.

charm A property of one of the six flavors of QUARK, which is expressed as a quantum number. Charm is conserved in strong and electromagnetic interactions but not in weak interactions.

charmonium A system consisting of a charm QUARK and an anticharm quark bound by the strong nuclear interaction. Charmonium is analogous to POSITRONIUM. The charmonium state with the lowest energy level is associated with the J/PSI PARTICLE.

charm quark See QUARK.

Charpak, Georges (1924–   ) Polish-born French physicist who worked at CERN and developed the DRIFT CHAMBER and the multiwire proportional chamber for detecting particles. Charpak was awarded the 1992 Nobel Prize for physics for his work on particle detectors.

chemical bond The attractive force that holds atoms together in molecules and solids. Chemical bonds between atoms are understood in terms of the ELECTRONIC STRUCTURE OF ATOMS and hence need to be described and calculated using quantum mechanics. The chemical bond is understood in terms of the shell structure of electrons in atoms, particularly the stability of electron configurations of the RARE GASES due to their shells having the maximum number of electrons allowed by the PAULI EXCLUSION PRINCIPLE.

One way in which it is possible for atoms to achieve the electronic configuration of rare gases is to share electrons between atoms. For example, single hydrogen atoms have one electron ($1s^1$) and needs one more electron to fill a shell. In a hydrogen molecule (H_2) each hydrogen atom shares two electrons, thus giving it the electron configuration of the rare gas helium. Similarly, a carbon atom has four electrons in its outer shell and needs four more electrons to have a full outer shell. In methane CH_4) the carbon atom gets a share in the four electrons of the hydrogen atoms, thus giving the carbon atom a full shell. Each hydrogen atom shares an electron from the carbon atom, thus giving the hydrogen atom a full shell. A chemical bond formed by the sharing of electrons is called a *covalent bond*. If one pair of electrons is shared the bond is called a *single bond*. It is also possible for bonds with two pairs of electrons (*double bonds*) and three pairs of electrons (*triple bonds*) to be formed.

Another type of chemical bond results from the transfer of electrons. For example, a sodium atom needs to lose one electron to achieve a rare-gas electron configuration. A chlorine atom is one electron short of a rare-gas electron configuration. If each sodium atom transfers one electron to a chlorine atom then both atoms have rare-gas electron configurations and become ions. The positively charged sodium ion (Na^+) and the negatively charged chloride ion (Cl^-) are held by electrostatic attraction between these ions. A bond formed in this way is called an *ionic bond* or *electrovalent bond*.

In a molecule such as the hydrogen molecule in which the pair of electrons is shared equally the bonding is said to be *homopolar*. In compounds such as sodium chloride where there is complete transfer of electrons to form an ionic bond the bonding is said to be *heteropolar*. There is a range of *intermediate bonds* between the two extremes of purely covalent bonding and purely ionic bonding. For example, in hydrogen chloride (HCl) there is sharing of a pair of electrons but since a chlorine atom is more electronegative than a hydrogen atom there is a greater probability of the shared electrons being near the chlorine atoms resulting in a net negative charge on the chlorine atom and net positive charge on the hydrogen.

Since quantum-mechanical systems with more than one electron cannot be solved exactly it is necessary to use approximations in quantitative descriptions of the chemical bond. The main ways for describing the chemical bond are MOLECULAR ORBITAL theory and VALENCE BOND THEORY.

chemical reaction A process in which one or more elements and/or compounds form new elements and/or compounds. Although all chemical reactions are reversible, in many cases the reverse reaction back to the starting point occurs only to a negligible extent, and the reaction is considered to be irreversible.

chirality 1. The property of 'handedness' of a particle that obeys the laws of quantum mechanics.
2. The property of certain molecules of existing in left- and right-handed forms.

chiral symmetry A symmetry that is associated with fermions in quantum field theory. It is an exact symmetry for massless fermions but only an approximate symmetry for fermions that have finite mass.

chiral symmetry breaking The situation that occurs when chiral symmetry is a BROKEN SYMMETRY. A notable example of chiral symmetry breaking occurs in QUANTUM CHROMODYNAMICS. If quarks were exactly massless then PIONS would be exactly massless GOLDSTONE BOSONS associated with chiral symmetry breaking. Since quarks are not exactly massless, then the bosons associated with chiral symmetry breaking are not exactly massless, thus explaining why pions are not massless. There have been many attempts to invoke chiral symmetry breaking in theories of elementary particles going beyond the STANDARD

MODEL but such theories have not been developed successfully.

chlorine A green reactive gaseous element belonging to the halogens; i.e. group 17 of the periodic table. It occurs in seawater, salt lakes, and underground deposits of halite, NaCl. It accounts for about 0.055% of the Earth's crust. The two main isotopes are ^{35}Cl (75.77%) and ^{37}Cl (24.23%). Several radioactive isotopes are also known, the most stable being ^{36}Cl (half-life 3.01×10^5 y).

Symbol: Cl. Melting pt.: −100.38°C. Boiling pt.: −33.97°C;. Density: 3.214 kg m^{-3} (0°C). Proton number: 17. Relative atomic mass: 35.4527. Electronic configuration: $[Ne]3s^23p^5$.

chromium A transition metal that occurs naturally as chromite $(FeO.Cr_2O_3)$, large deposits of which are found in Zimbabwe. Chromium is used in strong alloy steels and stainless steel and for plating articles. It is a hard silvery metal that resists corrosion at normal temperatures. There are four stable isotopes, the most abundant being ^{52}Cr (83.789%).

Symbol: Cr. Melting pt.: 1860±20°C. Boiling pt.: 2672°C. Relative density: 7.19 (20°C). Proton number: 24. Relative atomic mass: 51.9961. Electronic configuration: $[Ar]3d^54s^1$.

Chu, Steven (1948–) American physicist who was able to slow atoms down to extremely low temperatures, thereby enabling atomic states to be measured with great precision. Using six laser beams and a gas of sodium atoms Chu was able (in 1985) to cool the atoms to 240 microkelvins above absolute zero. He shared the 1997 Nobel Prize for physics with Claude COHEN-TANNOUDJI and William PHILLIPS.

classical electrodynamics The theory that describes electric and magnetic phenomena at the classical, i.e. nonquantum level. Classical electrodynamics is governed by the MAXWELL EQUATIONS.

classical field theory A theory that describes a field in classical rather than quantum-mechanical terms. Thus classical electrodynamics, which is described by the MAXWELL EQUATIONS and GENERAL RELATIVITY THEORY, which describes the classical gravitational field, are both examples of classical field theories. For a classical field theory to be applicable on a macroscopic scale it is necessary that the interactions are long-range, as is the case in electromagnetic and gravitational interactions, rather than short-range, as is the case for nuclear interactions. It is also mathematically convenient to use classical field theory to describe the physics of continuous media, such as fluids.

classical limit The limit of quantum mechanics which emerges as the PLANCK CONSTANT h tends to zero. This limit is analogous to geometrical optics being the limit of physical optics which emerges when the wavelength $\lambda \to 0$.

classical mechanics The term is used in two ways. One to mean NEWTONIAN MECHANICS, as opposed to mechanics involving the special and general theories of relativity. More commonly, it is taken to mean Newtonian mechanics modified by relativity theory, as opposed to quantum mechanics. In this sense, classical mechanics deals with quantities that can vary continuously whereas quantum mechanics deals with quantities that have discrete values.

Clausius, Rudolf Julius Emmanuel (1822–88) German theoretical physicist who was one of the founders of thermodynamics and made contributions to the kinetic theory of gases. In classic papers published in the 1850s he stated the second law of thermodynamics and introduced the concept of ENTROPY.

Clausius–Mossotti equation An equation which relates the POLARIZABILITY α of a molecule to the DIELECTRIC CONSTANT ε of a material composed of molecules with that polarizability. It is possible to write the Clausius–Mossotti equation in the form $\alpha = (3/4\pi N)/[(\varepsilon - 1)/(\varepsilon - 2)]$, where N is the number of molecules per unit volume. This equation was derived using electrostatics applied to macroscopic bodies by Ottaviano Fabrizio Mossotti in 1850 and independently by Rudolf CLAUSIUS in 1879. An important feature of the Clausius–Mossotti equation is that it links a microscopic quantity (the polarizability) and a macroscopic quantity (the dielectric constant). The equation works well for gases but holds only approximately for solids or liquids, especially if the dielectric constant is large.

Clifford, William Kingdon (1845– 79)
British mathematician who was one of the first people to speculate that RIEMANNIAN GEOMETRY might describe curved space. He also invented a type of abstract algebra, now called a *Clifford algebra*, which is used to analyze fermions in quantum field theory.

clock in a box experiment   A THOUGHT EXPERIMENT devised by Albert Einstein in the late 1920s in an attempt to show that it is possible to circumvent HEISENBERG'S UNCERTAINTY PRINCIPLE.

In one of his many debates with Niels Bohr on the foundations of quantum mechanics, Einstein invited Bohr to consider a box with a hole in the wall, which is opened and shut by a shutter operated by a clock in the box. This box is also full of electromagnetic radiation. At a specific time the clock causes the shutter to open long enough for one photon to escape before being shut again. Since mass and energy are equivalent the box should weigh less after the photon has escaped. Einstein argued that since it is possible to weigh the box before and after the photon has escaped and since the time it left the box was known both the exact energy of the photon and the exact time it left the box are known, thus circumventing the Heisenberg uncertainty principle. Bohr was able to refute the argument by showing that the act of weighing the box is itself subject to the uncertainty principle.

closed shell   An electron shell in an atom or molecule that is completely full, i.e. it contains the maximum number of electrons allowed by the PAULI EXCLUSION PRINCIPLE.

cloud chamber   An early type of detector for electrically charged particles. The *Wilson cloud chamber*, invented by Charles Wilson in 1911, consists of a container with air and ethanol vapor that was cooled suddenly by an adiabatic expansion, thus making the vapor supersaturated. Particles of ionizing radiation cause a trail of ions as they pass through the chamber and these induce condensation of droplets of liquid, forming a visible track. Tracks formed in this way can be photographed and magnetic fields used to provide information about the particles from the curvature of the tracks. This type of cloud chamber is also called an *expansion cloud chamber*. A later type, the *diffusion cloud chamber* was developed in 1950 by Cowan, Needels, and Nielsen. In this device strips of felt soaked in alcohol are placed in the top of the chamber and the bottom of the chamber is cooled in solid carbon dioxide. Vapor diffuses downward and there is a region in the center where the vapor is supersaturated and able to detect particles. This device thus acts as a continuous detector. Many important discoveries concerning nuclear physics, cosmic rays, and elementary particles were made using cloud chambers, including the discovery of the POSITRON.

cloudy crystal ball model (optical model)   A model of NUCLEAR REACTIONS in which the wave nature of the incident particle is taken into account. The nucleus is taken to be a semitransparent object with a complex refractive index. The real part of the refractive index is associated with refractive and diffractive effects while the imaginary part is associated with absorptive effects. The cloudy crystal ball model was put forward by Hans BOETHE in 1940 and has been extensively developed by many nuclear physicists since. It is very successful in describing nuclear reactions

in which it is necessary to take the wave nature of the incident particle into account.

cobalt A lustrous silvery-blue hard ferromagnetic transition metal occurring in association with nickel. It is used in alloys for magnets, cutting tools, and electrical heating elements and in catalysts and some paints. The main isotope is ^{59}Co. The radioisotope ^{60}Co (half-life 5.27 y) emits electrons and gamma rays and has a number of uses in medical diagnosis and radiotherapy.

Symbol: Co. Melting pt.: 1495°C. Boiling pt.: 2870°C. Relative density: 8.9 (20°C). Proton number: 27. Relative atomic mass: 58.93320. Electronic configuration: $[Ar]3d^74s^2$.

COBE (**Cosmic Background Explorer**) A NASA satellite that was launched in 1989. In 1992 it discovered ripples in the COSMIC MICROWAVE BACKGROUND providing impressive evidence in support of the big-bang theory of the Universe.

Cockcroft, Sir John Douglas (1897–1967) English physicist who, together with Ernest WALTON performed the first experiment in which one chemical element was changed into another element by artificial means. Cockcroft and Walton were awarded the 1951 Nobel Prize for physics for this work.

Cockcroft–Walton machine The first particle accelerator built in the early 1930s by Sir John COCKCROFT and Ernest WALTON at the Cavendish Laboratory, Cambridge. This device employed a potential difference of 800kV and was used to accelerate protons. In 1932 the Cockcroft–Walton machine was used to bombard lithium with protons, resulting in the production of helium.

$$^1_1H + ^7_3Li \rightarrow ^4_2He + ^4_2He$$

This was the first artificial transmutation from one chemical element to another element.

Cohen-Tannoudji, Paul (1933–) French physicist who developed techniques for cooling atoms down to extremely low temperatures. Following earlier work by William PHILLIPS and Steven CHU, Cohen-Tannoudji was able to cool helium atoms down to 0.18 microkelvin. He shared the 1997 Nobel Prize for physics with Chu and Phillips for this work.

coherent radiation Electromagnetic radiation in which there is a constant phase relation between two or more sets of waves. This means that the peaks and troughs are always similarly spaced.

cold dark matter *See* dark matter.

cold emission Emission of electrons from a solid by a process other than thermionic emission. The term is usually used to describe either field emission or secondary emission.

Coleman–Gross theorem The result that non-Abelian gauge theories with unbroken gauge symmetry are the only renormalizable quantum field theories that can have ASYMPTOTIC FREEDOM. It was stated by Sidney Coleman and David Gross in 1973.

collapse of the wave function *See* Copenhagen interpretation.

collective excitation A quantized mode in a system of many particles, such as a nucleus or a solid, in which there is cooperative motion of particles caused by interactions between the particles. PHONONS and PLASMONS in solids are examples.

collective nuclear model *See* nuclear structure.

collective oscillation A model of oscillation in a system of particles in which there is cooperative motion of particles because of interactions between them. Collective oscillations occur both in classical physics and in quantum mechanics. An example is a PLASMA OSCILLATION. *See also* collective excitation.

collimator A device for producing a parallel beam of radiation.

color *See* quantum chromodynamics; quark.

commute Two entities A and B are said to *commute* under some operation if the result of A followed by B is always the same as the result as B followed by A. For example, if A and B are ordinary numbers then A and B commute under the operations of addition and multiplication since A + B = B + A and A × B = B × A but A and B do not commute under the operations of subtraction and division.

complementarity The concept that an entity can have two distinct aspects which do not appear together in a single experiment. For example, both electrons and light behave as particles in some experiments and waves in other experiments. Complementarity is a characteristic feature of quantum mechanics. The concept was introduced by Niels BOHR in 1927.

complex conjugate If $z = x + iy$ is a complex number then the *complex conjugate* of z, denoted z^* is given by $z^* = x - iy$. The product of z and z^* satisfy $zz^* = x^2 + y^2$.

complexity The self-organization of a system that can arise from the interactions between the particles of that system. Complexity is frequently associated with BROKEN SYMMETRY and the existence of phase transitions between different states. SUPERCONDUCTIVITY, SUPERFLUIDITY, and FERROMAGNETISM, are examples of complexity.

complex number A number of the form $x + iy$ where x and y are real numbers and i is a quantity which is defined to be the square root of -1. x is called the *real part* and iy is called the *imaginary part* of the complex number. It is possible to represent a complex number by an *Argand diagram*, with the horizontal axis representing the real part and the vertical axis the imaginary part of the number.

complex variable A variable that is a function of complex numbers rather than real numbers. Complex variables have many applications in all branches of theoretical physics.

Compton, Arthur Holly (1892–1962) American physicist who discovered the COMPTON EFFECT in the early 1920s. Compton was awarded a share of the 1927 Nobel Prize for physics for this discovery.

Compton effect An increase in the wavelength of electromagnetic waves, particularly x-rays, when they are scattered by electrons. This effect was discovered by Arthur COMPTON in the early 1920s in the course of careful experiments on the scattering of x-rays. In 1923 Compton explained his result by using the quantum-mechanical idea that the x-rays behave like particles when they are scattered, losing some of their energy and thus increasing their wavelength. This explanation of the Compton effect was very influential in gaining acceptance for the idea that electromagnetic radiation has a dual wave–particle nature.

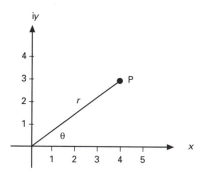

Complex number: the point P represents the number 4 + 3i. In polar form this is $r(\cos\theta + i\sin\theta)$.

Compton wavelength A length-scale that is characteristic of a particle in relativistic quantum mechanics. It is given by h/mc, where h is the Planck constant, m is the rest mass of the particle and c is the speed of light in a vacuum. The physical significance of the Compton wavelength is that it is not possible to locate a particle inside a distance similar to that of its Compton wavelength. The Compton wavelength of an electron is 2.426×10^{-12} m. The wavelength occurs in the theory of the COMPTON EFFECT.

conduction band *See* electronic structure of solids.

conductivity A measure of the ability of a material to conduct electricity (*electrical conductivity*) or to conduct heat (*thermal conductivity*).

configuration The arrangement of electrons in an atom or molecule. Configuration is determined by the AUFBAU PRINCIPLE and the PAULI EXCLUSION PRINCIPLE.

configuration interaction The mixing of wave functions of more than one electronic CONFIGURATION in a many-electron system such as an atom of molecule. The effects of electron CORRELATION can be taken into account using configuration interaction.

configuration space The n-dimensional space for a system with n DEGREES OF FREEDOM which has coordinates ($q_2, q_2, \dots , q_n$), where each q represents a degree of freedom. For example, if one has a gas of N atoms each atom needs three coordinates to specify its position. Thus, its configuration space is $3N$-dimensional. In the case of molecules the particles in a gas have internal degrees of freedom (vibration and rotation) and the configuration space has a higher dimension. The concept of configuration space is very useful in both classical mechanics and in statistical mechanics.

confinement *See* quantum chromodynamics.

Connes, Alain (1947–) French mathematician who invented NONCOMMUTATIVE GEOMETRY in the 1980s. Starting in the early 1990s Connes and his colleagues have applied noncommutative geometry to several topics in elementary particle theory including the STANDARD MODEL, SUPERSTRING THEORY, and RENORMALIZATION theory.

conservation law A law that states that the value of some quantity, such as mass, energy, or electric charge, is unchanged even though there is a series of changes in that system.

conservation of angular momentum The principle that in any process in a system the total angular momentum in the system is unchanged.

conservation of energy The principle that in any process in a system the total energy in the system is constant. This law is the first law of THERMODYNAMICS. It is of fundamental importance in physical science and has many manifestations and consequences. Because of the equivalence of mass and energy in special relativity theory it is necessary to include the rest masses of particles in the energy. *See also* conservation of mass.

conservation of mass The principle that in any process in a system the total mass in the system is constant. Because of the equivalence of mass and energy in SPECIAL RELATIVITY THEORY the law of conservation of mass is subsumed by the law of conservation of energy. In chemical reactions the mass change is so small that it cannot be detected. In nuclear reactions the mass change is large enough to be detected, with mass being converted into energy. In processes of PAIR CREATION or ANNIHILATION the changes in mass result in equivalent changes in energy.

conservation of momentum The principle that in any process in a system the total momentum in the system is conserved.

conservative system A closed system in which the mechanical energy is constant or an open system in which all external work is available as either kinetic energy or potential energy of the system.

consistent histories An interpretation of QUANTUM MECHANICS that asserts that the subset of all the possible ways in which an experimental result might have occurred that is consistent with the possible histories allowed by the laws of quantum mechanics is a small set and that predictions about the probabilities for these consistent histories can be made. This interpretation of quantum mechanics was put forward and developed by Robert Griffiths, Roland Omnes, James Hartle, and Murray GELL-MANN in the 1980s and 1990s. The consistent histories interpretation of quantum mechanics has attracted a great deal of interest.

containment 1. The prevention of the escape of radioactive matter from a container, such as a nuclear reactor.
2. The prevention of a plasma in a fusion reactor from touching the walls of the vessel. This is done using complicated configurations of electric and magnetic fields.

continuous group *See* group.

continuum A set of values of some quantity in which the values are not discrete but are continuous.

control rod A rod that can be moved into or out of the core of a NUCLEAR REACTOR to control the rate of the chain reaction. Control rods are made out of materials that absorb neutrons, such as boron or cadmium.

convergent series A series $a_1 + a_2 + ... + a_i$, for which the particle sum $Sn = a_1 + a_2 + ... + an$ tends to a finite limit as n tends to infinity. This finite limit is called the *sum* of the series. For example, the sum of the series $1 + 1/2 + 1/4 + 1/8 + ...$ is 2. It is a necessary condition for a series to be a convergent series that the individual terms ai tend to 0 as i tends to infinity but this is not a sufficient condition. For example, the series $1 + 1/2 + 1/3 + 1/4 + ...$ is a *divergent series*; i.e. a series which is not a convergent series. Many of the series encountered in physical science are divergent series.

Cooper, Leon Niels (1930–) American physicist who put forward the idea of Cooper pairs and developed the Bardeen–Cooper–Schrieffer (BCS) theory of SUPER-CONDUCTIVITY in the mid 1950s with John BARDEEN and Robert SCHRIEFFER. Cooper shared the 1972 Nobel Prize for physics with Bardeen and Schrieffer for this work.

Cooper pairs *See* superconductivity.

coordinate *See* coordinate system.

coordinate geometry A type of geometry in which points are located in space by a COORDINATE SYSTEM. This enables curves to be represented by an equation that describes the positions of all the points on the curve. This means that geometrical problems can be analyzed using algebra and vice versa.

coordinate system A system that specifies the position of points in space uniquely. The CARTESIAN COORDINATE system is the simplest coordinate system. Two coordinates are necessary to specify a point in a plane. Three coordinates are necessary to specify a point in three-dimensional space. In *n-dimensional* space n coordinates are necessary to specify a point. A point can be specified by many different coordinate systems and it is frequently the case that some coordinate systems are more convenient to use than others. For example, the SCHRÖDINGER EQUATION for the hydrogen atom cannot easily be treated using Cartesian coordinates but can be solved using spherical POLAR COORDINATES.

coordination number The number of atoms, ions, molecules, or functional groups surrounding a given (frequently central) atom or ion in a complex or crystal. For example, in an octahedral complex the coordination number of the central ion is eight.

Copenhagen interpretation The interpretation of quantum mechanics first put forward by Niels BOHR in 1927 and generally accepted as the conventional interpretation. It got its name because Bohr worked in Copenhagen and involves the interpretation of the wave function ψ in terms of probability (put forward by Max BORN), HEISENBERG'S UNCERTAINTY PRINCI-

PLE, and the importance of the process of measurement. The last aspect of the Copenhagen interpretation was emphasized by Bohr. He pointed out that the act of measuring the state of a system in quantum mechanics modifies that state. Bohr postulated that a quantum mechanical system exists in a superposition of states until a measurement is made. This measurement causes what is called a *collapse of the wave function*, formally known as *state-vector reduction*, forcing the system into one state, with that state being determined by probability. In other words, the system is in an indeterminate state until a measurement is made, at which point it assumes a 'real' state depending on probability.

Some scientists, including Albert EINSTEIN, were not satisfied with the Copenhagen interpretation of quantum mechanics and regarded it as provisional (*see* EPR experiment). It was to highlight what he considered to be the absurdities of the Copenhagen interpretation of quantum mechanics that Erwin SCHRÖDINGER put forward his paradox known as SCHRÖDINGER'S CAT in 1935. In the 1980s doubts about the Copenhagen interpretation became widespread, particularly in connection with QUANTUM COSMOLOGY, in which it is impossible to perform a measurement on the whole Universe. However, there is no consensus as to what should replace the Copenhagen interpretation as the correct interpretation of quantum mechanics.

copper A transition metal occurring in nature principally as the sulfide. Copper is used in electrical wires and in such alloys as brass and bronze. The two stable isotopes are ^{63}Cu (69.17%) and ^{65}Cu (30.83%).

Symbol: Cu. Melting pt.: 1083.5°C. Boiling pt.: 2567°C. Relative density: 8.96 (20°C). Proton number: 29. Relative atomic mass: 63.546. Electronic configuration: [Ar]$3d^{10}4s^1$.

core 1. A piece of iron or other magnetic material used to concentrate the magnetic lines of force in a transformer, electromagnet, or similar device.
2. The central part of a nuclear reactor in which the chain reaction occurs.

core orbital An atomic orbital that is associated with electrons in inner closed electron shells of an atom. Core orbitals of one atom have a small overlap with core orbitals on another atom; i.e. they play very little part in chemical bonding and are ignored in simple calculations of chemical bonds.

Cornell, Eric (1961–) American physicist who discovered a Bose–Einstein condensate in 1995 with Carl WIEMAN, in accord with a prediction made by Albert Einstein in 1925. The BOSE–EINSTEIN CONDENSATE found by Cornell and Wieman consisted of about 2000 rubidium atoms at 20 nanokelvin. Cornell, Wieman, and Wolfgang KETTERLE won the 2001 Nobel Prize for physics for their pioneering work on Bose–Einstein condensates.

correlation The mutual dependence of two or more events or phenomena. For example, in the free-electron theory of conductivity in metals the electrons are regarded as noninteracting. However, for a fully quantitative theory it is necessary to take into account that the motion of any one electron is affected by other electrons because of Coulomb repulsion (*see* Coulomb hole) and the Pauli exclusion principle (*see* Fermi hole). There are many phenomena in the theory of atoms, molecules, solids, and nuclei in which it is necessary to take the correlation between individual particles into account. *See also* correlation energy.

correlation diagram A diagram which shows how the energy levels of a system change as some parameter varies. Examples of varying parameters include the bond angle, the strength of a crystal field or ligand field around a central transition metal ion, and the distance between atoms. An important rule in constructing correlation diagrams is the *noncrossing rule*, which says that two electronic states do not cross if they have the same symmetry.

correlation energy The difference between the energy of a system given by HARTREE–FOCK THEORY and the true energy. In the case of helium, the correlation energy is 1.1eV. For systems with a very small number of electrons it is possible to incorporate the interelectronic distances into the wave functions to take electron correlation into account. This does not work for larger systems where CONFIGURATION INTERACTION is a more viable way of taking electron correlation into account.

correspondence principle The principle that the predictions of quantum theory must agree with the results of classical physics in the limits of particles with large masses and large quantum numbers. The correspondence principle was used by Niels BOHR in the development of the OLD QUANTUM THEORY and explicitly stated by him in 1923.

cosmic abundance The relative amount of a chemical element in the Universe, measured in terms of the number of atoms of each element. In the Solar System the abundance is 90.8% hydrogen, 9.1% helium, and 0.1% all other elements. *See also* abundance; nucleosynthesis.

cosmic censorship A hypothesis put forward by Sir Roger PENROSE in 1969 that a SINGULARITY of general relativity theory is always hidden behind an EVENT HORIZON of a BLACK HOLE. This hypothesis, known as the *cosmic censorship conjecture*, has not been proved in general, although there is some evidence in support of it. The cosmic censorship conjecture might not be relevant to observations since it may well be the case that singularities do not occur in QUANTUM GRAVITY.

cosmic microwave background A cosmic background of electromagnetic radiation in the microwave region discovered in 1965. It is thought to be a remnant of the big bang (see BIG-BANG THEORY). The existence of the cosmic microwave background provides important evidence in favor of the big-bang theory since its existence was predicted by George GAMOW and his colleagues in the 1940s as a consequence of the big-bang theory.

cosmic rays High-energy radiation that reaches the Earth from space. *Primary cosmic rays* mostly consist of nuclei, particularly protons, with electrons, positrons, neutrinos, and gamma rays also being present. These particles collide with oxygen and nitrogen nuclei in the atmosphere of the Earth, giving rise to *secondary cosmic rays*. The secondary cosmic rays mostly consist of protons, neutrons, pions, and gamma rays.

The primary cosmic rays have energies that range from 10^8 to 10^{20} eV. It is thought that the Sun is the main source of particles with energies less than 10^{10} eV and that all particles with energies less than 10^{18} eV originate within the Galaxy.

cosmic string A one-dimensional object that is predicted to occur in certain GRAND UNIFIED THEORIES. There is no evidence for the existence of cosmic strings. If cosmic strings exist they would have some important cosmological consequences.

cosmological constant A constant term that can be added to Einstein's field equations of GENERAL RELATIVITY THEORY. The cosmological constant was originally put forward by Albert EINSTEIN in 1917 to ensure that the application of general relativity theory to the Universe results in a static Universe rather than an expanding or contracting Universe. The discovery that the Universe is expanding removed the necessity for introducing the cosmological constant but cosmological models with a nonzero cosmological constant have been considered by theoreticians.

For many years it was thought that the value of the cosmological constant is exactly zero but, starting in the late 1990s, evidence began to accumulate that the cosmological constant has a small but nonzero value. This has the consequence that the expansion of the Universe is accelerating. There have been many attempts to show why the value of the cosmological constant is either zero or very small but there is no consensus as to why this should be the case.

cosmological principle The principle that the overall features of the Universe look the same anywhere in the Universe. The expansion of the Universe is an example of this principle since HUBBLE'S LAW will apply anywhere in the Universe. An extension of the cosmological principle put forward in the 1940s is the *perfect cosmological principle*, which asserts that the Universe looks the same overall at any time. However, there is substantial evidence that the Universe has changed with time, having started off with a big bang.

cosmology The study of the origin and evolution of the Universe. The theoretical study of cosmology requires a combination of general relativity theory, the theory of elementary particles, and nuclear physics.

Coulomb, Charles Augustin de (1736–1806) French physicist who established the law bearing his name for the force between two electrically charged particles (COULOMB'S LAW). This result was established independently by Henry CAVENDISH but Cavendish did not publish his findings whereas Coulomb did in 1785 and 1787.

Coulomb barrier The potential energy barrier that has to be overcome when one electrically charged body, such as a proton, approaches another body with the same sign of electric charge, such as a nucleus. It is possible for incoming particles that do not have sufficient energy to overcome the barrier to reach the target by the quantum-mechanical process of TUNNELING.

Coulomb blockade An effect that occurs in nanotechnology, in which a capacitor cannot be charged because the energy of getting an electron onto the capacitor is larger than the energies of the electrons in the rest of the circuit. The Coulomb blockade effect occurs because the electric charge is discrete and has a minimum size. When the coulomb blockade is combined with quantum-mechanical tunneling new types of quantum-mechanical electronic devices such as single electron transistor become possible.

Coulomb hole A region round an electron in an atom or molecule or in condensed matter, in which the probability of finding another electron is small because of the electrostatic Coulomb repulsion between electrons.

Coulomb's law The fundamental law of ELECTROSTATICS. It states that the force F between two particles with electrical charges Q_1 and Q_2 a distance d apart is given by $F = Q_1 Q_2 / (4\pi \varepsilon_0 d^2)$, where ε_0 is a constant, called the ELECTRIC CONSTANT or the permittivity of the vacuum (or free space). Coulomb's law is analogous to Newton's law of gravity, both being INVERSE SQUARE LAWS. The law is named after Charles de COULOMB who stated it in 1785.

coupling The interaction between two or more systems or between two different parts of one system. For example, in nuclear and atomic spectra effects occur as a result of RUSSELL–SAUNDERS COUPLING, J–J COUPLING, and SPIN–ORBIT COUPLING. In molecular spectra there are five HUND COUPLING CASES to consider. An example of coupling in solids is ELECTRON–PHONON COUPLING.

coupling constant A constant that gives a measure of the strength of the interaction between two or more systems or two different parts of one system. In FIELD THEORY the coupling constant is a measure of the force on a particle exerted by a field. An example of a coupling constant is the ELECTRIC CONSTANT.

covalent bond See CHEMICAL BOND.

covalent crystal A crystal in which the atoms of the crystal are held together by covalent bonds. Diamond is an example of a covalent crystal. Such crystals are frequently hard and have high melting points.

covalent radius An effective radius for an atom that is covalently bonded in a compound. Its value is taken to be half the distance between the nuclei of a diatomic molecule. For example, in Cl_2 the distance between the nuclei is 0.198 nm, which means that the covalent radius of a chlorine atom is 0.099 nm. It is frequently possible to obtain internuclear distances accurately by adding the covalent radii together.

Cowan, Clyde Lorrain (1919–74) American physicist who, together with Frederick REINES performed the first experiments that detected NEUTRINOS in 1956.

CPT theorem A theorem that states that the combination of charge conjugation C, parity P, and time reversal T, denoted CPT, is a symmetry of relativistic quantum field theory. The principles of relativistic quantum field theory are not affected when C, P, and T (or any two of them) are violated but are when CPT is violated. A violation of CPT has never been found experimentally.

CP violation The violation of the symmetry formed by the combination of charge conjugation C and parity P. CP violation is only observed to occur in weak interactions and has been extensively studied in KAONS. It is hoped that more information about CP violation will be obtained in future from experiments on B-MESONS. CP violation can be incorporated into the STANDARD MODEL of elementary particles, with a key feature of this description of CP violation being the existence of three GENERATIONS of elementary particles.

creation *See* pair creation.

creation operator An operator that is used to describe the creation of particles in the formalism of SECOND QUANTIZATION.

critical density *See* critical state.

critical mass The minimum mass of a particular fissile material that can sustain a nuclear CHAIN REACTION. In the case of uranium-235 the critical mass is 16 kg. A smaller amount of mass than the critical mass is called a *subcritical mass*. *See also* nuclear weapons.

critical pressure *See* critical state.

critical state The state of a fluid at which both the liquid and the gas phases have the same density. The density, pressure, and temperature of the fluid at the critical state are the *critical density*, *critical pressure*, and *critical temperature* respectively.

critical temperature 1. *See* critical state.
2. The temperature at which a PHASE TRANSITION occurs. The temperature at which the transition between the ferromagnetic state and the paramagnetic state occurs, known as the CURIE TEMPERATURE, is an example of this more general definition of a critical temperature. *See also* transition temperature.

Cronin, James Watson (1931–) American physicist who, together with Val FITCH, discovered CP VIOLATION in the weak interactions while investigating the decay of neutral KAONS. This unexpected discovery was published in 1964. Cronin and Fitch were awarded the Nobel Prize for physics in 1980.

cross product *See* vector product.

cross-section In a collision process – for example, the bombardment of nuclei by neutrons – the apparent area that particles present to the bombarding particles. This is not its 'true' cross-sectional area, but depends on the probability of a reaction occurring. In particular, it varies with the energy of the incident particles. The measurement is used for other types of collision reactions besides neutron absorption, including reactions of atoms, ions, molecules, electrons, etc. The unit is the square meter (m^2).

CRT *See* cathode-ray tube.

crystal A solid that has a regular geometrical shape resulting from the regular three-dimensional repeating arrangement in space of the particles (atoms, ions, or molecules) making up the crystal. This arrangement in space is called the *crystal structure*. The arrangement of the points in space at which the particles are located is called the *crystal lattice*.

crystal field theory A theory of the electronic structure of complexes consisting of a central metal atom or ion (frequently a TRANSITION METAL) surrounded by ions or molecules, called *ligands*. Crystal field theory is used to calculate how the energy levels of the central atom or ion are split by the ligands in terms of the electrostatic interactions between the central entity and the ligands. The theory has been used to explain the optical, spectroscopic, and magnetic properties of complexes with considerable success. It was initiated by Hans BETHE in 1929 and extensively developed in the 1930s. *See also* ligand field theory.

Curie, Marie (1867–1934), and **Curie, Pierre** (1859–1906) Marie Curie was a Polish-born French scientist who is best known for her pioneering studies of radioactivity with her husband Pierre. Together they discovered the radioactive elements polonium and radium. In addition to his work on radioactivity Pierre Curie made important contributions to our understanding of the electrical and magnetic properties of matter, notably the discovery of the CURIE TEMPERATURE. Pierre and Marie Curie shared the 1903 Nobel Prize for physics with Henri BECQUEREL for their work on radioactivity. Marie Curie also won the 1911 Nobel Prize for chemistry for her discovery of polonium and radium.

curie Symbol: Ci A former unit of radioactivity. It is equivalent to the amount of a radioactive substance in which there are 3.7×10^{10} disintegrations per second. This definition arises because it is the radioactivity of 1 gram of radium-226.

Curie temperature The temperature at which a ferromagnetic material loses its ferromagnetism and becomes paramagnetic. The Curie temperature is characteristic of the material; for example, it is

760°C for iron and 356°C for nickel. It is named for Pierre CURIE who discovered it in 1895.

curium A highly toxic radioactive silvery element of the actinoid series of metals. A transuranic element, it is not found naturally on Earth but is synthesized from plutonium. ^{244}Cm and ^{242}Cm have been used in thermoelectric power generators.

Symbol: Cm. Melting pt.: 1340±40°C. Boiling pt.: $\cong$ 3550°C. Relative density: 13.3 (20°C). Proton number: 96. Most stable isotope: ^{247}Cm (half-life 1.56 × 10^7 years). Electronic configuration: [Rn]$5f^7 6d^1 7s^2$.

curl (**rot**) An operator that gives the vector product of the gradient operator with a vector. If a vector V with components V_x, V_y and V_z in the x, y, and z directions, with respective unit vectors $\mathbf{i}$, $\mathbf{j}$ and $\mathbf{k}$ then curl V is given by:

$$\mathbf{i} \times \partial V_x/\partial x + \mathbf{j} \times \partial V_y/\partial y + \mathbf{k} \times \partial V_z/\partial z.$$

current The flow of electric charge.

curved space–time The combination of space and time that describes general relativity theory in which the geometry is locally that of MINKOWSKI SPACE–TIME but can be curved globally.

cyclic accelerator An accelerator that is circular in shape, such as a CYCLOTRON or SYNCHROTRON.

cyclotron A device for accelerating charged particles to high energies. Protons can be given energies of over 10 MeV, deuterons over 20 MeV, and alpha particles about 40 MeV. Particles are injected near the center of an evacuated space between two D-shaped boxes placed between the poles of a strong permanent magnet. Within each 'dee' the particles describe a semicircular orbit. An alternating voltage is applied to the dees at such a frequency that the particles are accelerated by the potential difference each time they reach the gap, causing them to increase their path radii and speeds in steps. Eventually, after

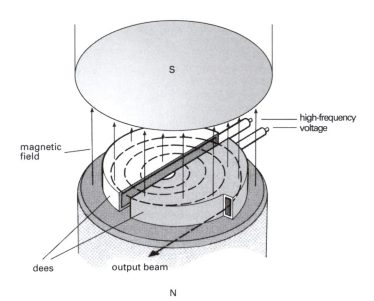

several thousand revolutions, they reach the perimeter of the dees at high speed, where a deflecting electric field directs them onto the target.

The energies that can be achieved are limited by the relativistic increase in mass of the particles. As their velocity approaches that of light, the increase in mass causes the period of rotation to increase so that they no longer reach the gaps in phase with the potential difference. *See also* synchrocyclotron.

Dalton, John (1766–1844) English chemist who put forward the idea that chemical reactions can be understood in terms of the concept of atoms, with the atoms of different chemical elements being characterized by different weights. He first stated his atomic theory in 1803 in his book *New System of Chemical Philosophy*, the first volume of which was published in 1808. Dalton proposed that matter is made up of atoms, that atoms cannot be created or destroyed, that atoms of different elements have different weights, and that chemical reactions are rearrangements of atoms. Dalton was able to explain the observed laws of chemical reactions, such as the conservation of mass, in terms of these postulates. He is also known for *Dalton's law of partial pressures*, which states that in a mixture of gases the total pressure is the sum of the pressures that each of the constituents of the mixture would exert if it was the only gas present in the container.

Dalton *See* atomic mass unit.

dark matter Matter in the Universe that cannot be detected directly because it does not emit or absorb electromagnetic radiation. There is a considerable amount of evidence that over 90% of the matter in the Universe consists of dark matter. Dark matter is detected by its gravitational effects. Some of this dark matter (sometimes called *missing matter*) may be made of BARYONS. However, the amount of helium in the Universe produced just after the big bang (*see* nucleosynthesis) indicates that most of the dark matter must be composed of particles that are not baryons. Many suggestions have been made as to what this nonbaryonic dark matter might be, but at present there is no established consensus on this topic. Candidates for dark matter have been divided into *cold dark matter*, sometimes known as WEAKLY INTERACTING MASSIVE PARTICLES (WIMPS), which consists of particles moving slowly compared to the speed of light, and *hot dark matter*, which consists of particles moving near to the speed of light. The discovery that neutrinos have mass means that they make up at least some of the hot dark matter.

dating The determination of the age of archeological or geological specimens. A number of techniques depend on measuring the presence of the decay product of some radioactive nuclide. Examples of radioactive dating include CARBON DATING, POTASSIUM–ARGON DATING, RUBIDIUM–STRONTIUM DATING, and URANIUM–LEAD DATING.

daughter A nuclide that is produced by the radioactive decay of another nuclide (called the *parent*).

Davisson, Clinton Joseph (1881–1958) American physicist who performed an experiment with Lester Germer demonstrating the wave nature of electrons. Davisson shared the Nobel Prize for physics in 1937 with George THOMSON.

Davisson–Germer experiment A classic experiment performed in 1927 with Lester Germer, in which a beam of electrons of known energy was directed at an angle onto a nickel surface. The angles at which electrons were reflected were measured and the results showed that the electrons were being diffracted by the regular surface lattice of the crystal. The experiment demonstrated that electrons could have wave properties, in accordance with the ideas of DE BROGLIE.

Davy, Sir Humphry (1778–1829) English chemist who discovered the elements sodium and potassium using electrolysis, thus demonstrating that there is a close connection between chemistry and electricity. He is also remembered for the *Davy lamp*, a safety lamp for coal miners. A major indirect contribution of Davy to science was his appointment of Michael FARADAY as his assistant in 1813.

d-block elements The block of elements in the PERIODIC TABLE consisting of three periods of elements: from scandium to zinc, from yttrium to cadmium and from lanthanum to mercury. These elements all have *d*-electrons in their penultimate shell and two outer *s*-electrons. See also TRANSITION METAL.

D-brane An extended object with D extended dimensions that arises in SUPERSTRING THEORY. D-branes can be regarded as SOLITON-like solutions of superstring theories with dimensions ranging from one to nine.

de Broglie, Prince Louis-Victor Pierre Raymond (1892–1987) French physicist who suggested in 1923 that particles of matter, such as electrons, could also behave like waves. This idea is fundamental in quantum mechanics and underlies electron diffraction (*see* Davisson–Germer experiment). De Broglie was awarded the Nobel Prize for physics in 1929.

de Broglie wavelength The wavelength associated with a particle in quantum mechanics. The wavelength λ is related to the momentum p of the particle by $\lambda = h/p$, where h is the PLANCK CONSTANT.

debye Symbol: D A unit of electric dipole moment equal to $3.335\,64 \times 10^{-30}$ coulomb meter. It is used in expressing the dipole moments of molecules.

Debye, Peter Joseph Willem (1884–1966) Dutch-born American chemist and physicist. Debye was a wide-ranging theoretician, with a common theme in his work being the use of physical techniques to elucidate the structure and properties of matter at the molecular level. He contributed to the theory of the specific heats of solids, x-ray diffraction, electrolytes, polymers, and dielectric constants and dipole moments. Debye won the 1936 Nobel Prize for chemistry for his work on molecular structure.

Debye–Hückel theory A theory of electrolytes in which it is postulated that ionization is complete and that each ion in the electrolyte is surrounded by a cluster of oppositely charged ions. The Debye–Hückel theory is based on a combination of electrostatics and statistical mechanics. It works well for very dilute electrolytes but is inadequate to describe strong electrolytes. The theory was put forward by Peter DEBYE and Erich Hückel in 1923.

Debye–Scherrer method A technique used in the determination of crystal structure by x-ray diffraction, in which a beam of x-rays is directed onto a powdered sample of the material. The powder contains crystals in all orientations and consequently the photographic plate that records the diffracted x-rays shows a pattern of concentric circles. The Debye–Scherrer method enables the dimensions of unit cells in crystals to be established very accurately. The technique, which is sometimes known as the *powder method*, was suggested by Peter DEBYE and Paul Scherrer in 1916.

Debye temperature A quantity with the dimensions of temperature that occurs in the DEBYE THEORY OF SPECIFIC HEAT. It is defined by $\theta = h\nu/k$, where h is the Planck constant, ν is the maximum frequency of the vibrations of the lattice under consideration and k is the Boltzmann constant. The Debye temperature is characteristic for a particular substance. For example, the Debye temperature for copper is 315 K.

Debye theory of specific heat A theory of the specific heat capacities of solids put forward by Peter DEBYE in 1912. He explained the specific heat capacity in terms of the vibrations of the lattice, with there being a distribution of frequencies for these vibrations up to some maximum frequency. The Debye theory gives reasonably good agreement between theory and experiment.

Debye–Waller factor A factor that occurs in determining the intensities of diffracted x-rays. The Debye–Waller factor is the effect on intensities resulting from thermal or quantum fluctuations in the crystals. It was introduced and calculated by Peter DEBYE in 1913 and elaborated on by Ivar Waller in 1923.

decay The process by which an entity such as a radioactive nucleus or an unstable particle transforms into another nucleus or particle, with the release of one or more particles and/or energy. It is frequently the case that decay is exponential with time, i.e. the number N of the original nuclei or particles at a time t is given by $N = N_0 \exp(-\gamma t)$, where N_0 is the original number and γ is a constant characterizing the time-scale of the decay, known as the *decay constant* or the *disintegration constant*. The decay constant is related to the HALF-LIFE $T_{1/2}$ of the decay by $T_{1/2} = 0.69315/\gamma$.

decoherence The process in which a quantum system interacts with its environment to become a system that has classical behavior. For example, consider a particle and a detector. According to the Copenhagen interpretation of quantum mechanics, the particle is in an indeterminate state until the measurement is made; i.e. the particle has a superposition of two quantum states. When it interacts with the detector, the combined system should also have a superposition of states with the detector indicating two different readings at the same time. The fact that this does not happen is ascribed to decoherence. Since the detector is a macroscopic object it interacts with its environment destroying the coherence of the superposed states. The concept of decoherence has been extensively discussed since the 1980s and has helped clarify the foundations of quantum mechanics. *See also* Schrödinger's cat.

deep inelastic scattering Inelastic scattering that occurs at very high energies. The analysis of deep inelastic scattering experiments led to the development of the PARTON model.

degeneracy The occurrence in a system of more than one quantum state with the same energy. For example, in an isolated atom, the three p-orbitals all have the same energy and are said to be *degenerate orbitals*. If an external magnetic field is applied, or if the atom is present in a compound, these orbitals may have different energies. The degeneracy is then said to be 'lifted'. The degeneracies of the energy levels of a quantum-mechanical system are determined by the symmetry of the system using GROUP THEORY.

degeneracy pressure The pressure that exists in a DEGENERATE GAS of fermions because of a combination of the Heisenberg uncertainty principle and the Pauli exclusion principle. The exclusion principle means that at high density two fermions close to each other must have different momenta, while the uncertainty principle means that the momentum difference is inversely proportional to the spacing. Thus, since a high-density gas has a small spacing the particles have large relative momenta and hence a much larger pressure than the classical pressure due to heat (*thermal pressure*). Degeneracy pressure prevents white dwarf stars and neutron stars from undergoing gravitational collapse to black holes.

degeneracy temperature The temperature below which the effects of degeneracy in a quantum-mechanical system manifest themselves. Above the degeneracy temperature these effects can be ignored and the system can be analyzed classically.

degenerate gas A gas made up of particles governed by quantum mechanics that is sufficiently dense that the behavior of the gas is governed by QUANTUM STATISTICS rather than by the classical MAXWELL–BOLTZMANN DISTRIBUTION. Examples of degenerate gases include the neutrons in a neutron star, the electrons in a white dwarf star, and the conduction electrons in a metal.

degenerate level An energy level in a system that is shared by two or more different quantum states.

degenerate states Quantum states of a system that all have the same energy level.

de Gennes, Pierre Gilles French theoretical physicist. By using concepts from the theory of phase transitions de Gennes was able to make important advances in our understanding of liquid crystals and polymers. He was awarded the 1991 Nobel Prize for physics.

degrees of freedom The set of independent coordinates of a dynamical system needed to specify the state of the system. Each degree of freedom is represented by one of these coordinates. For example, a single particle such as an atom has three degrees of freedom. These correspond to the three coordinates in space needed to specify its position (or to three modes of translational motion).

In general, a molecule with N atoms has $3N$ degrees of freedom. Thus, in a diatomic molecule there are six degrees of freedom. Three of these degrees of freedom correspond to different directions of translational motion. In addition, there are two degrees of freedom for rotation of the molecular axis and one degree of freedom for vibration along the bond between the atoms. A linear polyatomic molecule has three translational degrees of freedom, two rotational degrees of freedom and $(3N - 5)$ vibrational degrees of freedom. A nonlinear polyatomic molecule has three translational degrees of freedom, three rotational degrees of freedom and $(3N - 6)$ vibrational degrees of freedom.

The number of degrees of freedom determines the number of ways a system can take up energy. *See also* equipartition of energy.

delta bond A chemical bond that involves delta (δ) orbitals. A δ orbital is the analog for molecules of the d-orbitals for atoms, having two units of orbital angular momentum around an internuclear axis. Delta bonds are formed by the overlap of d-orbitals of different atoms and occur in certain cluster compounds of transition metals.

Democritus (*c.* 460–370 BC) Ancient Greek philosopher who developed atomic theory. He postulated that the world is made up of a very large number of very small uncuttable particles (atoms) moving in a void (the vacuum), with all the variety of matter that is observed in the world being the result of atoms combining in different ways.

density functional theory A method of calculating the energy of a many-electron system as a FUNCTIONAL of the electron density. Density functional theory has been used extensively to calculate properties of molecules and solids and has been extended to nuclei. Density functional theory was initiated by Walter KOHN and his colleagues in the mid 1960s.

density parameter Symbol: ω A key parameter in determining the fate of the Universe. It is defined as the ratio of the actual density of matter in the Universe to the critical density, where the critical density is the density of matter above which an expanding Universe will eventually recollapse. It is currently thought that the value of ω is about 0.3. Before 1980 it was one of the mysteries of cosmology as to why the value of ω is fairly close to 1. The theory of INFLATION provides an explanation of why this is the case.

Descartes, René (1596–1650) French philosopher, mathematician, and scientist who was one of the main pioneers of COORDINATE GEOMETRY. This branch of mathematics is also called *Cartesian geometry* in his honor. Descartes also made important contributions to optics. Much of the work of Descartes is contained in his classic book *A Discourse on the Method of Rightly Constructing Reason and Seeking Truth in Science*, usually known as *The Method*, which was published in 1637.

de Sitter, Willem (1872–1934) Dutch astronomer and cosmologist. He was one of the first astronomers to appreciate the relevance of both special and general relativity theory to the motions of planets and to use general relativity theory to construct a model of the Universe (*see* de Sitter space).

de Sitter space A model of the Universe constructed by Willem de Sitter based on GENERAL RELATIVITY THEORY. Originally it was thought that de Sitter space represented an empty static Universe. However, in the early 1920s it was realized that if matter was put into the model then the particles of matter would recede from each other very rapidly as the Universe expanded. When the actual expansion of the Universe was subsequently discovered it was found that the expansion is not as rapid as in de Sitter space. However, when INFLATION was proposed in the 1980s it was found that in the early Universe there was a period of very rapid expansion described by de Sitter space.

DESY (Deutsches Elektronen Synchrotron) A particle accelerator sited in a suburb of Hamburg. *See* DORIS.

detailed balance A state of equilibrium in which the effect of one process is cancelled by another process that operates with the opposite effect at the same time. For example, there is detailed balance for the chemical reaction A + B → C + D, where A, B, C, D are molecular species, if the rate at which the reaction occurs is equal to the rate at which the opposite reaction C + D → A + B occurs.

determinant A square ($n \times n$) array of numbers which can be derived from a square matrix. Determinants have the property that if two columns (or rows) of a square matrix are interchanged then the sign of the determinant of the matrix is changed. This property of determinants makes it convenient to use them to express wave functions in quantum mechanics.

determinism The view that all events are determined by preceding events. This philosophy underlies much of physics. However, both quantum mechanics and chaos theory make determinism less clearcut. *See also* Laplacian determinism.

deuterium Symbol: D The isotope of hydrogen in which the nucleus consists of one proton and one neutron. The mass of a deuterium atom is about twice the mass of the much more common form of hydrogen atom in which the hydrogen nucleus is one proton. For this reason deuterium is frequently known as *heavy hydrogen*. The chemical properties of deuterium are similar to hydrogen, although reactions involving deuterium compounds tend to be slower than with the analogous hydrogen compounds. The compound D_2O (deuterium oxide) is known as *heavy water*. Most physical properties of deuterium are similar to those of hydrogen.

deuteron The nucleus of a DEUTERIUM atom. It consists of one proton and one neutron.

diamagnetism See MAGNETISM.

diatomic molecule A molecule with two atoms. Gaseous elements such as hydrogen, oxygen, nitrogen, fluorine, and chlorine exist as diatomic molecules (H_2, O_2, etc.).

dichroism The property of certain materials of absorbing light that is polarized in one plane but transmitting light polarized in a perpendicular plane. A material with this property is said to be a *dichroic material*. Tourmaline is a naturally occurring dichroic mineral while Polaroid is an example of a synthetic dichroic material.

dielectric A name sometimes given to a nonconductor of electricity, i.e. an insulator.

dielectric constant *See* relative permittivity.

diffeomorphism A continuous transformation of a space that moves the points of space around but preserves the relationships between them that are used to define which points are close to one another.

diffeomorphism group The diffeomorphism group of a space described by Riemannian geometry can be regarded as the analog of the LORENTZ GROUP associated with MINKOWSKI SPACE–TIME, with a difformorphism being analogous to a LORENTZ TRANSFORMATION. Thus, the diffeomorphism group plays the same role in general relativity theory that the Lorentz group does in special relativity theory.

diffraction An effect occurring when light or other electromagnetic radiation passes an obstacle. The shadow edge is less sharp than expected because of the wave nature of the radiation. Some radiation bends round the obstacle into the shadow region and, in part of the nonshadow region, there is less intensity. If the obstacle is of similar size to the wavelength of the radiation or if radiation is passed through a narrow slit it is possible to get dark and bright fringes (*diffraction patterns*) very close to the edges of the shadow as a result of INTERFERENCE.

diffraction grating A device for producing diffraction of radiation. Typically it consists of a piece of glass on which are ruled a large number of closely spaced parallel lines using a diamond point (around 7000 per cm). If light is passed through the glass at right angles most is transmitted undeviated but INTERFERENCE also occurs for waves diffracted by the grooves, causing maxima and minima. The angles (θ) at which these occur depend on the wavelength λ according to the equation

$$n\lambda = d \sin \theta$$

where d is the distance between the lines of the grating and n is an integer (1, 2, 3, etc.).

diffusion The movement of matter as a result of the random thermal motion of the particles that make up the matter. Diffusion occurs quickly in gases, more slowly in liquids, and very slowly in solids at normal temperatures.

diode An electronic device with electrodes that only allows current to flow to any extent in one direction when a potential difference is applied. There are two main types of diode:
(1) In a *thermionic diode* a heated cathode emits electrons by THERMIONIC EMISSION to an anode when a positive potential is applied to it. This property of thermionic diodes was used in early electronic devices.
(2) In a *semiconductor diode* there is a single p–n junction (*see* transistor).

dipole Two equal and opposite charges that are separated by a distance. If the charges are electric charges the dipole is an *electric dipole*. If the charges are magnetic charges (i.e. magnetic poles) the dipole is a *magnetic dipole*.

dipole moment Symbol: μ The product of one of the charges of an electric dipole and the distance between the charges. Some molecules such as hydrogen chloride (HCl) act as dipoles because electric charge is attracted to the more electronegative atom (chlorine in this example). Thus the molecule has a permanent dipole moment, with a small positive charge on the hydrogen atom and a small negative charge on the chlorine. An *induced dipole moment* can be produced in a molecule by the application of an external electric field. The measurement of dipole moments is useful in the study of molecular structure.

dipole–dipole interaction The interaction between two molecules as a result of their dipole moments. The strength of this interaction depends on the size of the dipole moments, their distance apart, and relative orientation. Even if a molecule does not have a permanent dipole moment, dipole–dipole interaction is still possible because the presence of a nearby molecule can lead to an induced dipole moment in the first molecule and hence to *induced dipole–dipole interactions. See also* van der Waals' forces.

Dirac, Paul Adrien Maurice (1902–84)   British theoretical physicist who was one of the founders of quantum mechanics and quantum field theory. Between 1925 and 1927 he developed a formulation of quantum mechanics based on OPERATORS. He showed that the formulations of quantum mechanics in terms of MATRIX MECHANICS and WAVE MECHANICS could be derived from his operator formulation. In 1926 he worked out the rules governing many-fermion systems (*see* quantum statistics). In 1927 Dirac pioneered the quantum theory of radiation in two papers, these papers being among the first in QUANTUM FIELD THEORY.

In 1928 Dirac produced his greatest contribution to physics when he found an equation, now known as the DIRAC EQUATION, which combines quantum mechanics and special relativity theory. The equation had two sets of solutions, one of which corresponded to positive energy electrons, while the other corresponded to negative energy electrons. The physical significance of the negative energy electrons was not immediately apparent but in 1931 Dirac postulated that are positively charged ANTIPARTICLES of the electron. This antiparticle, called the POSITRON, was subsequently discovered.

Dirac made a number of subsequent contributions to theoretical physics. He shared the 1933 Nobel Prize for physics with Erwin SCHRÖDINGER for his contributions to quantum mechanics.

Dirac constant Symbol: $\hbar$;   The PLANCK CONSTANT (h) divided by 2π. The Dirac constant is sometimes called the *rationalized Planck constant* and has a value of 1.054589×10^{-34} Js. The Dirac constant is frequently a more convenient quantity to use than the Planck constant.

Dirac equation The equation that describes the RELATIVISTIC QUANTUM MECHANICS of spin-1/2 particles. The equation was put forward by Paul DIRAC in 1928. The existence of antiparticles and the occurrence of pair creation and annihilation of particle–antiparticle pairs as characteristic features of relativistic quantum mechanics are contained in the Dirac equation.

Dirac–Kapitza effect The scattering of particles by light. Since it is a feature of quantum mechanics that both matter and light have particle-like and wavelike properties Paul DIRAC and Pyotr KAPITZA predicted in 1933 that particles such as electrons should be scattered by light. This prediction was confirmed experimentally in 1986.

Dirac matrices A set of 4×4 matrices which is associated with the DIRAC EQUATION.

Dirac quantization condition A quantization condition put forward by Paul DIRAC in 1931 which allows the existence of magnetic monopoles in quantum electrodynamics (QED). Dirac showed that a magnetic monopole of strength g is allowed in QED if $eg = nhc/4\pi$ where e is the electron charge, h the Planck constant, c the speed of light in a vacuum, and n is an integer. The Dirac quantization condition was the first example of a TOPOLOGICAL QUANTIZATION CONDITION and can be generalized from QED to NON-ABELIAN GAUGE THEORIES.

Dirac spinor A four-component SPINOR that is a solution of the DIRAC EQUATION.

direct reaction *See* nuclear reaction.

discrete Describing a quantity in which only discontinuous values can occur.

discrete group See GROUP.

disintegration A process by which an atomic nucleus breaks up, either spontaneously or as a result of bombardment by another particle.

disordered system A material that does not have the order of a perfect crystal or a crystal with isolated defects. There are several types of disorder. In a *random alloy* the order of the different types of atom is random. Glass is an example of an *amorphous solid*, i.e. a random network of atoms. Other types of disordered solid have high concentrations of random defects. A liquid is another type of disordered system since the atoms in the liquid can move about and are not tied to a lattice.

dispersion The spreading out of light rays of different wavelengths when they undergo refraction. An example of dispersion is the production of colored light from the light of the Sun using a prism. Dispersion also causes chromatic aberration in lenses, i.e. an image is surrounded by colored images. Dispersion occurs because light of different wavelengths travels at different speeds in the refracting medium. Usually, the refractive index increases as the wavelength decreases. However, there may also be regions, known as regions of *anomalous dispersion*, in which refractive index *decreases* as the wavelength decreases.

dispersion forces *See* van der Waals' forces.

dissociation The splitting up of a molecule into smaller molecules or atoms or ions.

distribution function A function that gives the distribution of the values of some quantity using statistics. For example, distribution functions are used to describe the distribution of speeds of the molecules of a gas. Distribution functions are used to construct theories of the liquid state, starting from the intermolecular forces between the molecules of the liquid.

div *See* divergence.

divergence 1. (**div**) The scalar product of the gradient operator ∇ and a vector. If V is a vector with components V_x, V_y, and V_z in the x, y, and z directions, then:

$$\text{div } V = \partial Vx/\partial x + \partial Vy/y + \partial Vz/\partial z$$

Divergence is useful in many areas of modern physics.

2. An infinite result obtained when an integral is evaluated. Divergences of this sort occur both in the classical theory of the electron and in quantum field theory.

divergent series *See* convergent series.

Döbereiner, Johann Wolfgang (1780–1849) German chemist whose observations on triads of chemical elements, now known as DÖBEREINER TRIADS, helped to pave the way for the periodic table of the elements. Döbereiner was also a pioneer of the use of PLATINUM as a catalyst.

Döbereiner triads Combinations of three chemical elements in which one of the elements has chemical and physical properties which lie between those of the other two. In 1829 Döbereiner noticed that the properties of bromine, which had then been recently discovered are intermediate between those of chlorine and iodine. Similarly, he noticed that the properties of strontium are intermediate between those of calcium and barium and that sulfur, selenium, and tellurium also formed a triad. The concept of Döbereiner triads helped to pave the way for the development of the periodic table.

domain A small (about 1 millimeter across) region of a ferromagnetic material inside which all the atoms are magnetized in the same direction. In the absence of an external magnetic field the separate domains are magnetized in different directions. When a strong external magnetic field is applied the domains align themselves in the same direction, with domains which have magnetic axes in nearly the same direction as the external field growing at the expense of other domains.

domain wall The boundary of a DOMAIN. A domain wall is sometimes called a *Bloch wall* after the pioneering work of Felix BLOCH on magnetism.

donor *See* semiconductor.

Doppler effect An apparent change in the frequency of a wave due to relative motion between the source of the wave and an observer. The changes in pitch of an ambulance siren as it approaches and then passes an observer occur because of the Doppler effect. The Doppler effect is named after Christian Johann Doppler (1803–1853) who predicted it in 1842. He also predicted that the effect occurs for light waves. This is used in astronomy with a blue shift (to shorter wavelengths) being caused by approaching stars and a red shift (to longer wavelengths) being caused by receding stars. For electromagnetic waves it is necessary to use special relativity theory to describe the Doppler effect.

***d*-orbital** *See* orbital.

DORIS Double Ring Storage Facility; a particle accelerator built at DESY in the mid 1970s. DORIS consisted of two rings, with one ring on top of the other. Electrons were accelerated in one ring and positrons in the other. The accelerated electrons were then made to collide with the accelerated positrons. DORIS operated at an energy of 7 GeV, which enabled it to help to demonstrate the existence of the TAU particle.

dose The amount of energy from ionizing radiation that is absorbed by a unit mass of matter. People who are exposed to ionizing radiation, such as employees in the nuclear industry and radiologists, are subject to a *maximum permissible dose* in a specified period. This upper limit is determined by the International Commission on Radiological Protection.

double bond A chemical bond in which atoms share two pairs of electrons.

double refraction (birefringence) The optical property of certain crystals, such as calcite, of forming two refracted rays from one incident ray of light. One ray is called the *ordinary ray* while the other ray is called the *extraordinary ray*. Double refraction can be explained in terms of the in-

teraction of electromagnetic waves with the structure of the crystal.

double-slit experiment An experiment involving two slits which Richard FEYNMAN regarded as providing a clear demonstration of the mystery of quantum mechanics.

In the early nineteenth century Thomas YOUNG demonstrated the wave nature of light by producing an interference pattern. He shone a light onto a screen with a single slit in it. After going through the slit the light then reached a second screen with two slits. The light from the second screen falls on a third screen, on which the alternating bright and dark pattern characteristic of INTERFERENCE of waves is detected.

In the case of particles one would expect that this type of experiment would result in particles piling up behind each of the holes, with there being no interference pattern. However, when a double-slit experiment is performed with electrons it is found that an interference pattern also results. Nevertheless, when an electron arrives at a particular place on the detector screen it does so as a particle.

The quantum mechanical nature of this experimental result becomes even clearer when single electrons of photons go through the experimental apparatus one at a time, with the pattern on the detector screen emerging slowly. Each particle makes a single spot on the detector screen. However, over time an interference pattern emerges characteristic of wave behavior.

If the experiment is altered so that it is possible to detect which of the slits each

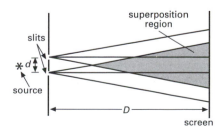

Double-slit experiment

particle goes then an interference pattern does not emerge and the particles finish in two piles behind the slits.

These experiments which are characteristic of quantum mechanics have been performed for protons, neutrons, and atoms as well as for electrons and photons.

doublet A pair of spectral lines in which the splitting between the lines occurs because of the spin of an electron. An example of a doublet is the pair of lines that make up the bright yellow sodium D line.

down quark *See* quark.

dressed charge *See* renormalization.

dressed mass *See* renormalization.

drift chamber A device for detecting particles, used with particle accelerators. A simple example has two wires, one positively charged and the other negatively charged, in a chamber containing gas. The wires are arranged so that there is a uniform electric field between them. A particle entering the chamber between the two wires causes ionization and the electrons produced 'drift' toward the anode (positive wire). The anode is a very thin wire, giving a high electric field in its vicinity, and as the electrons get very close they accelerate producing further ionizations. The resulting avalanche gives a measurable signal at the anode (as in a PROPORTIONAL COUNTER). The initial time of the event can be found by using a scintillation counter associated with the drift chamber and the time of drift can be used to find out where in the chamber the event occurred. Drift chambers can be quite complicated pieces of equipment with multiwire configurations. They were developed by Georges CHARPAK at CERN in the 1970s.

duality In quantum mechanics, the ability of an entity to exhibit both particle-like and wavelike behavior (*see* wave–particle duality). The idea of duality is also used in other branches of modern physics. For example, in certain models for magnetic phase transitions there is a duality relating quantities in the low-temperature phase to quantities in the high-temperature phase in such a way that it gives information about the transition point. In GAUGE THEORIES there are dualities, seen most clearly when SUPERSYMMETRY is present, between the formulation of the theories in terms of electric-like and magnetic-like variables. In SUPERSTRING THEORY it has been found that there are duality relations between the five possible superstring theories in 10 space–time dimensions and also SUPERGRAVITY THEORY in 11 space–time dimensions. This realization has led to the development of M-THEORY and speculations concerning duality relations between fundamental theories of particle physics, which describe the bulk of a system and theories which describe the surface of that system.

dubnium A radioactive synthetic element made by bombarding ^{249}Cf nuclei with ^{15}N atoms. It was first reported by workers at Dubna, a town near Moscow.

Symbol: Db. Proton number: 105. Most stable isotope: ^{262}Db (half-life 34 s). Electronic configuration: [Rn]$5f^{14}6d^37s^2$.

Dulong–Petit law The principle that for a chemical element in the solid state the product of the relative atomic mass and the specific heat capacity is a constant approximately equal to $3R$ (25 J mol^{-1} K^{-1}), where R is the gas constant. This law only applies to elements with simple crystal structures at normal temperatures. At low temperatures the specific heat capacity is temperature dependent and it is necessary to use quantum theory to get an accurate description of the behavior. The Dulong–Petit law was stated in 1819 by Pierre Dulong (1785–1838) and Alexis–Thérèse Petit (1791–1820).

dynamical system A system governed by some form of dynamics. A dynamical system can either be classical, as in the Solar System, or quantum mechanical, as in an atom. The behavior with time of dynamical systems, even those with a small number of DEGREES OF FREEDOM, can be complex. It is possible to study the evolution of dynamical systems using the concept of PHASE SPACE, with topology being a useful tool for this study. *See also* chaos theory.

dynamical symmetry breaking Symmetry breaking that occurs because of the dynamics of a system. Examples of dynamical symmetry breaking include the Bardeen–Cooper–Schrieffer theory of superconductivity and the breaking of chiral symmetry in quantum chromodynamics. It has frequently been suggested that the Higgs mechanism in the STANDARD MODEL of elementary particles should be implemented by a dynamical symmetry breaking mechanism involving bound states of fermions but this suggestion has never been implemented successfully.

dynamics The branch of mechanics that is concerned with how bodies move under the action of forces, i.e. with the relation between force and motion.

dyon A MAGNETIC MONOPOLE that also has an electric charge. There is no experimental evidence for the existence of dyons. They can occur in quantum electrodynamics, grand unified theories, and other theories which attempt to unify the fundamental interactions.

dysprosium A soft malleable silvery element of the lanthanoid series of metals. It occurs in association with other lanthanoids. One of its few uses is as a neutron absorber in nuclear reactors; it is also a constituent of certain magnetic alloys.

Symbol: Dy. Melting pt.: 1412°C. Boiling pt.: 22562°C. Relative density: 8.55 (20°C). Proton number: 66. Relative atomic mass: 162.50. Electronic configuration: $[Xe]4f^{10}6s^2$.

Dyson, Freeman John (1923–) British-born American physicist who made important contributions to clarifying the procedure of RENORMALIZATION of quantum electrodynamics in the late 1940s.

e A number that is the base of natural logarithms. It can be defined in several ways; one is as the sum of the series $1 + 1/1! + 1/2! + \dots + 1/n! + \dots$. The value of e is approximately 2.718 281 83. It has been proved that e is an irrational number and that it is also a transcendental number, i.e. a real number that is not a root of a polynomial equation with integer coefficients. The number e often occurs in mathematics and its physical applications. *See also* exponential function.

early Universe The very short period of time immediately following the BIG BANG. The theoretical study of cosmology in the early Universe combines general relativity theory with the theory of elementary particles. It has been mutually beneficial for both cosmology and elementary-particle theory, especially in work on GRAND UNIFIED THEORIES.

Because the temperatures in the early Universe were very high, symmetries that now appear as broken symmetries were unbroken at these temperatures. It is thought that, as the Universe cooled after the big bang, there was a sequence of phase transitions from which broken symmetry states emerged. It is also thought that the lightest chemical elements were formed by nuclear reactions soon after the big bang (*see* nucleosynthesis), and that as the Universe cooled heavier atoms were formed.

A particular area in which this synthesis of cosmology and particle physics has been productive is in the development of ideas about the problem of matter–antimatter asymmetry. Why does the present Universe appear to consist only of 'normal' matter and no antimatter (except for that produced artifically)? One approach is to combine cosmology with Grand Unified Theories, using SAKHAROV'S CONDITIONS. Another is that the matter–antimatter asymmetry in the early Universe may be explained using a combination of cosmology and the STANDARD MODEL of elementary particles. At present it is not known which (if either) of these approaches is correct.

A very important idea concerning the early Universe is that of INFLATION, in which it is postulated that the Universe underwent a period of very rapid expansion shortly after the big bang. This hypothesis explains a number of problems in cosmology and also provides a possible explanation for large-scale structure in the Universe – i.e. the formation of galaxies, clusters of galaxies, etc., starting from quantum fluctuations in the early Universe. It is not possible to give a complete description of the time immediately after the big bang because the physics of the very early Universe must necessarily be described by QUANTUM GRAVITY. No definitive theory of quantum gravity exists and consequently theories of the very early Universe up to the PLANCK TIME are speculative and incomplete. It is possible that SUPERSTRING THEORY will lead to a viable theory of quantum gravity and thence to a clearer understanding of the big bang and the very early Universe.

Earnshaw's theorem The result that a system of particles interacting via an inverse square force law cannot be in a stable state of static equilibrium. It was first stated by the Reverend Samuel Earnshaw (1805–88) in 1842. Since the interaction between electrically charged particles, such as electrons, is governed by COULOMB'S LAW which is an inverse square law. Earnshaw's theorem has the important consequence that any model of the atom in which the electrons are static cannot possibly be correct.

Eddington, Sir Arthur Stanley British physicist who was a pioneer in the study of stellar structure and general relativity. In 1919 Eddington led an expedition to observe a solar eclipse in order to test the prediction that light passing close to the Sun would be attracted. Albert EINSTEIN had made this prediction as a test of his general relativity theory. Eddington's expedition confirmed the prediction. Eddington was also one of the first people to try to understand stars in terms of processes going on inside them.

effective field theory A field theory that describes the interaction of particles at distances larger than the COMPTON WAVELENGTHS of these particles. The concept of effective field theory has been developed extensively since the late 1970s.

effective mass In solid state physics, the effective mass of an electron is an apparent mass that an electron has in a solid in responding to a force (e.g. in the presence of an electric field). The effective mass differs from the mass of a free electron because of the interaction of an electron with the crystal lattice. In relativistic quantum field theory the effective mass of a particle such as an electron can differ from that of the bare mass of the particle because of the 'mist' of VIRTUAL PARTICLES surrounding the particle.

Ehrenfest, Paul (1880–1933) Austrian physicist whose teaching and research was influential in the early development of quantum theory.

eigenfunction If Â is an OPERATOR then λ is an *eigenvalue* of Â if there is a function ψ such that $\hat{A}\psi = \lambda\psi$, with ψ being the *eigenfunction* corresponding to the eigenvalue λ. In the context of quantum mechanics an eigenfunction is a wave function for a quantum state of a system and an eigenvalue is the value of the energy of the energy level associated with that quantum state.

eigenvalue *See* eigenfunction.

eightfold way A classification of HADRONS proposed independently by Murray GELL-MANN and Yuval NE'EMAN in 1961. This classification was based on symmetry using a type of LIE GROUP known as SU(3). The eightfold way paved the way for the realization that all hadrons are composite particles of more fundamental particles called QUARKS.

Einstein, Albert (1879–1955) German-born theoretical physicist who formulated both the special and general theories of relativity and made major pioneering contributions to quantum mechanics.

In 1905 he published major papers on Brownian motion, the photoelectric effect, and special relativity theory. From his analysis of Brownian motion Einstein was able to give estimates of the sizes of molecules. This led to important experimental evidence in favor of the hypothesis that matter is made up of a very large number of atoms.

Special relativity enabled the inconsistency between classical electrodynamics and Newtonian mechanics to be resolved by showing that Newtonian mechanics requires modification for bodies that are moving extremely quickly. Einstein then showed that SPECIAL RELATIVITY THEORY has a number of important consequences, notably that no body can move faster than the speed of light in a vacuum and that mass can be regarded as a form of energy, as enshrined in his famous equation

$$E = mc^2.$$

Starting from the EQUIVALENCE PRINCIPLE in 1907, Einstein extended his analysis of relative motion in inertial frames of reference to noninertial frames of reference. This culminated in the modification of Newtonian gravity known as GENERAL RELATIVITY THEORY in 1915. Einstein made several important contributions to the development of general relativity theory such as the prediction of the existence of gravity waves in 1916 and the first application of general relativity theory to cosmology in 1917, including the introduction of the COSMOLOGICAL CONSTANT.

Einstein solved the problem of the photoelectric effect in 1905 by postulating that light has a particle-like aspect as well as a wavelike aspect. Einstein's other contributions to the development of quantum theory include his theory of the specific heat of solids in 1907, his theory of radiation in 1916–17 (which underpins the theory of LASERS), and BOSE–EINSTEIN STATISTICS in 1924–1925.

When quantum mechanics was formulated in the second half of the 1920s he did not accept the COPENHAGEN INTERPRETATION. In a prolonged correspondence with Niels BOHR, Einstein attempted to show that the formulation of quantum mechanics is incomplete. This led to the proposal of the EPR EXPERIMENT in 1935.

For the last thirty years of his life the main aim of Einstein's research was to find a UNIFIED FIELD THEORY to unify general relativity theory with electrodynamics, replace quantum mechanics with a deeper theory, and explain the elementary particles found in Nature. He was not successful in this bold quest.

An early success of general relativity was that Einstein showed in 1915 that it explained a long standing oddity in the motion of Mercury. However, he did not become a great international celebrity until 1919 when the observations of a solar eclipse by EDDINGTON showed that light is bent by the gravity of the Sun by the amount predicted by general relativity theory. Einstein was awarded the 1921 Nobel Prize for physics (which he received in 1922) for his explanation of the photoelectric effect. He is generally accepted as being the greatest physicist of the twentieth century and, along with Sir Isaac NEWTON, one of the greatest physicists of all time.

Einstein coefficients Two related quantities which occur in the QUANTUM THEORY OF RADIATION. They are related to the probability of a transition from the ground state to an excited state (or vice versa) occurring in the processes of INDUCED EMISSION and SPONTANEOUS EMISSION. The Einstein coefficients, denoted by A and B were put forward by Albert EINSTEIN in his 1916–17 papers on the quantum theory of radiation.

Einstein–de Sitter Universe A model of the Universe based on general relativity theory put forward by Albert EINSTEIN and Willem DE SITTER in 1932. It is an expanding Universe in which space–time is flat. The Einstein–de Sitter Universe is a realistic model of the Universe after INFLATION has occurred.

Einstein equation 1. The relation $E = mc^2$ deduced by Albert EINSTEIN in 1905 as a consequence of his special relativity theory, where E is an amount of an energy, m is the amount of mass corresponding to that energy, and c is the speed of light in a vacuum. The physical significance of this equation is that it shows that mass can be regarded as a form of energy. 2. The equation $E_{max} = h\nu - W$, which describes the PHOTOELECTRIC EFFECT in terms of photons of light, where E_{max} is the maximum kinetic energy of an electron ejected in the photoelectric effect, h is the Planck constant, ν is the frequency of the incident electromagnetic radiation, and W is the WORK FUNCTION of the solid being bombarded by light. This equation was first put forward by Albert Einstein in his 1905 paper on the photoelectric effect.

einsteinium A radioactive transuranic element of the actinoid series, not found naturally on Earth. It can be produced in milligram quantities by bombarding ^{239}Pu with neutrons to give ^{253}Es (half-life 20.47 days). Several other short-lived isotopes have been synthesized.

Symbol: Es. Melting pt.: 860±30°C. Proton number: 99. Most stable isotope: ^{254}Es (half-life 276 days). Electronic configuration: [Rn]5f^{11}7s^2.

Einstein–Podolsky–Rosen experiment *See* EPR experiment.

Einstein–Rosen bridge *See* wormhole.

Einstein's field equation The fundamental equation of GENERAL RELATIVITY THEORY, which relates the energy (mass) density of matter to the curvature of space–time. It is a generalization of Newton's law of gravity. However, since it is a much more complicated equation, or, to be more precise, set of coupled equations, than Newton's law of gravity it is much more difficult to solve. Nevertheless, some exact solutions of Einstein's field equation exist. Einstein put forward his equation in 1915.

Einstein theory of specific heat A theory of the specific heat of solids put forward by Albert EINSTEIN in 1907. In this work, which initiated the quantum theory of solids, Einstein calculated the specific heat capacity of solids by making the assumption that all the atoms in the solid undergo simple harmonic motion with the same frequency. This theory gives the DULONG–PETIT LAW for high temperatures and shows that the specific heat capacity tends to zero as the absolute temperature goes to zero. However, this theory does not give an accurate quantitative description of the specific heat at low temperatures. A better description is given by the DEBYE THEORY OF SPECIFIC HEAT, which assumes that the atoms in the crystals oscillate with a range of frequencies.

elastic collision A collision in which the kinetic energy of the colliding bodies is conserved. Collisions between atoms are elastic collisions. Collisions between molecules are only approximately elastic since some of the kinetic energy of the molecules can be converted into vibrational and rotational energy.

electric constant (**permittivity of free space**) Symbol: ε_0 A quantity that occurs in the quantitative statement of COULOMB'S LAW. If two particles with charges Q_1 and Q_2 are placed a distance r apart then the force F between the charges is given by $F = Q_1Q_2/(4\pi\varepsilon_0 r^2)$. Thus, $1/(4\pi\varepsilon_0)$ is the analog for electrostatics of the GRAVITATIONAL CONSTANT in Newton's law of gravity. In SI units ε_0 has the value 8.854×10^{-12} farad per meter.

electric dipole moment An electrical analog of the MAGNETIC MOMENT of a particle. If the center of the distribution of electric charge of a particle were different from the center of mass of a particle then there would be a twisting effect as the particle moved through an electric field. The STANDARD MODEL of elementary particles predicts that the electric dipole moment of an electron is zero. This result has been established experimentally to a high degree of accuracy.

electric field The field associated with an electric charge. An electric field can be represented by lines of force. The strength of an electric field is characterized quantitatively by the ELECTRIC FIELD STRENGTH.

electric field strength (**electric intensity**) Symbol: E A measure of the strength of an electric field at a point. It is defined as the force per unit charge experienced by a small electric charge at that point. Thus, $E = F/Q$. The electric field strength is also equal to the gradient of the electric potential at a point. The unit of electric field strength is the volt per meter (Vm^{-1}).

electric flux The electric flux through an area perpendicular to the lines of force of an electric field is the quantity defined by the product $E \times A$, where E is the electric field strength normal to the area A.

electricity Any of various phenomena associated with a property of certain particles called *electric charge*. There are two types of electric charge, denoted positive and negative. See also COULOMB'S LAW.

electric potential Symbol: V The electric potential at a point is the energy required to bring a unit electric charge from infinity to that point. The unit of electric potential is the volt.

electric susceptibility Symbol: χ_e The parameter that relates the POLARIZATION of a material to the external electric field applied to the material. It is given by $P/(\varepsilon_0 E)$, where P is the polarization, E is the electric field strength causing it and ε_0 is the ELEC-TRIC CONSTANT. The electric susceptibility is a dimensionless quantity which is equal to $\varepsilon_r - 1$, where ε_r is the relative permittivity of the medium.

electrochemical equivalent Symbol: z The mass of a chemical element released from a solution of one of its ions in electrolysis by one coulomb of electric charge.

electrochemistry The branch of chemistry concerned with ions in solutions or molten salts. It includes the study of electrolysis and of electric cells.

electrode A part of an electrical system that emits or collects electrically charged particles such as electrons. *See* anode; cathode.

electrode potential A quantity that gives a measure of the ability of a metal to lose electrons to a surrounding solution of its ions. It is the potential difference between a metal and such a solution. It is not possible to measure the electrode potential directly since it is necessary to have another electrode to complete the electrical circuit. For that reason, standard *electrode potentials* are defined, which give the potential relative to that of a standard hydrogen electrode.

electrodynamics The study of the forces and dynamics associated with electric and magnetic fields.

electrolysis The production of chemical change by passing charge through certain conducting liquids (electrolytes). The current is conducted by migration of ions – positive ones (cations) to the cathode (negative electrode), and negative ones (anions) to the anode (positive electrode). Reactions take place at the electrodes by transfer of electrons to or from them.

In the electrolysis of water (containing a small amount of acid to make it conduct adequately) hydrogen gas is given off at the cathode and oxygen is evolved at the anode. At the cathode the reaction is:
$$H^+ + e^- \rightarrow H$$

$$2H \rightarrow H_2$$
At the anode:
$$OH^- \rightarrow e^- + OH$$
$$2OH \rightarrow H_2O + O$$
$$2O \rightarrow O_2$$
In certain cases the electrode material may dissolve. For instance in the electrolysis of copper(II) sulfate solution with copper electrodes, copper atoms of the anode dissolve as copper ions
$$Cu \rightarrow 2e^- + Cu^{2+}$$
See also Faraday's laws (of electrolysis).

electrolyte A liquid that conducts electricity because of the presence of positive and negative ions. Electrolytes are molten ionic salts or solutions of acids, bases, or salts.

electromagnetic induction The production of an electromotive force in a conductor resulting from a change in an external magnetic field. Electromagnetic induction was discovered by Michael FARADAY in 1831 and is described by FARADAY'S LAWS OF MAGNETIC INDUCTION and LENZ'S LAW.

electromagnetic interaction *See* fundamental interactions.

electromagnetic pump A pump with no moving parts, used for circulating electrically conducting fluids. A direct current is passed between two electrodes inserted in the fluid circuit and a constant magnetic field is directed perpendicular to this current. There is a force on the fluid mutually perpendicular to these two directions, and therefore parallel to the direction of flow, thus propelling the fluid. One of its applications is in pumping liquid sodium coolant through nuclear reactors.

electromagnetic radiation Radiation consisting of ELECTROMAGNETIC WAVES.

electromagnetic spectrum The range of wavelengths of electromagnetic waves. Visible light is a very small part of this spectrum. The other waves in the electromagnetic spectrum show the same wave

ELECTROMAGNETIC SPECTRUM (note: the figures are only approximate)		
Radiation	*Wavelength (m)*	*Frequency (Hz)*
gamma radiation	-10^{-10}	10^{19-}
x-rays	$10^{-12} - 10^{-9}$	$10^{17} - 10^{20}$
ultraviolet radiation	$10^{-9} - 10^{-7}$	$10^{15} - 10^{17}$
visible radiation	$10^{-7} - 10^{-6}$	$10^{14} - 10^{15}$
infrared radiation	$10^{-6} - 10^{-4}$	$10^{12} - 10^{14}$
microwaves	$10^{-4} - 1$	$10^{9} - 10^{13}$
radio waves	$1 -$	-10^{9}

properties such as light waves, diffraction, interference, and polarization.

The electromagnetic spectrum can be divided into various types of electromagnetic radiation according to their wavelengths. These divisions are not sharp boundaries but convenient labels. There is a gradual change in behavior as the wavelength changes. The shortest waves are gamma rays (below 10^{-10} m), followed by x-rays (10^{-10}–10^{-9} m), ultraviolet radiation (10^{-9}–10^{-7} m), visible light (10^{-7} – 10^{-6} m), infrared radiation (10^{-6}–10^{-4} m), microwaves (10^{-4}–1), with radio waves having the longest wavelengths (greater than 1 m). The figures quoted are approximate and different authors quote slightly different figures.

Since all electromagnetic waves travel at the same speed in a vacuum, denoted c, there is an inverse relationship between the wavelength λ and the frequency v of a wave in the electromagnetic spectrum given by $\lambda v = c$.

electromagnetic wave A type of transverse wave involving oscillating electric and magnetic fields that are perpendicular to each other and to the direction of propagation. Light waves and all the other waves in the ELECTROMAGNETIC SPECTRUM are electromagnetic waves. The existence of these waves was predicted by James Clerk MAXWELL in the 1860s in work that unified electricity and magnetism. This prediction was confirmed experimentally by Heinrich HERTZ in 1888. Electromagnetic waves do not require a material medium to support their propagation. They transport energy and in a vacuum they all move at the speed of 2.9979×10^8 meters per second. These waves can be generated by accelerating electrically charged particles. *See also* photon.

electromagnetism The phenomena associated with the relations between electric and magnetic fields.

electromotive force (e.m.f.) Symbol: E If a source of electrical power (such as a cell) is able to drive electric current round a circuit then the source is said to have an electromotive force. The unit of e.m.f. is the same as that of potential difference, i.e. the volt. In a voltaic cell the potential difference between the terminals is less than the e.m.f. when current is flowing because of the internal resistance of the cell. It is equal to the e.m.f. only when there is no current flowing.

electromotive series (electrochemical series) A series of chemical elements arranged in order of their standard electrode potentials. If an element has a greater tendency to lose electrons from its atoms to its ionic solutions than hydrogen then it is said to be *electropositive*; if it has a lesser tendency then it is *electronegative*. Some of the main elements of the series are: potassium, calcium, sodium, magnesium, aluminum, zinc, cadmium, iron, nickel, tin, lead, hydrogen, copper, mercury, silver, platinum, gold. Since elements higher in the series tend to lose electrons and become positive ions more readily than those lower in the series they tend to displace lower elements from ionic solutions. For example, all the electropositive elements, i.e. all the elements before hydrogen, replace hydrogen from acids. Similarly, if a piece of zinc is placed in a copper sulfate solution then copper is deposited and zinc ions dissolve in the solution. *See also* electrode potential.

electron   A stable elementary particle with a negative electric charge. The electron is the lightest charged lepton. It has a rest mass (denoted m_e) of $9.1093897 \times 10^{-31}$ kg and a spin of ½. Electrons occur in atoms (*see* electronic structure of atoms). When an electron has been detached from an atom it is called a *free electron*. Electrons are also ejected in the process of BETA DECAY. The electron has an antiparticle called the POSITRON.

The electron was discovered by Sir Joseph John (J.J.) THOMSON in 1897. There is no experimental evidence to suggest that the particle has an internal structure. If an electron is regarded as a point charge then its SELF-ENERGY is infinite. Serious difficulties also arise for an equation describing the motion of an electron called the LORENTZ–DIRAC EQUATION. On the other hand, postulating that the electron has a nonzero size also has serious difficulties since it becomes necessary to postulate the existence of POINCARÉ STRESSES to explain the stability of electrons. It is now thought that problems associated with any structure of the electron cannot be analyzed in terms of classical electrodynamics and that it is necessary to take quantum mechanics into account.

electron affinity   The energy released when an electron is added to an atom (or molecule) to form a negative ion.

electron capture   1. The formation of a negative ion as a result of the capture of an electron by an atom or molecule or the formation of an atom or molecule by the capture of an electron by a positive ion.
2. The transformation of a proton into a neutron in a nucleus by the capture of an inner orbital electron of the atom. A neutrino is emitted in this process. The emission of an x-ray photon is also associated with this process as a result of an outer electron falling into the vacancy created by the capture of the inner electron. The wavelength of the emitted photon is char-

acteristic of the daughter element. *See also* capture.

electron configuration   *See* electronic structure of atoms.

electron-deficient compound   A compound in which there are fewer electrons taking part in the chemical bonds than take part in normal covalent bonding involving pairs of electrons. Boranes, i.e. boron hydrides such as diborane (B_2H_6), are examples of electron deficient compounds. It is necessary to explain the bonding in such compounds in terms of MULTI-CENTER BONDS.

electron diffraction   The diffraction of a beam of electrons by, for example, the atoms of a crystal. Electron diffraction occurs because of the quantum mechanical feature of electrons that they can behave like waves as well as particles. The phenomenon was discovered by Clinton DAVISSON and Lester Germer and by Sir George THOMSON in 1927, thus giving important confirmation of the idea of Louis DE BROGLIE on the dual wave–particle nature of electrons in quantum theory.

Electron diffraction is used to study the structure of solid surfaces. It is also a technique for determining the structure of isolated molecules in the gas phase.

electronegative   Describing chemical elements that tend to gain electrons and become negative ions (*see* electromotive series). The halogens are examples of electronegative elements. In molecules electronegative atoms tend to attract electrons in chemical bonds. This can be seen by considering hydrogen chloride (HCl). Chlorine is more electronegative than hydrogen and, as a result, the molecule is polar, with negative charge on the chlorine atom and positive charge on the hydrogen. *See also* electronegativity.

electronegativity   A measure of how ELECTRONEGATIVE a chemical element is. There are several definitions of electronegativity. The *Mulliken electronegativity E* is defined by $E = (I + A)/2$, where I is the ionization potential and A is the electron affinity. The *Pauling electronegativity* is based on a scale using bond dissociation energies. Some of the values on this scale are fluorine 4, carbon 2.5, nitrogen 3.0, oxygen 3.5, chlorine 3.0, and bromine 2.8.

electron gas A model of electrons in metals or plasmas in which the electrons are regarded as a gas, with there being a uniform distribution of positive electric charge to make the system electrically neutral. This system is analyzed theoretically either using classical or quantum statistical mechanics, including kinetic theory. The electron-gas model gives a good qualitative and approximate quantitative description of many properties of metals and plasmas but cannot give an accurate quantitative description of these systems since it does not take either lattice structure or motions of the positive ions into account.

electron gun A device that generates a beam of electrons. An electron gun usually consists of a heated cathode in an evacuated tube, with the electrons being emitted from the cathode by THERMIONIC EMISSION. The electrons are accelerated towards an anode by a high voltage and are focussed by other electrodes to form a narrow beam. Electron guns are used in cathode-ray tubes and in experiments involving electron impact.

electronics The study and design of devices and circuits that make use of the movement of electrons.

electronic structure of atoms The arrangement of electrons about the nucleus of an atom and the energy levels they occupy. The electrons move in ORBITALS round the nuclei. Each electron in an atom is characterized by a set of four quantum numbers:
(1) The *principal quantum number*, denoted n, which has the values 1, 2, 3, etc, with the higher the value of n the further the electron is from the nucleus on average. An *electron shell* is the set of electrons in an atom with the same value of n. The maximum number of electrons a shell charac-

terized by n can have is $2n^2$. For historic reasons associated with spectroscopy the shells are denoted by letters K, L, M, etc., with $n = 1$ for the K shell, $n = 2$ for the L shell, etc. The principal quantum number has the largest effect on the energies of the quantum states.
(2) the *orbital quantum number*, denoted l, which specifies the orbital angular momentum of an electron. For a given value of n, the possible values of l are $(n-1)$, $(n-2)$, ... , 2, 1, 0, with each of the possible values of l being a *subshell* of the shell characterized by n. For example, the L shell ($n=2$) has two subshells, the possible values of l being $l=1$ and $l=0$. Similarly, the M shell ($n=3$) has three subshells, the possible values of l being $l=2$, $l=1$ and $l=0$. The subshells of a given shell have slightly different energies, with higher values of l giving rise to higher energy quantum states. For historic reasons associated with spectroscopy the subshells are denoted by letters s ($l=0$), p ($l=1$), d ($l=2$), f ($l=3$). For given values of n and l, the maximum number of electrons in a subshell is $2(2l+1)$. Thus, an s-subshell ($l = 0$) can contain 2 electrons. A p-subshell ($l = 1$) can contain 6 electrons, a d-subshell ($l = 2$) can contain 10 electrons, and an f-subshell ($l = 3$) can contain up to 14 electrons. The orbital quantum number is sometimes known as the *azimuthal quantum number*.
(3) The *magnetic quantum number*, denoted m, which characterizes the energy levels of electrons when the atom is in an external magnetic field. For a given value of n and l the possible values of m are $+l$, $(+l-1)$, ..., 1, 0, -1, ..., $-(l-1)$, $-l$. For example, the p-subshell ($l=1$) can have values of m of $m=1$, $m=0$, and $m=-1$, corresponding to three p ORBITALS. In the absence of an external magnetic field all the quantum states with the same values of n and l but different values of m have the same energy (i.e. they are degenerate). However, in the presence of an external magnetic field this energy level splits, with there being a small energy difference between each state characterized by a different value of m (*see* Zeeman effect).

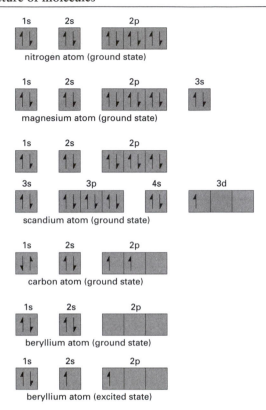

1s 2s 2p
nitrogen atom (ground state)

1s 2s 2p 3s
magnesium atom (ground state)

1s 2s 2p
3s 3p 4s 3d
scandium atom (ground state)

1s 2s 2p
carbon atom (ground state)

1s 2s 2p
beryllium atom (ground state)

1s 2s 2p
beryllium atom (excited state)

Electronic configurations of typical atoms indicated using box-and-arrow diagrams

(4) The *spin quantum number*, denoted M_s, which characterizes the intrinsic angular momentum (spin) of an electron. The possible values are $+\frac{1}{2}$ and $-\frac{1}{2}$. The existence of spin is responsible for the pattern of splitting of energy levels of atoms found in FINE STRUCTURE of atomic spectra and the ANOMALOUS ZEEMAN EFFECT.

The existence of these four quantum numbers for electrons in atoms combined with the PAULI EXCLUSION PRINCIPLE explains the electronic structure of atoms and why the PERIODIC TABLE of chemical elements occurs.

Each chemical element is specified by its *electronic configuration*, which is the way electrons fill the energy levels of the atom. The principal quantum numbers are indicated by their values and the orbital quantum numbers by the letters associated with them, with the number of electrons in a subshell written as a superscript. For example, an oxygen atom has two *s*-electrons which have principal quantum numbers with value 1, two *s*-electrons which have principal quantum numbers with value 2, and four *p*-electrons which have principal quantum numbers with value 2. Thus, the electron configuration of oxygen is written $1s^2 2s^2 2p^4$. It is also conventional to use a shorthand involving the configuration of a noble gas. For instance, Neon has the electronic configuration $1s^2 2s^2 2p^6$ and calcium has the configuration $1s^2 2s^2 2p^6 3s^2$. The configuration of calcium can be written $[Ne]3s^2$. It is necessary to use the AUFBAU PRINCIPLE to determine the electron configuration of an element.

electronic structure of molecules   The arrangement of electrons in molecules. It is possible to determine what electronic

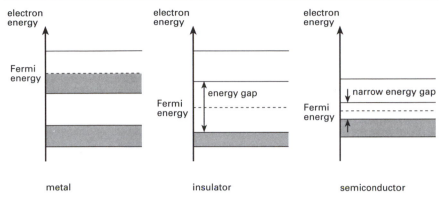

The energy bands in solids

states can arise from the atoms of the molecule by applying the WIGNER–WITNER RULES or their generalization from diatomic molecules to polyatomic molecules. The quantum numbers of a molecule are associated with the representations of the POINT GROUP of that molecule. The existence of the CHEMICAL BOND is understood in terms of the electronic structure of molecules. It is not possible to solve the Schrödinger equation for molecules with more than one electron. As with atoms, it is necessary to use approximation techniques and/or powerful computers. The two main approaches to understand the electronic structure of molecules are MOLECULAR ORBITAL THEORY and VALENCE BOND THEORY.

electronic structure of solids The arrangement of electrons in solids. The problem of electrons in crystals is much easier to analyze then electrons in disordered solids. A fundamental result for electrons in crystals is BLOCH'S THEOREM which explains how electrons can propagate through crystals, and hence explains why it is sometimes a good approximation to regard electrons in metals as free electrons.

In a crystal the sharply defined energy levels of atoms become *energy bands* of allowed energy. In each band there is a large number of allowed quantum states. The gaps between the energy bands are called *forbidden gaps* since quantum states for electrons in crystals cannot exist in the gaps. A full energy band in a crystal is the analog of a full shell in an isolated atom.

The fully occupied band with the highest energy is called the *valence band* of the solid since it is formed by the outer electrons of the atoms making up the solid (i.e. the ones that take part in chemical bonding and are responsible for valence). The band above the valence band in energy is called the *conduction band*.

The optical and electrical properties of solids are accounted for by their band structure. In particular, the existence of energy bands enables the distinction between metals and insulators to be understood. In order for electrical conduction to occur it is necessary that an electron can increase its energy, in which case it needs to make the transition from one quantum state to another quantum state with almost the same energy. If the valence band is full then, by the PAULI EXCLUSION PRINCIPLE, the electron cannot make a transition to another quantum state in the band. Electrical conduction can occur if the electron with the highest energy is in the conduction band. A solid is a metal either if the conduction band is not completely full or the conduction band overlaps with the valence band.

If the gap between the valence band and the conduction band is wide the material is an insulator. At normal temperatues, thermal energies are insufficient to cause electrons to go from the valence band to the conduction band. However, if the gap between a full valence band and an empty conduction band is small then, thermal energies at normal temperatures can cause electrons to move from the valence band to the conduction band. Thus, although such a solid is an insulator at absolute zero, it is a conductor at nonzero temperature, with the electrical conductivity increasing with temperature. This type of solid is called an *intrinsic* semiconductor. *See* semiconductor. When semiconductors are 'doped' by adding small amounts of impurity atoms then energy levels appear in the gaps between the bands.

Disordered solids also have energy bands. However, in disordered solids the phenomenon of *localization* of electrons can occur. A localized electron is one that cannot contribute to the electrical conductivity at absolute zero temperature because it does not propagate through the solid. If there is only a small amount of disorder the localized electron states are in 'tails' at the top and bottom of the bands, with there being boundary energies called *mobility edges* which separate the localized states from *extended states*, i.e. states that can propagate through the solid and hence contribute to electrical conductivity at absolute zero temperature. If the highest occupied energy level of a disordered solid is a localized state then at absolute zero temperature, the solid is an insulator. In 1958 Philip ANDERSON showed that for sufficiently great disorder all the electronic states are localized. For that reason, electron localization is frequently referred to as *Anderson localization*. The problem of electron localization in disordered systems may be analyzed theoretically by methods analogous to those used in the theory of phase transitions, with the mobility edge being analogous to the transition temperature.

electron microscope A type of microscope that uses a beam of electrons rather than a beam of light to produce a magnified image of a small object. The electron microscope makes use of the fact that electrons behave as waves as well as particles and that the wavelength of electrons at suitable energies is smaller than that of light. Since the resolution of a microscope is limited by the wavelength, electron microscopes can produce much higher resolution than optical microscopes.

In a *transmission electron microscope* the electron beam passes through a very thin sample and is focussed onto a fluorescent screen where a visual image is produced. It is possible to obtain magnifications of 200 000 in this way. In a *scanning electron microscope* thicker samples can be used. A beam of electrons scans the sample. The reflected electrons, and any secondary electrons emitted, are focussed onto a screen. Both the magnification and resolution are lower than in the transmission microscope.

electron optics The use of electric and magnetic fields to focus beams of electrons. Electron optics is used in the design of cathode-ray tubes and electron microscopes.

electron probe microanalysis (EPM) A method of quantitative analysis used for small samples in which a beam of electrons is focused onto the surface of the sample so that x-rays of characteristic intensities are emitted.

electron shell *See* electronic structure of atoms.

electron-spin resonance (ESR) A method of spectroscopy that provides information about the chemical bonds and structure of paramagnetic molecules (*see* magnetism) and free radicals. This is possible because there is a small energy difference in energy between the two possible alignments of the spin of an unpaired electron in an external magnetic field. The transition between these states can be investigated using microwave radiation. The value of the energy difference between the two states depends on the surrounding electrons in the molecule. Electron-spin resonance for electrons is analogous to nuclear magnetic resonance in nuclei.

electronvolt Symbol: eV A unit of energy defined as the energy needed to move an electron across a potential difference of one volt. It is equal to 1.602×10^{-19} joule. It is often used to measure the energy of particles.

electropositive Describing chemical elements that tend to lose electrons and become positive ions (*see also* electromotive series). The alkali metals, such as lithium, sodium, and potassium, are examples of electropositive elements.

electroscope A device for detecting electrical charge and electrically charged particles. The most common type is the *gold leaf electroscope* in which a pair of rectangular gold leaves are suspended side by side from a rod, which is insulated, with a plate on top of the rod. When a charge is applied to the plate the leaves move apart because of the mutual repulsion between the charges they have received.

electrostatics The study of electric charges at rest. Electrostatics is governed by COULOMB'S LAW.

electroweak interaction The combination of the electromagnetic and weak interactions into a single unified mathematical description. The WEINBERG–SALAM MODEL model is a successful electroweak theory.

element A substance that cannot be decomposed into simpler substances by chemical means. In a sample of a given element all the atoms in the sample have the same number of protons (and electrons), although it is possible for the number of neutrons to vary (*see* isotope).

There are 92 elements that occur naturally on the Earth and several other elements known as TRANSURANIC ELEMENTS which have been produced artificially. The light elements originated soon after the big bang while all the other elements were produced in stars (*see* nucleosynthesis).

The elements have physical and chemical regularities (*see* periodic table). These regularities were understood after the explanation of the ELECTRONIC STRUCTURE OF ATOMS in terms of quantum mechanics.

elementary particle One of the fundamental constituents of matter. Our view of what elementary particles are has changed with time. At one time it was thought that atoms were elementary particles. However,

the discovery of the *electron* in 1897 showed that this cannot be the case. The discovery of the nuclear structure of the atom and the subsequent discoveries of the *proton* and *neutron* led to the idea that the atom has an internal structure consisting of a nucleus made up on protons and neutrons surrounded by a number of electrons equal to the number of protons. In addition, it was postulated that particles called *neutrinos* are emitted in beta decay and that the particles inside the nucleus are held together by short-lived particles called *mesons*. After World War II many short-lived particles were discovered.

Between the 1950s and the mid 1970s a great deal of theoretical and experimental work led to our present state of knowledge, which is that the only particles not known to be made up of other particles are *quarks* and *leptons*. It is thought that there are six types of quark, called up, down, strange, charm, top, and bottom, and that there are also six types of lepton; the electron, the electron neutrino, the muon, the muon neutrino, the tau, and the tau neutrino. All these particles have spin $\frac{1}{2}$. Some quarks have electric charges of +2/3 while some have charges of –1/3. The electron, muon and tau all have a charge of –1. All the neutrinos are electrically neutral. These quarks and leptons are divided into three *families* (*generations*), each of which contains two quarks, an electrically charged lepton, and the neutrino associated with that lepton.

Although it is known that *hadrons* are not elementary particles, with *baryons* being composed of three quarks, and *mesons* being composed of two quarks, it is frequently convenient to regard hadrons as elementary particles, particularly as isolated quarks are not observed.

Elementary particles can be classified by the FUNDAMENTAL INTERACTIONS they are subject to. Quarks are subject to all four fundamental interactions: strong, weak, electromagnetic, and gravitational. Electrically charged leptons are subject to weak, electromagnetic, and gravitational interactions but not strong interactions. Neutrinos are subject to weak and gravitational interactions but not to strong and electromagnetic interactions. In accord with RELATIVISTIC QUANTUM MECHANICS all the quarks and leptons have ANTIPARTICLES. A list of the elementary particles and some of their properties is given in the Appendix.

In addition to the spin ½ particles of matter there are also bosons which mediate the fundamental interactions between the spin ½ particles of matter. These bosons are *photons*, which mediate electromagnetic interactions, *gluons*, which mediate strong interactions, W *bosons* and Z *bosons*, which mediate weak interactions, and *gravitons*, which mediate gravitational interactions. Photons, gluons, W-bosons, and Z bosons are spin-1 particles while gravitons are spin-2 particles. In addition, there is a spin-zero particle called the *Higgs boson* associated with the W and Z bosons, which has been predicted to exist but has not been found experimentally.

The family structure of elementary particles suggests that the quarks and leptons are not the ultimate elementary particles. It is not known why this family structure arises or why there are three families. There have been many attempts to understand the pattern of quarks and leptons. At the present time it is thought that an understanding of elementary particles requires a theory that unifies all four fundamental interactions. This may well also require a theory that successfully combines quantum mechanics and general relativity. *Superstring theory* appears to be a promising step in this direction, with the different particles being different modes of excitation of the string. However, superstring theory has not been formulated definitively yet and it is unclear how exactly quarks and leptons emerge from the theory. *See also* standard model.

ellipse   *See* conic sections.

emanation   The name formerly given to certain isotopes of radon, which diffused out of the parent material containing radium, from which they are formed by alpha decay. ^{220}Radon is a member of the ^{232}thorium decay series and it was therefore known as *thoron* or thorium emanation. The radon isotope that is a member of the actinium series was known as actinium emanation or *actinon*. Summarizing: radium emanation is now known as ^{222}Rn; thoron emanation is now known as ^{220}Rn; actinium emanation is now known as ^{219}Rn.

e.m.f.   *See* electromotive force.

emission spectrum   *See* spectrum.

Empedocles (*c.* 494–434 BC)   Ancient Greek thinker who was one of the first people to propose that the four elements are air, earth, fire, and water.

energy   A measure of the ability of a system to do work. Energy can be classified into *kinetic energy*, which is the energy a body has as a result of its motion, and *potential energy*, which is the energy stored in a system because of its position or state. Energy has many forms such as gravitational energy, electrical energy, and nuclear energy and special relativity shows that mass is a form of energy. It is a fundamental law of physical science, known as the *law of conservation of energy*, also known as the *first law of thermodynamics*, that energy cannot be created or destroyed in any process but can only be converted from one form into another.

energy band   *See* electronic structure of solids.

energy level   A stationary state of energy of a system governed by quantum mechanics. For example, an electron in an atom has a definitie energy associated with the ORBITAL of the electron. In quantum mechanics a particle that is in a bound state, such as an electron in an atom, can only have a discrete set of energy levels. The energy level with the lowest possible allowed energy is called the *ground state* of the system, with all the other states being called *excited states*.

enrichment   The process of increasing the fraction of a particular ISOTOPE in a mixture of isotopes of the same element. The word is usually applied to the enrichment of uranium by increasing the proportion of U-235. It is also used for the process of obtaining heavy water from naturally occurring water.

entanglement The feature of a quantum-mechanical system that two particles which have separated and are a long way apart are still part of the same system. *See also* EPR experiment.

entropy Symbol: S A measure of the disorder of a system or, equivalently, the availability of the energy of a system to do work. These definitions are equivalent because if some of the energy of a system becomes heat, which is the most disordered form of energy, then the amount of energy that is able to do work decreases. The second law of thermodynamics states that in a closed system any real (irreversible) process leads to an increase in entropy.

Epicurus (*c.* 342–271 BC) Ancient Greek thinker who was one of the earliest people known to have put forward the idea of atoms. Epicurus was a pupil of LEUCIPPUS. Like him, he postulated that matter is made up of a very large number of small indestructible particles called atoms, which differ from each other in size, shape, and position and move randomly in an infinite void.

EPM *See* electron probe microanalysis.

EPR experiment (Einstein–Podolsky–Rosen experiment) A thought experiment proposed by Albert EINSTEIN, Boris Podolsky, and Nathan Rosen in 1935 to illustrate what they regarded as the incorrectness of Niels Bohr's view of quantum mechanics. In this experiment they postulated a composite particle that separates into two different particles which move apart to a large distance. Einstein analyzed the momentum of each particle in terms of quantum theory and argued that Bohr's interpretation of quantum mechanics led to the conclusion that there must be instantaneous communication between the particles no matter how far apart they are.

To see this it is instructive to consider a modified version of the EPR experiment put forward by David BOHM in 1951. Bohm considered a spin-zero particle decaying into two particles with equal but opposite spin (by the conservation of spin) which fly apart. Once the particles are far away the spin of one of the particles is measured. It is either up or down. By the conservation of spin this means that measuring one spin instantaneously determines the spin of the other particle since it must necessarily be opposite to that of the first particle. However, this is not as trivial as it sounds since Niels Bohr's interpretation of quantum mechanics asserts that the spin of the particles is not up or down but exists as a superposition of up and down states until the measurement is made. Einstein rejected the possibility that measurement of the state of one particle could instantaneously determine the state of another particle in this way and concluded that Bohr's interpretation of quantum mechanics was incorrect. Einstein believed that the states of the particles were determined locally at the point of separation by as yet undiscovered factors – so-called *hidden variables*.

In the 1960s John BELL considerably sharpened the analyses of Einstein and Bohm by proving a result, known as *Bell's theorem*, concerning measurements of the spins of particles in an EPR-like experiment. Bell showed that a set of inequalities, known as *Bell's inequalities*, would be satisfied by a hidden variable theory obeying locality but would be violated if Bohr was correct.

By the 1980s technology had advanced to the stage where actual experiments to test the theories could be performed. Alain Aspect and his colleagues in Paris performed experiments, known collectively as the *Aspect experiment*, to test Bell's inequalities using photons. These experiments showed that Bell's inequalities are violated and hence that Bohr was right and Einstein was wrong and that measurement of one part of a system instantaneously determines the state of another part of the system, even if the other part is a long way away. This is a phenomenon known as *quantum entanglement* and shows that

NONLOCALITY is an essential feature of quantum mechanics. Careful analysis by a number of authors in the 1980s has shown that this nonlocality in quantum mechanics does not violate special relativity theory since it is not possible to send information faster than the speed of light in a vacuum using this feature of quantum mechanics.

equation of state An equation that relates the pressure p, the volume V and absolute temperature T of a substance. The simplest equation of state is the *ideal gas law*: $pV = nRT$, where n is the number of moles of the substance and R is the universal gas constant. This equation is only applicable to ideal gases, which means that neither the volume of the gas molecules or intermolecular forces between them are taken into account. A number of other equations of state take these features into account, notably the VAN DER WAALS EQUATION.

equilibrium statistical mechanics *See* statistical mechanics.

equipartition of energy The principle that the energy in a system is divided equally among the degrees of freedom of that system. This principle was proposed by Ludwig Boltzmann and given some theoretical support by James Clerk Maxwell. At thermal equilibrium the average energy for each degree of freedom is kT, where k is the Boltzmann constant and T is the absolute temperature. The principle is a good approximation at high temperatures but not at low temperatures, where it is necessary to take quantum effects into account.

equivalence principle A fundamental principle underlying general relativity theory, which can be stated in several equivalent ways. One statement of the equivalence principle is that the effects of acceleration are indistinguishable from those of a uniform gravitational field. Another statement is that the gravitational mass of a body is equal to its inertial mass. This equality has been shown to be true experimentally with great accuracy by Roland Eötvös in the nineteenth century

and various investigators in the twentieth century.

erbium A soft malleable silvery element of the lanthanoid series of metals. It occurs in association with other lanthanoids. Erbium has uses in the metallurgical and nuclear industries and in making glass for absorbing infrared radiation.

Symbol: Er. Melting pt.: 1529°C. Boiling pt.: 2863°C. Relative density: 9.066 (25°C). Proton number: 68. Relative atomic mass: 167.26. Electronic configuration: $[Xe]4f^{12}6s$.

Erlangen Programme The name given to the sugestion of Felix KLEIN in 1872, made in his inaugural lecture at the University of Erlangen, that a type of geometry can be characterized by a group of transformations. Frequently, this group is the group of symmetry transformations. For example, in 1891 Arthur Schoenflies derived the 230 SPACE GROUPS of crystals as an application of the Erlangen programme. In 1910–1911 Klein himself showed that the LORENTZ GROUP is the group that characterizes MINOKOWSKI SPACE–TIME. As Klein already realized in 1872, RIEMANNIAN GEOMETRY does not readily fit into the Erlangen Programme. In 1926 Jan Schouten suggested that the DIFFEOMORPHISM GROUP characterizes Riemannian geometry. This suggestion means that the diffeomorphism group plays the same role in general relativity theory that the Lorentz group plays in special relativity theory. More generally, in 1939 Hermann WEYL proposed that the appropriate group characterizing a geometry is the set of transformations which preserves the mathematical structure of the geometry. Weyl's suggestion may very well be relevant for finding the group that characterizes a theory unifying general relativity, quantum mechanics, elementary particles, and the nongravitational forces.

Esaki diode (**tunnel diode**) A highly doped p–n semiconductor junction diode with an unusual voltage–current characteristic due to electrons tunneling from the valence band to the conduction band. The

Esaki diode was discovered by Leo ESAKI in 1957.

ESCA *See* photoelectron spectroscopy.

escape velocity The minimum velocity a body must have to escape from the gravitational attraction of a celestial body such as a star, planet, or moon. For a body to escape, its kinetic energy must be greater than the gravitational potential energy. This means that the escape velocity must be greater than $\sqrt{(2GM/r)}$, where G is the gravitational constant, and M and r are the mass and radius respectively of the celestial body. The escape velocity from the Earth is 11 200 ms^{-1} and from the Moon is 2370 ms^{-1}.

ESR *See* electron-spin resonance.

eta particle Symbol: η An electrically neutral meson made up of a down quark and an anti-down quark. It has a mass of 548.8 MeV and a lifetime of 2.5×10^{-19} s, with pions and photons being its main decay products.

ether (aether) A hypothetical medium that was once thought to permeate all space and to be the medium through which electromagnetic waves and interactions are propagated. The failure of the MICHELSON–MORLEY EXPERIMENT cast doubt on the existence of the ether. In special relativity theory it is not necessary for the ether to exist because electromagnetic waves can be propagated through empty space.

Euclid (*c.* 330 BC–*c.* 260 BC) Ancient Greek mathematician. Euclid is best known as the author of a mathematical treatise, mostly concerned with geometry, called the *Elements* which consisted of 13 volumes. In this very influential book the mathematics of the time was presented in a systematic way, starting from a set of basic axioms.

Euclidean field theory A formulation of quantum field theory initiated by Julian SCHWINGER in 1958 in which the time coordinate in Minkowski space–time is treated like a fourth space dimension. This procedure, which can be rigorously justified, makes the analysis of quantum field theory much simpler. For example, Euclidean quantum field theory enables powerful methods from STATISTICAL MECHANICS to be used in quantum field theory.

Euclidean geometry The type of geometry that describes flat surfaces and their generalizations to three and higher dimensions. In Euclidean geometry the sum of the three angles of a triangle is always 180°. In the nineteenth century several types of NONEUCLIDEAN GEOMETRY were found.

Euclidean quantum gravity The formulation of quantum gravity as a EUCLIDEAN FIELD THEORY. The Euclidean formulation of quantum gravity, proposed by Stephen HAWKING in the second half of the 1970s, is controversial since it is much trickier technically and much more difficult to justify theoretically for quantum gravity than for quantum field theory in flat space–time. Nevertheless, Hawking and his colleagues have obtained many significant results using this formulation, particularly in QUANTUM COSMOLOGY.

Euler, Leonhard (1707–1783) Swiss-born mathematician. Euler was one of the greatest and most prolific mathematicians of all time. He made many major contributions to many branches of mathematics and physics including calculus, mechanics, optics, and fluid mechanics. In addition, he introduced many mathematical notations which have become standard, including e, i, and π.

europium A silvery element of the lanthanoid series of metals. It occurs in association with other lanthanoids. Its main use is in a mixture of europium and yttrium oxides widely employed as the red phosphor in television screens. The metal is used in some superconducting alloys.

Symbol: Eu. Melting pt.: 822°C. Boiling pt.: 1597°C. Relative density: 5.23 (25°C). Proton number: 63. Relative atomic mass: 151.965. Electronic configuration: [Xe]4f^{7}6s^2.

even–even nucleus A nucleus that consists of an even number of protons and an even number of neutrons. Even–even nuclei are more stable, and hence more common, than EVEN–ODD NUCLEI, ODD–EVEN NUCLEI, and ODD–ODD NUCLEI.

even–odd nucleus An atomic nucleus that contains an even number of protons and an odd number of neutrons.

event horizon *See* black hole.

Everett, Hugh (1930–1982) American physicist who developed the MANY WORLDS INTERPRETATION of quantum mechanics while he was a graduate student at Princeton University in the 1950s.

exceptional groups *See* Lie groups.

exchange energy An effect that arises because of the feature of quantum mechanics that identical particles are indistinguishable. This means, for example, that if two electrons are close together then it would not be possible to tell the difference if the electrons exchanged their quantum states. As a consequence, it is necessary to describe the two-electron system with a wave function in which the two electrons have changed places. There are two combined states which are possible: the symmetric state and the antisymmetric state. The energy of each of these states is different to that of the original state. Since the energy differences arise from the exchange of particles this type of energy is called *exchange energy*. The covalent bond, ferromagnetism, and antiferromagnetism are examples of phenomena which occur because of exchange energy.

excitation A process that changes a quantum mechanical system such as a nucleus, atom, or molecule from one quantum state to another quantum state with a higher energy. Frequently the quantum state with the lower energy is the ground state. The difference in energy between the two quantum states is called the *excitation energy*.

excited state *See* energy level.

exclusion principle *See* Pauli exclusion principle.

exotic atom An atom in which either an electron is replaced by another negatively charged particle such as a muon or a proton is replaced by another positively charged particle such as a positron. Usually exotic atoms are the analogs of the hydrogen atom. Examples include positronium (e^+e^-), muonium (μ^+e^-), pionic hydrogen ($p^+\pi^-$), and muonic hydrogen ($p^+\mu^-$). Exotic atoms are unstable. However, many of them have sufficiently long lifetimes for them to have well defined spectra.

expectation value The value of a quantity that is obtained when the value of a quantity has been determined experimentally many times and the average is taken of these experimentally determined values.

exponential decay A function of the form e^{-x}, sometimes written as $\exp(-x)$. In exponential decay the value of $y = \exp(-x)$ goes down very rapidly as x increases.

exponential function In general, a function that varies as the power of some other quantity. Usually, one is concerned with the function e^x, which is sometimes written as $\exp(x)$. *See also* e.

exponential growth The very rapid growth of the EXPONENTIAL FUNCTION as the value of x increases.

extrinsic semiconductor *See* semiconductor.

Eyring, Henry (1901–81) American chemist who made many contributions to the application of quantum mechanics and statistical mechanics to chemistry, particularly the theory of the rates of chemical reactions. Eyring was a prolific author of scientific papers and co-authored several classic books, notably *The Theory of Rate Processes* with Samuel Glasstone and Keith Laidler which was published in 1941.

F

Fajans, Kasimir (1887–1975) Polish-born American chemist who is best known for his work on isotopes and for FAJANS' RULES of chemical bonding. In 1913 Fajans suggested, independently of Frederick SODDY that isotopes are atoms of elements that have the same proton number but different atomic weights.

Fajans' rules Rules that determine how readily covalent bonding rather than ionic bonding occurs. The first rule states that covalent bonding is more likely to occur if the number of electrons to be removed or donated increases, i.e. if the charge of the ions would be high. Fajans' second rule states that covalent bonding is favored by small positive ions and large negative ions.

fall-out Radioactive particles deposited on the Earth from the atmosphere. This usually originates from nuclear explosions but can also occur because of a nuclear accident. It consists of fission fragments, the most dangerous of which are iodine-131, which accumulates in the thyroid gland, and strontium-90, which accumulates in the bones. These isotopes can be passed on to the human population by milk, milk products, and meat from grazing animals.

false vacuum A vacuum state in a quantum field theory that is a local minimum but not the state of the system with the minimum energy overall (called the *true vacuum*). It is possible for tunneling from the false vacuum to the true vacuum to occur. This tunneling can be calculated using INSTANTON techniques. The false vacuum has been predicted to exist in certain quantum field theories of interest in the theory of elementary particles and cosmol-

ogy but has never been observed experimentally.

family *See* elementary particle.

farad Symbol: F   The SI unit of capacitance. When the plates of a capacitor are charged by one coulomb and there is a potential difference of one volt between them, then the capacitor has a capacitance of one farad. 1 farad = 1 coulomb volt^{-1} (C V^{-1}).

Faraday, Michael (1791–1867) English chemist and physicist who is best remembered for his fundamental contributions to electricity and magnetism. Faraday discovered the phenomenon of ELECTROMAGNETIC INDUCTION in 1831. Faraday demonstrated a connection between magnetism and optics when he discovered the FARADAY EFFECT in 1845. An important feature of Faraday's work on electricity and magnetism was that it was expressed in terms of fields and lines of force. Faraday also made many important contributions to chemistry.

Faraday constant Symbol: *F*   The electric charge carried by one mole of electrons or singly ionized ions. It is equal to the product of the AVOGADRO CONSTANT and the charge of an electron and has the value 9.648×10^4 coulombs per mole.

Faraday effect The rotation of the plane of polarization of electromagnetic radiation in certain substances (including quartz and water) in a magnetic field. The angle of rotation is proportional to both the magnetic field strength and the length of the path of the radiation in the medium.

Faraday's laws of electrolysis Two

laws found by Michael FARADAY in his experiments on electrolysis: (1) The amount of chemical change produced during electrolysis is proportional to the charge passed. (2) The charge Q needed to deposit a mass of m grams of an ion of relative ionic mass M is given by $Q = FmZ/M$, where F is the FARADAY CONSTANT and Z is the charge of the ion.

Faraday's laws of electromagnetic induction

Three laws which describe the phenomenon of electromagnetic induction: (1) When a magnetic field surrounding a conductor changes then an e.m.f. is induced in the conductor. (2) The size of the induced e.m.f. is proportional to the rate of change of the magnetic field. (3) The direction of an induced e.m.f. is such that it opposes the change producing it (LENZ'S LAW).

far infrared The longer-wavelength part of the infrared region of the electromagnetic spectrum; i.e. the part farthest in wavelength from the visible region and nearest to the radio-wave region.

fast neutron A term for rapidly-moving neutrons used in two senses. One sense of the term refers to neutrons that have kinetic energy greater than 0.1 MeV. More often the term means neutrons that have sufficient energy to start fission in uranium-238. This means an energy of more than 1.5 MeV.

fast reaction *See* nuclear reactor.

***f*-block elements** The block of elements of the periodic table consisting of the LANTHANIDES (from cerium to lutetium) and the ACTINIDES (from thorium to lawrencium). All these elements have two *s*-electrons in their outer shell (n) and *f*-electrons in their next to outer shell ($n-1$).

feedback *See* self-organization.

FEM *See* field-emission microscope.

femtochemistry The study of chemical processes occurring on the time-scale of a

femtosecond, i.e. 10^{-15} s. Femtochemistry became possible due to the development of lasers that could be pulsed in femtoseconds and it has enabled the molecular processes in a chemical reaction to be studied in a detailed way. In femtochemistry a femtosecond pulse results in dissociation of a molecule. Then, a further series of femtosecond pulses is released, with the frequency of the pulses being that of an absorption of a product of the dissociation process. The abundance of the dissociation product can be determined by the amount of absorption.

Fermat, Pierre de (1601–65) French mathematician and physicist. Fermat was a pioneer of both coordinate geometry and probability theory. His investigations of finding maxima and minima of curves and finding tangents to them anticipated some of the subsequent work on differential calculus by Sir Isaac NEWTON and Gottfried LEIBNIZ. He also worked in optics, making the fundamental discovery known as FERMAT'S PRINCIPLE OF LEAST TIME. He is perhaps best known for *Fermat's last theorem*. This theorem states that the equation
$$a^n + b^n = c^n$$
has no solutions for n greater than 2 if a, b, and c are all integers. Fermat's last theorem was proved in 1995.

Fermat's principle of least time The governing principle of geometrical optics. It was put forward by Pierre de FERMAT in letters in 1657 and 1662 and states that the path of a light ray is the one that takes the least time. Fermat showed that this principle leads to the law of rectilinear propagation of light and the laws of reflection and refraction.

fermi A unit of length, named after Enrico FERMI, which was formerly used in atomic and nuclear physics. It is equal to 10^{-15} m.

Fermi, Enrico (1901–1954) Italian physicist who made many important contributions to physics both as a theoretician and an experimentalist. In the mid 1920s he showed that a gas of particles that obey the PAULI EXCLUSION PRINCIPLE also obey what is now known as FERMI–DIRAC STATISTICS. In 1933 Fermi proposed his theory of beta decay, based on the realization that a new type of force, known as the *weak nuclear force*, is involved and that this force needs to be described by a relativistic quantum field theory.

In the mid 1930s, following the discovery of artificial radioactivity by Irène and Frédéric JOLIOT-CURIE in 1934, Fermi and his colleagues, performed experiments to generate new isotopes using slow neutrons. Fermi was awarded the 1938 Nobel Prize for physics for this work. Fermi and his colleagues constructed the first nuclear reactor, then known as an atomic pile, at the University of Chicago in 1942.

Fermi constant Symbol: G (sometimes written G_F or G_W) The constant characterizing the weak interactions, i.e. the analog for weak interactions of the ELECTRIC CONSTANT and the GRAVITATIONAL CONSTANT.

Fermi–Dirac distribution The distribution function for a gas of particles that are subject to the PAULI EXCLUSION PRINCIPLE.

Fermi–Dirac statistics *See* quantum statistics.

Fermi hole A region around a fermion, such as an electron, in a many-fermion system in which there is a low probability of finding another fermion of the same type with the same spin as the first fermion. The concept of a Fermi hole is useful for analyzing electron correlation in atoms and molecules.

Fermilab A laboratory with a particle accelerator located at Batavia, near Chicago, which was opened in 1972. The accelerator, known as the *Tevatron*, consists of two rings that accelerate protons to very high energies. The top quark was discovered at Fermilab in the 1990s.

Fermi level The energy level with the highest occupied quantum state at absolute zero temperature in a many-fermion system. This concept is particularly useful in the electronic structure of solids.

Fermi liquid A many-fermion system in which there is continuity between the 'bare' fermions (such as electrons) in the absence of interactions with other particles and the QUASI-PARTICLES in the interacting fermion system. The electrons in a normal metal constitute a Fermi liquid but it appears that the electrons in a high-temperature superconductor do not make up a Fermi liquid.

fermion A particle that is subject to the PAULI EXCLUSION PRINCIPLE. The wave function for a system containing more than one identical fermion is *antisymmetric*; i.e. it changes sign when the positions of any two of the fermions are interchanged. By the SPIN–STATISTICS THEOREM all fermions have ½-odd integer spin. Composite systems that obey the Pauli exclusion principle are also fermions (*see* Wigner's rule).

Fermi theory of weak interactions A RELATIVISTIC QUANTUM FIELD THEORY proposed by Enrico FERMI in 1933 to describe beta decay. The theory successfully describes beta decay at low energies but breaks down at high energies due to the fact that it is nonrenormalizable (*see* renormalization). The Fermi theory of weak interactions has been incorporated into the WEINBERG–SALAM MODEL, which gives a unified description of the weak and electromagnetic interactions.

fermium A radioactive transuranic element of the actinoid series, not found naturally on Earth. It is produced in very small quantities by bombarding ^{239}Pu with neutrons to give ^{253}Fm (half-life 3 days). Several other short-lived isotopes have been synthesized.

Symbol: Fm. Proton number: 100. Most stable isotope: ^{257}Fm (half-life 100.5 days). Electronic configuration: [Rn]$5f^{12}7s^2$.

ferromagnetism *See* magnetism.

fertile material Material that can absorb a neutron to form a FISSILE MATERIAL. For example, uranium-238 is a fertile material able to absorb a neutron to form uranium-239, which then decays to plutonium-239. The conversion of fertile material into fissile material occurs in a breeder reactor. *See* nuclear reactor.

Feynman, Richard Phillips (1918–1988) American theoretical physicist who made many important contributions to theoretical physics, particularly to the theory of elementary particles.

In his PhD thesis of 1942 and in a paper published in 1948, Feynman reformulated quantum mechanics in terms of what he called PATH INTEGRALS. In the late 1940s he showed that the difficulties of the infinities that occur in calculations of QUANTUM ELECTRODYNAMICS (QED) could be removed by a procedure called RENORMALIZATION. In his work on QED Feynman introduced what are now called FEYNMAN DIAGRAMS to aid and systematize the calculations. In the late 1960s Feynman developed the PARTON MODEL, an important milestone in the theory of strong interactions, which paved the way for QUANTUM CHROMODYNAMICS. In the early 1960s he elucidated some important features of QUANTUM GRAVITY by analyzing quantum gravity as a quantum field theory.

Feynman shared the 1965 Nobel Prize for physics with Julian SCHWINGER and Sin-Itiro TOMONAGA for their work on QED.

Feynman diagram A pictorial representation of the way that particles interact with each other, which is used extensively in calculations of these interactions. For example, a Feynman diagram can be produced for the interaction of two particles (represented by straight lines) by the exchange of a third particle (represented by a wavy line). The two particles could be electrons, with the third particle being a photon. Each Feynman diagram represents a term in a perturbation series, which can be calculated using certain rules for the diagram. Feynman diagrams originated with work on quantum electrodynamics. They are also used in other quantum field theories and in the quantum theory of many-body systems. Similar diagrams are also used in statistical mechanics.

field Formally, a field is a quantity whose value varies continuously. Usually, a field is taken to be a region in space in which a particle or body experiences a force due to some other particle or body. For example, there is an *electric field* around an electrically charged particle and a *gravitational field* around a body that has mass. The concept was introduced by Michael FARADAY in the 1840s and subsequently expressed mathematically by Lord KELVIN and James Clerk MAXWELL.

The strength of a field at a point is defined as the ratio of the force experienced by a small appropriate specimen to the relevant property of that specimen. For example, the gravitational field strength is the ratio force/mass and the electric field strength is the ratio force/electric charge. *See also* field theory.

field emission (cold emission) The emission of electrons from a cold metal by a strong external electric field. To obtain sufficiently high electric fields the metal is shaped to a sharp point. The electric field distorts the potential barrier holding the electrons in the metal. At sufficiently high fields the barrier is narrow enough for electrons to tunnel through. *See also* field-emission microscope.

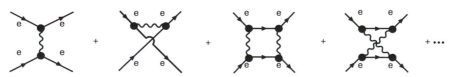

Feynman diagrams for electron–electron scattering

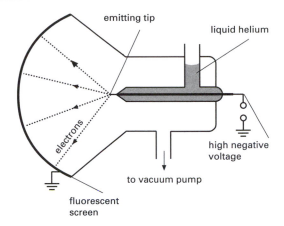

emitting tip

liquid helium

electrons

high negative
voltage

to vacuum pump

fluorescent
screen

Field-emission microscope

field-emission microscope (FEM) A device in which FIELD EMISSION from a sharp metal point is used to investigate atoms on the surface of a material. The emitted electrons are accelerated to a fluorescent screen, which gives a magnified image of the metal tip. The device is not able to resolve individual ions, but a pattern is obtained showing different crystal planes on the surface of the tip.

field ionization Ionization of atoms or molecules at the surface of a metal under the influence of a strong electric field. The metal is in the form of a very sharp point with a high positive potential. The process is similar to that in FIELD EMISSION with the difference that an electron moves from the atom or molecule absorbed on the surface into the metal by tunneling through the potential barrier. The resulting positive ion is accelerated away from the surface.

FIM *See* field-ion microscope.

field-ion microscope (FIM) A device similar to a FIELD-EMISSION MICROSCOPE, using field ionization to produce an image of the surface. Usually helium is the gas used (at very low pressure) and the device is cooled in liquid helium. It is possible to resolve the pattern of individual atoms on the metal tip and to identify adsorbed atoms.

field theory A theory of interactions in terms of fields. If the interaction is long range, as in the case of electromagnetic and gravitational interactions, then the interaction can be described by a CLASSICAL FIELD THEORY. Classical electrodynamics describes the electromagnetic interactions and general relativity theory describes the gravitational interactions. For interactions at the subatomic level, including the strong and weak nuclear interactions it is necessary to use QUANTUM FIELD THEORY.

fine structure Closely spaced lines in a spectrum. Fine structure in atomic spectra arises because of the existence of energy levels with almost exactly the same energy, with the energy level splitting being due to effects of electron spin. In molecular spectra, fine structure also occurs because of rotational and vibrational motion of the molecules. *See also* hyperfine structure.

fine structure constant Symbol: α A dimensionless number that occurs in calculations of atomic fine structure and, more generally, defines the strength of the electromagnetic interaction in quantum electrodynamics. It is defined by $e^2/\hbar c$, where e is the charge of an electron, $\hbar$ is the Dirac constant and c is the speed of light in a vacuum.

fissile material Material that undergoes nuclear fission. Sometimes the fission occurs spontaneously but, more often, it is induced when the material is irradiated by neutrons. Uranium-235 and plutonium-239 are examples of fissile nuclides. Fissile material is used in NUCLEAR REACTORS and nuclear weapons.

fission *See* nuclear fission.

fission-track dating A method of measuring the age of glasses and other minerals that depends on the tracks made in these solids by fragments from the spontaneous fission of contained uranium. The density of tracks in the material depends on the uranium content, the age of the material, and any fading of the tracks. After counting the tracks, the uranium content is estimated by irradiating the materials with neutrons to induce fissions. The number of extra tracks produced depends on the uranium content, which can then be estimated, giving the age since solidification.

Fitch, Val Logsdon (1923–) American physicist who discovered CP VIOLATION with James CRONIN in 1964 when investigating the decay of neutral KAONS. Cronin and Fitch shared the 1980 Nobel Prize for physics for this discovery.

Fizeau, Armand Hippolyte Louis (1819–1896) French physicist who is best remembered for determining the speed of light in air reasonably accurately. He obtained this result in 1849. In 1850 he showed, independently of Léon FOUCAULT, that light travels faster in air than in water, in accord with the wave theory of light.

flatness problem *See* inflation.

flavor *See* quark.

fluctuation A random deviation in the value of a quantity from its average value. In any system governed by quantum mechanics there are *quantum fluctuations*, even at the absolute zero temperature, because of the HEISENBERG UNCERTAINTY PRINCIPLE. Above the absolute zero temperature *thermal fluctuations* occur. It is thought that quantum fluctuations in the EARLY

UNIVERSE were responsible for large scale structures such as galaxies. It is necessary to take fluctuations into account to get a more accurate description of phase transitions than that given by MEAN FIELD THEORY.

fluorescence The emission of electromagnetic radiation, frequently in the form of light, from a substance after the substance has been stimulated in some way other than by heat. The stimulation can come from bombardment by electrons or absorption of electromagnetic radiation. For electromagnetic radiation the emitted radiation often has a longer wavelength than the absorbed radiation. An example of this is that certain materials absorb ultraviolet radiation, which is followed by the emission of visible light. This is the mechanism producing light in *fluorescent tubes*. Fluorescence occurs because of the decay of excited electronic states in molecules. *See also* luminescence; phosphorescence.

fluorine A greenish-yellow highly reactive gaseous element belonging to the halogens (group 17 of the periodic table). The main isotope is ^{19}F. A number of radioactive isotopes are known, the most stable being ^{18}F (half-life 1.83 h).
  Symbol: F. Boiling pt.: −188.14°C. Melting pt.: −219.62°C. Density: 1.696 kg m^{-3} (0°C). Proton number: 9. Relative atomic mass: 18.99840.32. Electronic configuration: [He]$2s^2 2p^5$.

fluoroscope An instrument that makes x-rays, or other non-visible radition, visible by means of a fluorescent screen.

flux In general, some type of flow. This flow can be a flow out of or into some area of either matter or energy or of the lines of force of a field such as an electric or magnetic field.

flux quantization The quantization of flux, particularly of magnetic flux and its analogs in other theories. Flux quantization occurs in superconductors and has been postulated to occur in QUANTUM

CHROMODYNAMICS and QUANTUM GRAVITY. It is an example of quantization that occurs because of TOPOLOGY rather than symmetry.

Fock, Vladimir Alexandrovich (1898–1974)   Russian theoretical physicist who made many important contributions to theoretical physics, particularly quantum mechanics and quantum field theory. In 1926 Fock was one of several physicists who independently found the equation of relativistic quantum mechanics now known as the KLEIN–GORDON EQUATION. In the late 1920s he was one of the first people to appreciate the importance of gauge invariance in quantum theory. In 1930 he put forward what is now known as HARTREE–FOCK THEORY to calculate many-electron systems.

forbidden transition   A transition between energy levels that is not allowed to occur because of a SELECTION RULE. It is possible for forbidden transitions to occur in practice, but with a much lower probability than for an allowed transition. Forbidden transitions are responsible for the existence of weak lines in spectra. A forbidden transition can occur for three reasons: (1) In the actual situation the selection rule that forbids the transition is only an approximate rule. (2) The selection rule applies only to dipole radiation; i.e. only the electric dipole part of the interaction between a system such as an atom and an electromagnetic field is considered. In reality, magnetic dipole radiation and quadrupole radiation may be involved in the transitions. (3) Although the selection rule is correct for atoms or molecules in isolation it does not necessarily hold in external fields or if collisions or other interactions can occur.

f-orbital   See orbital.

force   Symbol: F The vector quantity that tends to change the momentum of a body.

forces of Nature   See fundamental interactions.

Foucault, Jean Bernard Léon (1819–1868)   French physicist who made several important contributions to physics. In 1850, independently of Armand FIZEAU, he showed that the speed of light in air is greater than it is in water, in accord with the wave theory of light. In 1851 he invented what is now known as the *Foucault pendulum* to demonstrate the rotation of the Earth. In 1852 he invented the gyroscope. In 1862 he determined the speed of light to an accuracy within one per cent of the best current determination.

Fowler, William Alfred (1911–95) American physicist who collaborated with Sir Fred HOYLE on theories of the production of chemical elements inside stars (*see* nucleosynthesis). Fowler, Hoyle, and Geoffrey and Margaret Burbidge (who were also involved in this work) gave a detailed exposition of nucleosynthesis in a classic paper published in 1957. He shared the 1983 Nobel Prize for physics with Subrahmanyan CHANDRASEKHAR.

fractal   A figure that is produced by indefinite successive subdivision. For example, to produce a *snowflake curve* one can start with an equilateral triangle. Each side is then divided into three segments and the midle segments are replaced by two equal segments which would form a smaller equilateral triangle. A star-shaped figure with 12 sides is the result of this process. The sides of the resulting figure are subdivided in the same way and this process is repeated indefinitely. The resulting snowflake-like figure has a *fractional dimensional*, i.e. a dimension intermediate between that of a line (1) and a surface (2). The snowflake curve has a dimension 1.26. Fractals arise in several branches of physics including CHAOS THEORY and turbulence.

frame of reference   A set of coordinate axes that enables the position of any object in space to be defined at any instant in time. In four-dimensional space–time the frame of reference has four coordinate axes, three of space and one of time.

francium A radioactive element of group 1 of the periodic table. It is found on Earth only as a short-lived product of radioactive decay, occurring in uranium ores in minute quantities. A large number of radioisotopes of francium are known.

Symbol: Fr. Melting pt.: 27°C. Boiling pt.: 677°C. Proton number: 87. Most stable isotope: ^{223}Fr (half-life 21.8 minutes). Electronic configuration: [Rn]7s^1.

Franck, James (1882–1964) German-born American physicist who together with Gustav HERTZ provided the first experimental support of the theory of atomic energy levels put forward by Niels BOHR in 1913. Franck and Hertz bombarded mercury atoms with electrons. They found that the mercury atoms absorb exactly 4.9 electron volts (eV) of energy, thus showing that energy is only absorbed in atoms in definite, precise amounts. Franck and Hertz won the 1925 Nobel Prize for physics for this work.

Frank, Ilya Mikhailovitch (1908–90) Russian physicist who together with Igor TAMM gave the correct theoretical interpretation of CERENKOV RADIATION in 1937. Cerenkov, Frank, and Tamm shared the 1958 Nobel Prize for physics.

Frankland, Sir Edward (1825–99) British chemist who originated the concept of VALENCE in chemistry in 1852 and elaborated on it in 1866.

Fraunhofer, Josef von (1787–1826) German physicist and optician who developed the spectroscope, studied diffraction (*see* Fraunhofer diffraction), and found what are now known as FRAUNHOFER LINES in the spectrum of the Sun.

Fraunhofer diffraction The type of diffraction that occurs when the wave fronts are parallel when they reach the diffracting object. Usually Fraunhofer diffraction is concerned with light waves. It corresponds to a light source that is an infinite distance away from the diffracting object. Fraunhofer diffraction is a special, limiting case of FRESNEL DIFFRACTION but is easier to analyze and is more useful practically.

Fraunhofer lines Dark lines found in the spectrum of the Sun. They were discovered by William Wollaston in 1802 but not studied in detail until Josef FRAUNHOFER started doing so in 1814. The correct explanation of Fraunhofer lines – that they arise because of the absorption of some of the visible radiation emitted by the hot interior of the Sun by elements in its cooler outer regions – was put forward by Gustav KIRCHOFF in 1859.

frequency Symbol: f or ν The number of times a regular event is repeated per unit time. Usually, this refers to the number of cycles of a wave, or some other oscillation or vibration, per second. The unit of frequency is the hertz, denoted Hz, with one hertz being one cycle per second.

Fresnel, Augustin Jean (1788–1827) French physicist who made major contributions to establishing the wave theory of light. He pointed out that the phenomenon of light polarization requires that light waves are transverse rather than longitudinal.

Fresnel diffraction The type of diffraction that occurs when the wave fronts cannot be regarded as parallel plane waves as in FRAUNHOFER DIFFRACTION. This is the case when the source of the waves is fairly close to (in principle, not an infinite distance from) the diffracting object. Usually, Fresnel diffraction is concerned with light waves. Fresnel diffraction is more difficult to analyze than Fraunhofer diffraction since it is necessary to take account of the curvature of the wave fronts.

Friedman, Jerome Isaac (1928–) American physicist who took part in experiments that conclusively showed that there are particles inside protons and neutrons. Working with Henry KENDALL and Richard TAYLOR at SLAC in the late 1960s he investigated the internal structure of nucleons by scattering beams of high-energy electrons. The analysis of these experiments indicated that there are pointlike spin-½ particles inside nucleons. Friedman, Kendall, and Taylor shared the 1990 Nobel Prize for physics for this work.

Friedmann, Aleksandr Alexandrovich (1888–1925) Russian astronomer who is best known for finding solutions to Einstein's field equation of general relativity theory that correspond to an expanding Universe. *See also* Universe.

Friedmann–Lemaître–Robertson–Walker Universe A solution of the Einstein field equation of general relativity theory that corresponds to a homogeneous, isotropic expanding Universe. This model of the Universe was found by Alexander FRIEDMANN in 1922 and independently by Georges LEMAÎTRE in 1927 and expressed in a modified way by Howard Robertson and Arthur Walker in the mid 1930s. It is a good model of the expanding Universe in accord with the big-bang theory.

Frisch, Otto Robert (1904–1979) Austrian-born British physicist who worked with his aunt Lise MEITNER on the theory of nuclear fission. In 1939 Frisch and Meitner realized that the results of Meitner's former colleague Otto HAHN – that when uranium is bombarded by neutrons the lighter element barium is one of the products – meant that fission had occurred. In 1940 Frisch together with Rudolf PEIERLS wrote a report to the British Government indicating that an explosive nuclear chain reaction could be produced using uranium-235.

frontier orbital A molecular ORBITAL that is either the highest occupied molecular orbital or the lowest unoccupied molecular orbital. The chemical properties and the spectrum of the molecule are usually determined by these two orbitals.

Fukui, Kenichi (1918–98) Japanese physical and theoretical chemist who developed the FRONTIER ORBITAL concept and applied it to chemical reactions. Fukui shared the 1981 Nobel Prize for chemistry with Roald HOFFMANN.

functional A function of a function. Functionals are used extensively in quantum field theory, the many-body problem in quantum mechanics, and statistical mechanics.

fundamental constants of Nature Certain parameters that have the same size throughout the Universe. Examples include the Planck constant, the speed of light in a vacuum, the electric constant, the magnetic constant, the gravitational constant, the Fermi constant, and the charge on an electron. It used to be thought that the values of these constants did not change with time but evidence has begun to emerge that this may not necessarily be the case. It has been observed that if the values of several of the fundamental constants were different then life as we know it would not be possible. This has given rise to the MULTIVERSE idea – that there are many Universes, each of which has its own set of fundamental constants.

fundamental interactions The four different types of interaction that can occur between particles or bodies: the *gravitational, electromagnetic, strong,* and *weak nuclear interactions* (usually referred to as the *strong interactions* and the *weak interactions*). The gravitational and electromagnetic interactions are long-range interactions. This means that they manifest themselves on scales much larger than the atomic scale, with classical field theories giving suitable descriptions of phenomena on these scales. The strong and weak interactions are short-range interactions, with their range being similar to that of an atomic nucleus. Indeed, the size of atomic nuclei is due to the short range of these interactions.

The gravitational interaction is the weakest type of interaction. Nevertheless, because it acts between all massive bodies and is always attractive it is able to overcome all the other interactions for a sufficiently large amount of matter. A very good description of the familiar manifestations of gravity, such as an apple falling from a tree and planets orbiting the Sun, is given by NEWTON'S LAW OF GRAVITY in which the strength of the force between two bodies falls off as the square of the distance between the bodies. A more accurate theory of gravity is provided by the Einstein field equation of general relativity theory, in which gravity arises because of the curvature of space–time rather than occurring because of action at a distance as in Newtonian gravity. It is necessary to use general relativity theory to describe very dense bodies with strong gravitational fields, such as neutron stars and black holes, and to describe the evolution of the whole Universe. For systems such as the Solar System the effects of general relativity theory are minor corrections to Newtonian gravity. Although it has not been found possible to construct a quantum field theory of the gravitational interactions it is frequently useful to regard gravitational interactions between bodies as being mediated by particles known as GRAVITONS. Very promising descriptions of gravitational interactions at the microscopic level include SUPERSTRING THEORY and SPIN NETWORKS. It has been suggested that the BRANE WORLD picture provides an explanation of why the gravitational interaction is much weaker than the other interactions, with the nongravitational interactions living in the 'surface' of the brane while the gravitational interactions live in the 'bulk'.

Electric and magnetic forces are different aspects of one type of interaction called the *electromagnetic interaction*. All electrically charged particles are subject to electromagnetic interactions. Unlike gravitational interactions, electromagnetic interactions can be either attractive or repulsive. Since the electromagnetic interaction is long range it has a well-defined classical field theory called CLASSICAL ELECTRODYNAMICS, which is governed by the MAXWELL EQUATIONS. The properties of atoms and molecules, and hence chemistry and biology, are dominated by electromagnetic interactions they are responsible for holding electrons and nuclei together and for the chemical bonds between atoms in molecules. At the microscopic level electromagnetic interactions are governed by a quantum field theory called QUANTUM ELECTRODYNAMICS, with the electromagnetic interactions being mediated by PHOTONS.

The strong interactions that hold protons and neutrons together in nuclei, in spite of the electrostatic repulsion between protons, are a vestige of the fundamental strong interactions that hold quarks together in nucleons. All quarks, and hence all hadrons, are subject to the strong interaction. The strong interaction is described by a quantum field theory called QUANTUM CHROMODYNAMICS, with the interactions between the quarks being mediated by GLUONS. Because of ASYMPTOTIC FREEDOM the strong nuclear force has the property that, in complete contrast to the gravitational

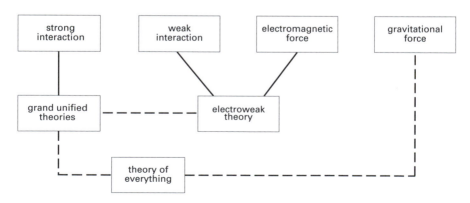

The relationship between the four fundamental interactions

and electromagnetic interactions, it becomes stronger as quarks are further apart.

The weak interaction is responsible for BETA DECAY and other decay processes in which one type of quark changes into another type of quark. All the LEPTONS and quarks take part in weak interaction processes. The weak interactions are described by a theory called QUANTUM FLAVORDYNAMICS (the WEINBERG–SALAM MODEL) which gives a unified description of the weak and electromagnetic interactions, with the weak interactions being mediated by W BOSONS and Z BOSONS.

The strong interaction is about 100 times as strong as the electromagnetic interaction (which is why the heaviest stable nucleus has an proton number of about 100). The weak interaction is about 10^{10} times weaker than the electromagnetic interaction. The gravitational interaction is roughly 10^{40} times weaker than the electromagnetic interaction. It is not known how this pattern of interaction strengths emerges and, in particular, why the gravitational interaction is so much weaker than the nongravitational interactions.

Another difference between the gravitational and the nongravitational interactions is that the bosons which mediate the nongravitational interactions have spin 1 whereas gravitons have spin 2.

The unification of all the interactions into one model remains one of the fundamental aims of physics. The unification of the weak and electromagnetic interactions has been achieved with considerable success but none of the many attempts to produce further unification can be considered to be definitive at the time of writing. *See also* grand unified theories; M-theory; superstring theory; unified field theory.

fusion   *See* nuclear fusion.

G

Gabor, Dennis (1900–79) Hungarian-born British physicist who invented HOLOGRAPHY in 1948 while working on improving the resolution of the electron microscope. The importance of his work became apparent after the invention of the laser. Gabor was awarded the 1971 Nobel Prize for physics.

gadolinium A ductile malleable silvery element of the lanthanoid series of metals. It occurs in association with other lanthanoids. Gadolinium is used in alloys, magnets, and in the electronics industry.
Symbol: Gd. Melting pt.: 1313°C. Boiling pt.: 3266°C. Relative density: 7.9 (25°C). Proton number: 64. Relative atomic mass: 157.25. Electronic configuration: $[Xe]4f^75d^16s^2$.

galaxy A very large collection of stars and their satellites, dust, and gas held together by mutual gravitational attraction. The Sun belongs to a galaxy usually known as the *Galaxy* or the *Milky Way*. There are about a hundred billion stars in our Galaxy.

galaxy formation *See* structure formation.

Galilean transformations A set of transformations that describes the change in the parameters of motion and position from a frame of reference with the origin at O and coordinates (x, y, z) to a frame with the origin at O' and coordinates at (x', y', z') which is moving in the x' direction with a constant velocity v relative to the first frame. The set is: $x' = x - vt$, $y' = y$, $z' = z$, $t' = t$, where O and O' coincide at $t = 0$. If motion is invariant under Galilean transformations then the motion is said to be characterized by *Galilean invariance*. Newtonian mechanics is characterized by Galilean invariance. *See also* Lorentz transformations.

Galileo Galilei (1564–1642) Italian astronomer and physicist who was one of the founders of the scientific study of motion. He paved the way for the EQUIVALENCE PRINCIPLE underlying general relativity theory about 300 years later when he showed that bodies with different weights all fall at the same speed (in the absence of air resistance). He was an early pioneer of the telescope, with which he made many important discoveries. For example, his discovery of the moons of Jupiter refuted the old motion that all heavenly bodies had to orbit the Earth. His discoveries led to a conflict with the Roman Catholic Church.

gallium A soft silvery low-melting metallic element belonging to group 3 of the periodic table. It is found in minute quantities in several ores, including zinc blende (ZnS) and bauxite ($Al_2O_3.H_2O$). Gallium is used in low-melting alloys, high-temperature thermometers, and as a doping impurity in semiconductors. Gallium arsenide is a semiconductor used in light-emitting diodes and in microwave apparatus.
Symbol: Ga. Melting pt.: 29.78°C. Boiling pt.: 2403°C. Relative density: 5.907 (solid at 20°C), 6.114 (liquid). Proton number: 31. Relative atomic mass: 69.723. Electronic configuration: $[Ar]3d^{10}4s^24p^1$.

gamma decay A type of radioactive decay in which gamma rays are emitted by the specimen. Gamma decay occurs when a nuclide is produced in an excited state, gamma emission occurring by transition to a lower energy state. It can occur in association with alpha decay and beta decay.

gamma radiation Electromagnetic radiation in which the wavelength of the electromagnetic waves is less than 10^{-10} m. This type of radiation is emitted by an atomic nucleus when it returns to the ground state from an excited state. A nucleus is left in an excited state by the emission of alpha particles or beta decay or by bombardment by particles such as neutrons. There are also astronomical sources of gamma radiation.

gamma rays Streams of gamma radiation.

Gamow, George (1904–68) Russian-born American physicist who made important contributions to nuclear physics and cosmology. In 1928 he showed that alpha decay occurs because of quantum mechanical tunneling. Perhaps his most significant contribution to physics was his development of the BIG-BANG THEORY in the late 1940s and early 1950s, in which he combined the idea of an expanding Universe starting with a big bang found by his old teacher Aleksandr FRIEDMANN with nuclear physics. This led to the prediction of the existence of the COSMIC MICROWAVE BACKGROUND radiation and the realization that the light elements such as helium were mostly formed in the early Universe rather than in stars.

gas constant Symbol: R The constant which appears in the EQUATION OF STATE for an ideal gas. It has the value 8.314 J K^{-1} mol^{-1}.

gas-cooled reactor *See* nuclear reactor.

gauge boson A boson that mediates the interactions between particles in a GAUGE THEORY. Examples of gauge bosons include the photon in quantum electrodynamics, gluons for the strong interactions in quantum chromodynamics, and W- and Z-bosons for the weak interactions in the Weinberg–Salam model. All these particles are spin-one vector bosons. The spin-two

GRAVITON can be regarded as a gauge boson that mediates the gravitational interactions if general relativity theory is regarded as a gauge theory.

gauge group *See* gauge theory.

gauge theory A type of field theory that involves a symmetry group (called the *gauge group*) for the fields and potentials. In the case of electrodynamics the group is the Abelian group U(1), whereas the gauge theories for the strong interactions and the theory unifying the weak interactions with the electromagnetic interactions use non-Abelian groups. This is why QUANTUM ELECTRODYNAMICS is a much simpler theory than QUANTUM CHROMODYNAMICS (QCD), which describes the strong interactions, and the WEINBERG–SALAM MODEL, which unifies the weak and electromagnetic interactions. In the case of QCD the group is SU(3) while for the Weinberg–Salam model the group is the product SU(2) × U(1) (*see* Lie groups). GRAND UNIFIED THEORIES, which attempt to unify the strong, weak, and electromagnetic interactions, are gauge theories using groups such as SU(5) and SO(10). In the gauge theories mentioned so far the interactions are mediated by spin-1 GAUGE BOSONS. Gravity can also be regarded as a gauge theory but the gauge group is more complicated than the groups used to describe the nongravitational interactions and the gravitational interactions are mediated by spin-2 GRAVITONS.

Gauss, Karl Friedrich (1778–1855) German mathematician who made major contributions to many branches of mathematics. He is generally regarded as being one of the greatest mathematicians of all time. Among his mathematical contributions of relevance to physics were his study of non-Euclidean geometry and complex numbers. In physics itself he contributed to the development of celestial mechanics, optics, electricity, and magnetism.

Gay-Lussac, Joseph-Louis (1778–1850) French chemist who made many contributions to chemistry, particularly the properties of gases.

Gay-Lussac's law 1. An alternative name sometimes given to CHARLES' LAW. 2. The principle that when gases combine chemically they do so in simple whole number proportions by volume. For example, one volume of oxygen combines with two volumes of hydrogen to form water. The statement of the law by Joseph-Louis Gay-Lussac in 1808 was an important step in the development of the theory of atoms and molecules.

gedanken experiment *See* thought experiment.

Geiger, Hans Wilhelm (1882–1945) German physicist who developed a counter to measure radioactivity. Geiger developed the first version of the device with Ernest RUTHERFORD in 1908 to count alpha particles. In addition to this work the investigations of Geiger and Ernest Marsden on the scattering of alpha particles in 1909 led Rutherford to propose the nuclear structure of the atom.

Geiger counter (Geiger–Müller counter) A device used to detect ionizing radiation and hence to count the particles and photons producing the ionization. It consists of a tube containing a cylindrical metal cathode with a wire anode at its center and a low-pressure gas, which is frequently a mixture of argon and methane. There is a potential difference of about 1000 volts between the electrodes. When an electrically charged particle entering the counter causes an ion to be produced the ion is accelerated toward an electrode. The energy of the ion causes further ionization by collision in the gas. This produces a measurable pulse.

Geiger–Nuttall law A law put forward as an empirical observation by Hans Wilhelm Geiger and John Mitchell Nuttall in 1912 stating that the energy E of an alpha particle is directly proportional to log γ, where γ is the DECAY CONSTANT of the alpha emitter. The Geiger–Nuttall law was ex-plained theoretically in 1928 by George Gamow and independently by Ronald Gurney and Edward Condon in terms of their theory of alpha decay as a quantum mechanical tunneling effect.

Gell-Mann, Murray (1929–) American physicist who made major contributions to the classification of elementary particles. In 1953 Gell-Mann, and independently Kazuhiko Nishijima, explained the unexpectedly long lifetimes of KAONS by postulating that a quantity, which he called STRANGENESS, is conserved. In 1961 Gell-Mann, and independently Yuval NE'E-MAN, postulated that hadrons can be classified in a pattern called the EIGHTFOLD WAY based on the group SU(3). This led to the successful prediction of the existence of the OMEGA MINUS PARTICLE. In 1964 Gell-Mann, and independently George ZWEIG, explained this pattern by postulating that hadrons are composite particles of more fundamental particles, which Gell-Mann called QUARKS. In the early 1970s Gell-Mann was one of the founders of QUANTUM CHROMODYNAMICS. In the 1980s Gell-Mann collaborated with James Hartle in developing the CONSISTENT HISTORIES interpretation of quantum mechanics. Gell-Mann was awarded the 1961 Nobel Prize for physics.

general relativity theory A classical field theory that describes gravitation and extends special relativity theory to noninertial frames of reference. It was put forward by Albert EINSTEIN in 1915. General relativity theory reduces to NEWTONIAN GRAVITY in the limit at which the escape velocity from a massive body is much smaller than the speed of light in a vacuum. It is necessary to use general relativity rather than Newtonian gravity to describe very dense bodies such as neutron stars and black holes and to give an account of the evolution of the Universe as a whole. In general relativity theory the gravitational field arises because of the curvature of space–time caused by the presence of mass. The equivalence between inertial mass and gravitational mass is central to general relativity rather than being an additional postulate as in Newtonian gravity.

There are several differences between general relativity theory and Newtonian gravity that allow the two theories to be distinguished experimentally. For example, general relativity theory predicts that light is bent by massive bodies such as the Sun. Experiments performed during solar eclipses have confirmed this expectation. There are small differences between general relativity theory and Newtonian gravity in the motions of planets, particularly in the orbit of Mercury. Very accurate measurements have confirmed that the motions of planets are in very good accord with the predictions of the theory. An important prediction of general relativity is that accelerating bodies with mass radiate GRAVITATIONAL WAVES just as accelerating charged bodies radiate electromagnetic waves. Although gravitational waves have not been directly detected their existence has been inferred from observations of the BINARY PULSAR. There is extremely good agreement between the predictions of general relativity theory and all known observations.

The basic equation of general relativity theory is the EINSTEIN FIELD EQUATION. The curvature of space–time in general relativity theory is described using RIEMANNIAN GEOMETRY. The occurrence of SINGULARITIES in general relativity theory is frequently regarded as indicating that general relativity theory breaks down at extremely short distances and that a theory of QUANTUM GRAVITY is required at such short distances.

generation *See* elementary particles.

geometrical optics See OPTICS.

germanium A hard brittle gray metalloid element belonging to group 14 of the periodic table. It is found in sulfide ores and in zinc ores and coal. Most germanium is recovered during zinc or copper refining as a by-product. Germanium was extensively used in early semiconductor devices but has now been largely superseded by silicon. It is used as an alloying agent, catalyst, phosphor, and in infrared equipment.

Symbol: Ge. Melting pt.: 937.45°C. Boiling pt.: 2830°C. Relative density: 5.323 (20°C). Proton number: 32. Relative atomic mass: 72.61. Electronic configuration: $[Ar]3d^{10}4s^24p^2$.

Germer, Lester Halbert (1896–1971) American physicist who participated with Clinton Davisson on the experiments on electron diffraction. *See* Davisson–Germer experiment.

Giaever, Ivar (1929–) Norwegian-born American physicist who did important research on the tunnel effect in superconductors for which he shared the Nobel Prize for physics with Leo ESAKI and Brian JOSEPHSON.

Gibbs, Josiah Willard (1839–1903) American theoretical physicist who made major contributions to thermodynamics and statistical mechanics. In the 1870s he pioneered the application of thermodynamics to chemistry. He was also one of the pioneers of STATISTICAL MECHANICS, showing how thermodynamic properties are related to statistical properties. In mathematics, Gibbs was one of the founders of VECTOR theory.

Glaser, Donald Arthur (1926–) American physicist who built the first BUBBLE CHAMBER in 1952, for which he was awarded the 1960 Nobel Prize for physics.

Glashow, Sheldon Lee (1932–) American physicist who paved the way for the WEINBERG–SALAM MODEL by suggesting that the weak and electromagnetic interactions are unified by a gauge theory associated with the group $SU(2) \times U(1)$. He also predicted the existence of charm quarks. Together with Howard GEORGI, Glashow constructed one of the first GRAND UNIFIED THEORIES in 1974. Glashow, Weinberg and Salam shared the 1979 Nobel Prize for physics.

global symmetry See SYMMETRY.

glueball An electrically neutral meson consisting of two or more gluons bound together by the strong interaction. Glueballs are predicted to exist by QUANTUM CHROMODYNAMICS and to decay very rapidly into other particles. This makes it very difficult to detect them; nevertheless, evidence was found in the mid 1990s for glueballs with masses of 1500 MeV and 1700 MeV. Although these masses are in accord with the theoretical expectations the evidence for the existence of glueballs is not yet generally regarded as conclusive.

gluon The GAUGE BOSON that mediates strong interactions. Gluons have never been observed directly and, as with QUARKS, it is thought that gluons can never be seen in isolation. The most direct evidence for the existence of gluons comes from JET events.

Goeppert-Mayer, Maria (1906–72) German-born American physicist who developed the nuclear SHELL MODEL. In 1948 she showed that the shell model explains the MAGIC NUMBERS of particularly stable nuclei. Goeppert-Mayer and Hans JENSEN, who developed the shell model independently, were awarded the 1963 Nobel Prize for physics (which they shared with Eugene WIGNER).

Goeppert-Mayer's rule A rule stating that in the nuclear shell model the state with the lowest energy is the state with the lowest spin. This rule, which was put forward in 1950, is useful for predicting the spin of nuclei. It is analogous to (but opposite to) HUND'S RULE in atomic shell theory.

gold A transition metal that occurs native. It is unreactive and is very ductile and malleable. Gold is used in jewelry, often alloyed with copper, and in electronics and colored glass. Pure gold is 24 carat; 9 carat indicates that 9 parts in 24 consist of gold.

Symbol: Au. Melting pt.: 1064.43°C. Boiling pt.: 2807°C. Relative density: 19.320 (20°C). Proton number: 79. Relative atomic mass: 196.96654. Electronic configuration: [Xe]$4f^{14}5d^{10}6s^1$.

Gold, Thomas Austrian-born American astronomer who is best known as one of the originators of the STEADY STATE THEORY of the Universe. Soon after the discovery of PULSARS in 1968 Gold put forward the correct explanation, independently of Franco Pacini, that they are rapidly rotating NEUTRON STARS. Gold correctly predicted that pulsars should very gradually slow down.

goldstino (Goldstone fermion) A fermion that is the analog of the GOLDSTONE BOSON in SUPERSYMMETRY.

Goldstone, Jeffrey (1933–) British-born American physicist who was one of the first people to appreciate the importance of SPONTANEOUS SYMMETRY BREAKING in quantum field theory. Goldstone has also made important contributions in a number of other areas.

Goldstone boson *See* Goldstone's theorem.

Goldstone's theorem A theorem in relativistic quantum field theory stating that if a continuous symmetry is a spontaneously broken symmetry then there must be a spin-zero, massless particle. This particle is called a *Goldstone boson*, which can be either an elementary particle or a composite particle.

Goudsmit, Samuel Abraham (1902–78) Dutch-born American physicist who put forward the idea of electron spin with George UHLENBECK in 1925 to explain FINE STRUCTURE and the ANOMALOUS ZEEMAN EFFECT in atomic spectra.

grad *See* gradient operator.

gradient operator (grad) .

grand unified theories (GUTs) Theories that attempt to unify the strong, weak, and electromagnetic interactions into a single theory. Several different GUTs have been proposed based on gauge groups such as SU(5) and SO(10). These theories postulate that at high energies the interactions merge into a single interaction, with the standard model emerging from the GUT as a result of spontaneous symmetry breaking associated with the HIGGS MECHANISM. The energies at which the interactions come together in GUTs has been calculated to be about 10^{15} GeV. This is much higher than any conceivable accelerator of the future but was attained in the very EARLY UNIVERSE. For that reason GUTs are of relevance to the early Universe; for example, they provide a solution to the problem of matter–antimatter asymmetry in the Universe.

GUTs suggest that very massive gauge bosons called X-BOSONS mediate interactions between quarks and leptons. This leads to the main prediction of GUTs, which is that protons decay to positrons and pions, with baryon number ceasing to be a conserved quantity. The lifetime of the proton is predicted to be much larger than the age of the Universe. Nevertheless, the probabilistic nature of quantum mechanics means that if sufficiently many protons are present in an observed sample then some should decay each year. Extensive experiments have been performed in attempts to detect proton decay. These have indicated that if proton decay does occur then the lifetime of the proton must be longer than the predictions of GUTs. This finding rules out many GUTs. However, if GUTs is combined with SUPERSYMMETRY to give *supersymmetric grand unified theories (SUSY GUTs)*, then such theories predict a lifetime about a thousand times longer than the predictions of nonsupersymmetric GUTs. It is hoped that experiments in the next few years will be able to test this prediction of SUSY GUTs.

Most GUTs correctly predict that neutrinos have nonzero masses. Many also predict the existence of MAGNETIC MONOPOLES and/or COSMIC STRINGS for topological reasons but there is no evidence for such entities.

GUTs were first proposed in the mid 1970s by Howard Georgi, Sheldon GLASHOW, and others and combined with supersymmetry in the early 1980s.

gravitino A spin 3/2 particle that is predicted to exist in SUPERGRAVITY as the supersymmetric partner of the GRAVITON.

The gravitino is described by the RARITA–SCHWINGER EQUATION. There is no evidence for the existence of the gravitino at present.

gravitational collapse A phenomenon predicted in general relativity theory in which a sufficiently large concentration of mass in a body results in the collapse of the body to a BLACK HOLE due to the mutual gravitational attraction of the matter being large enough to overcome the thermal pressure or DEGENERACY PRESSURE resisting the collapse. Gravitational collapse to a black hole is thought to occur to stars whose mass is greater than the OPPENHEIMER–VOLKOFF LIMIT after their nuclear 'fuel' has been exhausted.

gravitational constant Symbol: G The constant that appears in Newton's law of gravity. For that reason it is sometimes known as the *Newton constant*. It has the value 6.725985×10^{-11} Nm^2 kg^{-2}.

gravitational field The region of space surrounding a MASSIVE BODY. In this region there is a force of attraction on any other massive body. The *gravitational field strength* at a point in the gravitational field around a body of mass M is given by the ratio of the force to a test mass, i.e. GM/R^2, where G is the gravitational constant and R is the distance between the body and the test mass.

gravitational interaction *See* fundamental interactions.

gravitational lens A massive object that deflects light in a way that is analogous to a lens in optics. Gravitational lenses were predicted to exist in 1936 by Albert EINSTEIN using general relativity theory. In 1979 a 'double' quasar was observed as a result of a galaxy, or cluster of galaxies, between the quasar and the Earth acting as a gravitational lens to produce the multiple image. The images produced by gravitational lenses are used in astronomy to obtain information about the mass distribution of galaxies or clusters of galaxies.

gravitational mass   *See* mass.

gravitational potential   The gravitational potential at a point in a gravitational field is the energy needed to bring unit mass from infinity to that point.

gravitational radiation   The emission of *gravitational waves* as a result of the acceleration of massive bodies. This is analogous to ELECTROMAGNETIC RADIATION being the emission of ELECTROMAGNETIC WAVES as a consequence of the acceleration of electrically charged bodies. However, gravitational radiation is very weak (only 10^{-40} as strong as electromagnetic radiation). For example, the amount of gravitational radiation produced by a planet such as the Earth orbiting the Sun is about the same as the energy produced by an ordinary electric light bulb. Because it is very weak gravitational radiation is very difficult to detect directly and no direct evidence exists. It has been detected indirectly in observations of the BINARY PULSAR, with these observations being in extremely good accord with the expectation that the motion produces gravitational radiation.

Gravitational radiation sets matter vibrating just as electromagnetic radiation sets charged particles vibrating. This is the principle behind *gravitational wave detectors*, which have a large mass kept under conditions in which other sources of vibration are removed. Sensitive detectors are attached to the mass in attempts to detect vibrations caused by gravitational waves. Typically, a number of widely spaced detectors are used, and observations are made to look for coincident events in two or more detectors. If they could be made sufficiently accurate, such detectors might be able to register waves from violent events, such as the collapse of a star to form a neutron star or black hole.

Gravitational waves move at the same speed as light in a vacuum. They were predicted as consequences of general relativity theory by Albert EINSTEIN in 1916.

gravitational wave   *See* gravitational radiation.

graviton   A spin-2 particle that is thought to mediate the gravitational interaction. The graviton is the analog in quantum gravity of the photon in quantum electrodynamics. The particle has not been detected in any experiments.

gravity   The force associated with the GRAVITATIONAL INTERACTION. Gravity acts on all objects that have mass.

Green, Michael Boris (1946–) British physicist who developed SUPERSTRING THEORY with John SCHWARZ in the early 1980s.

Grotrian diagram   A diagram that shows the energy levels and allowed transitions in an atom. Each energy level is represented by a horizontal line. It is possible to represent the intensity of a transition in a Grotrian diagram by making the thickness of the line representing the transition to be proportional to the intensity. This type of diagram was invented by the German spectroscopist W. Grotrian in 1928.

ground state   *See* energy level.

group   A mathematical structure that is used to study symmetry. Formally, a group G is a collection of *elements a*, *b*, *c*, etc., with the elements being related to each other by certain rules.
(1) For any two members *a* and *b* of G the product *ab* must also be a member of G.
(2) For any three elements *a*, *b*, *c*, there is an associative relation $a(bc) = (ab)c$.
(3) There is an element called the *identity element* (sometimes called the *unit element*), usually denoted *e*, such that $ae = ea = a$ for all elements *a*.
(4) Each element *a* has an inverse, denoted a^{-1}, which is also a member of G.

Two elements *a*, *b* of a group G are said to *commute* if $ab = ba$. If $ab = ba$ for all pairs *a*, *b* of elements of G then G is said to be an *Abelian group*. If this is not the case then G is said to be a *non-Abelian group*. If a group has a finite number of elements then the group is said to be a *discrete group*. For example, the set of reflection and rotation symmetries of shapes that molecules can have, such as triangles and tetrahedra, gives discrete point groups. If the group has an infinite number of elements with the elements being continuous then the group is said to be a *continuous group*. For example, the set of rotations about a fixed axis forms a continuous group called the *rotation group*. A particular type of continuous group that is of great importance in elementary particle theory is a LIE GROUP.

Because of the connection between symmetries and conservation laws expressed by NOETHER'S THEOREM there is a close relation between group theory and conserved quantities. For example, the rotation group is closely associated with angular momentum. Group theory is of particular importance in quantum mechanics. Group theory enables the quantum numbers which characterize the energy levels and the selection rules between these levels to be understood for atoms, molecules, solids, nuclei, and elementary particles. *See also* irreducible representation; Lie group; Lorentz group; permutation group; point group; rotation group; space group.

group 0 elements *See* group 18 elements.

group 1 elements (alkali metals) A group of elements in the PERIODIC TABLE: lithium, sodium, potassium, rubidium, cesium, francium.

group 2 elements (alkaline-earth metals) A group of elements in the PERIODIC TABLE: beryllium, magnesium, calcium, strontium, barium, and radium.

groups 3–12 *See* TRANSITION METALS.

group 13 elements A group of elements in the PERIODIC TABLE: boron, aluminum, gallium, indium, and thallium.

group 14 elements A group of elements in the PERIODIC TABLE: carbon, silicon, germanium, tin, and lead.

group 15 elements A group of elements in the PERIODIC TABLE: nitrogen, phosphorus, arsenic, antimony, and bismuth.

group 16 elements (chaleogens) A group of elements in the PERIODIC TABLE: oxygen, sulfur, selenium, tellurium, and polonium.

group 17 elements (halogens) A group of elements in the PERIODIC TABLE: fluorine, chlorine, bromine, iodine, and astatine.

group 18 elements (noble gases; rare gases; inert gases; group 0 elements) A group of elements in the PERIODIC TABLE: helium, neon, argon, krypton, xenon, and radon.

Guth, Alan Harvey (1947–) American physicist who proposed the concept of INFLATION in 1980 to solve some problems in cosmology that could not be solved by the BIG-BANG THEORY. Although Guth's original mechanism was soon realized to be incorrect, his general idea that inflation is necessary to understand the evolution of the Universe is correct.

GUTs *See* grand unified theories.

gyromagnetic ratio The ratio of the magnetic moment of a system, such as an electron or a nucleus, to its angular momentum. For the orbital angular momentum of an electron the gyromagnetic ratio has the value $-e/2mc$ and for the spin $-e/mc$, where e and m are the charge and mass of an electron respectively and c is the speed of light in a vacuum.

H

hadron A type of composite particle made up of QUARKS held together by strong interactions. Hadrons can be divided into two groups: *baryons*, which are made up of three quarks, and *mesons*, which are made up of a quark and an antiquark. The most familiar examples of baryons are the proton and the neutron.

hafnium A transition metal found in zirconium ores. Hafnium is difficult to work and can burn in air. It is used in control rods for nuclear reactors and in certain specialized alloys and ceramics.

Symbol: Hf. Melting pt.: 2230°C. Boiling pt.: 5197°C. Relative density: 13.31 (20°C). Proton number: 72. Relative atomic mass: 178.49. Electronic configuration: [Xe]$4f^{14}5d^26s^2$.

Hahn, Otto (1879–1968) German chemist who discovered nuclear fission. Hahn devoted his career to the study of radioactivity, mostly in collaboration with Lise MEITNER. In 1917 Hahn and Meitner discovered the element protactinium. In 1921 Hahn defined nuclear isomerism. In 1938 Hahn and Fritz STRASSMANN discovered that when uranium is bombarded by slow neutrons it is changed into the much lighter element barium. Meitner and Otto FRISCH subsequently showed that Hahn

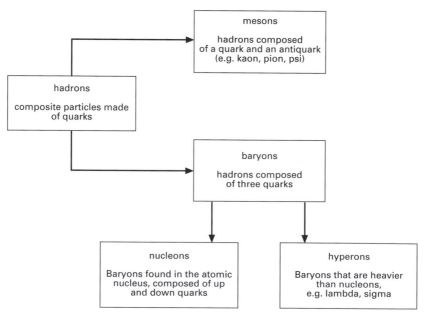

Classification of hadrons

and Strassmann had discovered NUCLEAR FISSION. Hahn was awarded the 1944 Nobel Prize for chemistry.

hahnium   *See* transactinide elements.

half-life Symbol: $T_{1/2}$ The time it takes for half the nuclei in a sample of radioactive material to decay. The stability of a nucleus is indicated by its half-life. If nuclei are stable their half-lives can be regarded as infinite. There is enormous variety in the half-lives of radioactive nuclei, ranging from picoseconds to billions of years.

Hall effect   The production of an electric potential when an electric current flows through a conductor with a magnetic field applied at right angles to the conductor. The electric potential is perpendicular to both the direction of the current and the direction of the magnetic field. It occurs because the LORENTZ FORCE acting on the moving charged particles causes them to be deflected. The direction of the potential can provide information about the type of charge carriers (for example, whether they are electrons or positive holes). The effect was discovered by Edwin Herbert Hall in 1879. *See also* quantum Hall effect.

halogen   *See* group 17 elements.

Hamilton, Sir William Rowan (1805–1865)   Irish mathematician who formulated geometrical optics and Newtonian mechanics in closely related ways and who invented QUATERNIONS. In the mid 1830s he showed that Newtonian mechanics could be formulated in a related way, now known as the HAMILTONIAN formulation. He also found a minimum principle for Newtonian mechanics known as *Hamilton's principle* (*see* least action).

Hamiltonian Symbol: H A function used to express the energy of a system in terms of position and momentum. Newtonian mechanics can be formulated in terms of the Hamiltonian, giving rise to a set of equations called *Hamilton's equations*. Field theories, including relativistic field theories such as general relativity

theory can also be expressed in Hamiltonian form. In the WAVE MECHANICS formulation of quantum mechanics the Hamiltonian is an operator. *See also* Schrödinger equation.

harmonic   An oscillation that has a frequency that is a simple integer multiple of sinusoidal oscillation called the *fundamental*. The fundamental frequency, denoted f, is called the *first harmonic*. The *second harmonic* has a frequency $2f$ which is twice the fundamental, the *third harmonic* has a frequency $3f$, etc. Harmonics above the first are called *overtones*; the second harmonic is the *first overtone*; the third harmonic is the *second overtone*, etc.

harmonic oscillator   A system that executes SIMPLE HARMONIC MOTION. The equation of a harmonic oscillator can be solved exactly both in classical mechanics and in quantum mechanics. Many real systems behave like harmonic oscillators to a good approximation. A simple pendulum is an example in classical mechanics. In quantum mechanics the harmonic oscillator is a good approximation for an atom oscillating about its mean position in a molecule. The solution of the harmonic oscillator is useful, even for systems that are not exactly harmonic oscillators (*anharmonic oscillators*), because it is a starting point for finding an approximation solution using PERTURBATION THEORY.

Hartree–Fock theory   An approximation technique used to calculate wave functions and energy levels in many-fermion systems. It is used particularly in calculations for atoms with many electrons. The technique involves considering a single particle (e.g. one electron) moving in an averaged-out potential caused by all the other particles (e.g. all the remaining electrons in the atom). Douglas Hartree introduced this technique in 1928, with Vladimir FOCK and John SLATER improving it in 1930 by taking account of the Pauli exclusion principle.

hassium   A transactinide element that is formed artificially by bombarding lead-208 with iron-58.

Symbol: Hs. Proton number: 108. Most stable isotope: ^{265}Hs (half-life 2×10^{-3}s). Electronic configuration: $[Rn]5f^{14}6d^67s^2$.

Hawking, Stephen William (1942–)
British physicist who has made major contributions to cosmology, particularly the theory of BLACK HOLES. After Roger PENROSE showed in 1965 that there is a SINGULARITY at the center of a black hole formed by the gravitational collapse of a large star, Hawking realized that, by reversing the time in Penrose's result, it is possible to show that the BIG BANG at the beginning of the Universe was also a singularity. Since the mid 1970s Hawking and his colleagues have developed EUCLIDEAN QUANTUM GRAVITY and applied it to many problems in cosmology such as the beginning of the Universe.

Hawking's area theorem
A result of general relativity theory proved by Stephen HAWKING in papers published in 1971 and 1972, stating that in any process the area of a black hole can only increase. This increase in area is analogous to the increase in entropy in the second law of thermodynamics. The analogy led Jacob BEKENSTEIN to postulate that a black hole has an entropy associated with its area.

Hawking radiation
The emission of particles from a BLACK HOLE as a result of quantum-mechanical effects. Particle–antiparticle pairs are produced near the surface of a BLACK HOLE. It is possible for one member of the pair (either the particle or the antiparticle) to fall into the black hole while the other escapes. The stream of escaping particles produced in this way is called *Hawking radiation*. The particle that escapes carries an amount of energy that is exactly equal to the negative energy of the particle falling into the black hole. The net effect is that, since the negative energy reduces the mass of the black hole, the particles given off appear to be carrying away the mass of the black hole. The existence of Hawking radiation means that a black hole has a characteristic temperature, known as the *Hawking temperature*, with this temperature being inversely pro-

portional to the mass of the hole. The existence of this temperature explains why the laws of thermodynamics apply to black holes. The Hawking temperature of a black hole with the mass of the Sun is about 10^{-7} K. At this very low temperature Hawking radiation is negligible. The Hawking temperature of a mini black hole of about 10^{12} kg, which might have been formed in the early Universe, is about 10^{11}K. Such a black hole would radiate gamma rays, neutrinos, and electron–positron pairs copiously. However, observations of cosmic gamma rays put strong constraints on the possible number of such mini black holes that could exist. Hawking radiation has not been observed at the present time.

heat A form of energy associated with the random motion of atoms and molecules. If energy is transferred to a body in the form of heat then the average kinetic energy of the molecules making up the body increases.

heat capacity *See* specific heat capacity.

heat of atomization The energy that is required to dissociate one mole of a given substance into atoms.

Heaviside, Oliver (1850–1925)
British electrical engineer and physicist who made important contributions to electromagnetic theory. After the transmission of radio waves across the Atlantic Ocean in 1901 he and Arthur Kennelly independently suggested in 1902 that there is an electrically charged layer in the atmosphere that reflected the waves. This layer, sometimes called the *Heavisde layer*, was detected in 1924 and exists because of ionization of the molecules in this layer by radiation from the Sun.

heavy-fermion system A solid in which electrons have very high effective masses of several hundred times the mass of a free electron as a result of complicated many-body interactions. The cerium–copper–silicon compound $CeCuSi_2$ is an example of a heavy-fermion system.

heavy hydrogen   *See* deuterium.

heavy water (**deuterium oxide**) Symbol: D_2O Water in which hydrogen atoms ^1H are replaced by DEUTERIUM atoms ^2H (D). The physical and chemical properties of heavy water are similar to but not identical with those of H_2O. Heavy water is used as a moderator in some nuclear reactors. It is also used for 'labeling' in studies of chemical reactions.

Heisenberg, Werner (1901–1976) German physicist who was one of the founders of quantum mechanics and quantum field theory. In 1925 he was the first person to formulate quantum mechanics in a consistent way. He developed this formulation of quantum mechanics, known as MATRIX MECHANICS, with Max BORN and Pascual JORDAN. In 1927 he showed that quantum mechanics imposes a fundamental limitation, now known as the HEISENBERG UNCERTAINTY PRINCIPLE, on what can be known about particles. In 1928 he showed that the quantum-mechanical concept of EXCHANGE explains the phenomenon of FERROMAGNETISM. After the discovery of the neutron in 1932, Heisenberg proposed that an atomic nucleus is composed of protons and neutrons and that the two particles are two different states of one entity, with the two particles being distinguished by what he called ISOTOPIC SPIN. In 1929–1930 Heisenberg wrote two papers with Wolfgang PAULI that laid the foundations of quantum field theory.

Heisenberg model   A model for ferromagnetism proposed by Werner HEISENBERG in 1928. Although the model is not exactly soluble Heisenberg was able to show that quantum mechanical EXCHANGE involving electrons could explain why ferromagnetism occurs. The Heisenberg model has been investigated extensively by many workers.

Heisenberg uncertainty principle   *See* uncertainty principle.

Heitler, Walter (1904–1981)   German physicist who is best known for his paper with Fritz LONDON in 1927 explaining the chemical bond in the hydrogen molecule in terms of quantum mechanics. London also applied group theory to the problem of chemical bonds in more complicated molecules.

helium   A colorless monatomic gas; the first member of the rare gases (group 18 of the periodic table). Helium consists of a nucleus of two protons and two neutrons (equivalent to an α-particle) with two extra-nuclear electrons. It has an extremely high ionization potential and is completely resistant to chemical attack of any sort. The gas accounts for only $5.2 \times 10^{-4}\%$ of the atmosphere; up to 7% occurs in some natural gas deposits. Helium is the second most abundant element in the universe, the primary process on the Sun being nuclear fusion of hydrogen to give helium.

Helium is unusual in that it is the only known substance for which there is no triple point (i.e., no combination of pressure and temperature at which all three phases can co-exist). This is because the interatomic forces, which normally participate in the formation of solids, are so weak that they are of the same order as the zero-point energy. At 2.2 K helium undergoes a transition from liquid helium I to liquid helium II, the latter being a true liquid but exhibiting superconductivity and an immeasurably low viscosity (*superfluidity*). The low viscosity allows the liquid to spread in layers a few atoms thick, described by some as 'flowing uphill'.

Helium also has an isotope. ^3He is formed in nuclear reactions and by decay of tritium. This also undergoes a phase change at temperatures close to absolute zero..

Symbol: He. Melting pt.: 0.95 K (pressure). Boiling pt.: 4.216 K. Density: 0.1785 kg m^{-3} (0°C). Proton number: 2. Relative atomic mass: 4.002602. Electronic configuration: $1s^2$.

Helmholtz, Hermann Ludwig Ferdinand von (1821–94)   German physicist and physiologist with wide-ranging interests in physics. In 1847 he became one of the co-discoverers of the law of conservation of energy. He also made other important contributions to thermodynamics. In 1881 he suggested that FARADAY'S LAWS OF ELECTROLYSIS imply the existence of an 'atom of electricity', i.e. the electron.

henry Symbol: H   The SI unit of inductance, equal to the inductance of a closed circuit that has a magnetic flux of one weber per ampere of current in the circuit. 1 H = 1 Wb A^{-1}.

Herman, Robert (1922–97)   Physicist who worked with Ralph ALPHER and George GAMOW on the big bang model.

Hero (*c.* first century AD)   Ancient Greek mathematician, sometimes known as *Heron*, who postulated that light rays travel in straight lines. This was the first VARIATIONAL PRINCIPLE in science.

hertz Symbol: Hz   A standard unit of frequency equal to one cycle per second. It was named after Heinrich HERTZ.

Hertz, Gustav Ludwig (1887–1975)   German physicist who participated with James FRANCK in the experiments which showed that atoms only absorb energy in discrete amounts. Franck and Hertz shared the 1925 Nobel Prize for physics for this work. Gustav Hertz was the nephew of Heinrich HERTZ.

Hertz, Heinrich Rudolph (1857–1894)   German physicist who demonstrated the existence of electromagnetic waves when he produced radio waves in 1888. His discovery was a major boost to the theory of electrodynamics put forward by James Clerk Maxwell, which had predicted the existence of such waves from the MAXWELL EQUATIONS. In 1887 Hertz discovered the PHOTOELECTRIC EFFECT.

Herzberg, Gerhard (1904–99)   German-born Canadian spectroscopist who made extensive contributions to molecular spectra and the electronic structure of molecules. Herzberg wrote a classic series of books on atomic and molecular spectra be-

tween 1937 and 1979. He won the 1971 Nobel Prize for chemistry.

Hess, Victor Francis (1883–1964)   Austrian-born American physicist who discovered COSMIC RAYS in a series of balloon flights in 1911–12. Hess shared the 1936 Nobel Prize for physics with Carl ANDERSON.

Hess's law   A law which states that the overall heat of a chemical reaction is the same no matter what route is taken from reactants to products. This law was stated by Germain Henri Hess in 1840. It is a consequence of the principle of the CONSERVATION OF ENERGY.

heterotic string   *See* superstring theory.

Hevesy, George Charles von (1885–1966)   Hungarian-born Swedish chemist who developed the use of radioactive isotopes for LABELING for investigating complex systems, including biological systems. He was awarded the 1943 Nobel Prize for chemistry for this work.

hidden-variables theory   A type of theory that regards quantum mechanics as incomplete, with there being a deeper level of reality described by variables known as *hidden variables*. For many years it was thought that hidden variables could not exist because of a theorem by John VON NEUMANN in his book *The Mathematical Foundations of Quantum Mechanics*, which appeared to demonstrate the impossibility of hidden variables. However, von Neumann made an incorrect assumption, a mistake that was pointed out by Grete Hermann in 1933. This refutation of the result was largely ignored until the work of John BELL in the mid 1960s.

David BOHM attempted to construct a hidden variables theory of quantum mechanics. In the 1960s Bell took Bohm's work further by proposing a precise experimental test that could distinguish between a hidden variables theory based on LOCALITY and the conventional expectations of quantum mechanics (*see* Bell's inequality). It was suggested in the 1990s that a successful hidden variables theory could be constructed using ideas from CHAOS THEORY but at present no successful hidden variables theory has been constructed in this way, or any other way.

Higgs, Peter Ware (1929–) British physicist who was one of the physicists who discovered what is now known as the HIGGS MECHANISM in the mid 1960s.

Higgs boson *See* Higgs mechanism.

Higgsino A fermion that is the partner of the Higgs boson in SUPERSYMMETRY.

Higgs mechanism The mechanism by which the W BOSON and the Z BOSON and, more generally, all massive particles, acquire their mass. In the mid 1960s Peter HIGGS, and independently Robert Brout and François Englert, showed that if the gauge symmetry in a gauge theory is a SPONTANEOUSLY BROKEN SYMMETRY then there is an exception to GOLDSTONE'S THEOREM that means that this type of broken symmetry is associated with a massive spin-zero particle. This particle is now known as the *Higgs boson*. In addition, Higgs *et al.* showed that the gauge bosons which were massless in the case of unbroken symmetry became massive in the case of broken symmetry. Since massless bosons are associated with long-range interactions, gauge bosons that acquire mass by the Higgs mechanism are associated with short-range interactions. The Higgs boson is associated with a scalar field, now known as the *Higgs field*.

In the WEINBERG–SALAM MODEL the W boson and the Z boson which mediate the weak interaction acquire their masses by the Higgs mechanism. In 1971 Gerardus 'T HOOFT showed that the procedure of RENORMALIZATION can be applied to gauge theories in which the Higgs mechanism operates. It is thought that the Higgs mechanism is the only way in which gauge bosons can acquire mass. It is possible to calculate the masses of the W and Z bosons in the Weinberg–Salam model. However, it is not possible to calculate the mass of the Higgs boson in the Weinberg–Salam model, al-

though various bounds on its value exist. Higgs bosons have not been found experimentally, although thorough searches up to 130 GeV have been conducted. It is thought that fermions in the STANDARD MODEL acquire their masses by the Higgs mechanism but it has not been possible to implement this suggestion.

high-temperature superconductivity *See* superconductivity.

Hilbert, David (1862–1943) German mathematician with wide-ranging interests in mathematics and its physical applications. His early work on the theory of *invariants* is of relevance to physics because of the relations between symmetry, invariance, and conservation laws. Hilbert also made important contributions to the mathematical formulation of the kinetic theory of gases. In 1915 he expressed general relativity theory in Lagrangian form. He also took an interest in the development of quantum mechanics and his work on what came to be known as a HILBERT SPACE was very useful in quantum mechanics.

Hilbert space A VECTOR SPACE that can have an infinite number of dimensions. The state of a system in quantum mechanics can be represented by a vector in Hilbert space. The Hilbert space formulation of quantum mechanics was developed by John VON NEUMANN in the late 1920s. Other formulations of quantum mechanics, including MATRIX MECHANICS and WAVE MECHANICS, can be derived from this formulation.

Hoffmann, Roald (1937–) Polish-born American chemist who formulated what are now known as the WOODWARD–HOFFMANN RULES with Robert Burns Woodward in 1965. This work led to Hoffmann sharing the 1981 Nobel Prize for chemistry with Kenichi FUKUI. (Woodward did not share this prize because he died in 1979).

Hofstadter, Robert (1915–1990) American physicist who conducted experiments that showed that the proton and neutron are not pointlike objects but have definite sizes and shapes. Starting at Stanford University in 1950, Hofstadter investigated atomic nuclei by studying the way in which electrons are scattered by nuclei. He shared the 1961 Nobel Prize for physics with Rudolph MÖSSBAUER.

hole   *See* semiconductor.

holmium   A soft malleable silvery element of the lanthanoid series of metals. It occurs in association with other lanthanoids. It is used in some magnetic devices but has few other applications.
  Symbol: Ho. Melting pt.: 1474°C. Boiling pt.: 2695°C. Relative density: 8.795 (25°C). Proton number: 67. Relative atomic mass: 164.93032. Electronic configuration: $[Xe]4f^{11}6s^2$.

holographic hypothesis   A general principle of QUANTUM GRAVITY that information about the bulk of a system can be inferred from information about the surface of that system. The holographic hypothesis was put forward by Gerardus 'T HOOFT and Leonard Susskind in the mid 1990s. There are several specific versions of the holographic hypothesis. In the context of black holes it is related to the BEKENSTEIN BOUND relating information to the area of a black hole. In 1998 Juan Maldacena conjectured that there is a realization of the holographic hypothesis in SUPERSTRING THEORY. There is a substantial amount of evidence in support of Maldacena's conjecture. At present, there is a great deal of research devoted to the holographic hypothesis and its implications.

holography   A technique for recording a three-dimensional image of an object. The object is illuminated with light from some source. The reflected light from the object is combined with direct light from the source to give an interference pattern on a photographic plate. When the plate is developed the pattern is called a *hologram*. A three-dimensional image of the object is produced by illuminating the interference pattern with the original light. Holography

was invented by Dennis Gabor in 1948. Frequently the light source is a laser.

homogeneous   Describing a material or system that has the same properties in any direction; i.e. uniform without irregularities.

horizon   A surface in space–time from beyond which an observer cannot see or receive any signal. An example is the *event horizon* of a black hole.

horizon problem   *See* inflation.

hot dark matter   Dark matter that consists of particles moving at speeds near to the speed of light. Since neutrinos have nonzero masses it is thought that they may contribute much of the hot dark matter in the Universe.

Houtermans, Fritz (1903–)   German physicist who performed calculations with Robert ATKINSON showing that nuclear fusion could occur in stars as a consequence of quantum mechanical tunneling by protons.

Hoyle, Sir Fred (1915–2001)   British astronomer who explained how the chemical elements are produced inside stars and in the early Universe (*see* nucleosynthesis). Hoyle also, together with Herman BONDI and Thomas GOLD, proposed the STEADY STATE THEORY of the Universe. In 1957 Hoyle summed up his work on the production of elements heavier than helium in the stars in a classic comprehensive paper with his colleagues William FOWLER and Geoffrey and Margaret Burbidge. In spite of his strong preference for the steady state theory over the BIG-BANG THEORY (a name he invented), he performed calculations with Fowler and with Robert Wagoner in the 1960s on the production of helium from hydrogen in the big-bang theory, which give strong evidence in support of the theory. Hoyle and his colleagues put forward many stimulating but controversial ideas such as the idea that complex molecules of the type needed for life exist in INTERSTELLAR MATTER.

Hubble, Edwin Powell (1859–1953) American astronomer who showed in the 1920s that the Milky Way is one galaxy amongst many others and that the galaxies are moving apart, i.e. the Universe is expanding. *See also* Hubble's law.

Hubble constant A quantity that gives information about the rate at which the Universe is expanding. It is not strictly a 'constant', because it changes with time as a result of the expansion rate slowing down because of the mutual gravitational attraction of all the matter in the Universe. The reciprocal of the Hubble constant is called the *Hubble time*. If the Hubble constant was actually constant, the Hubble time would be a measure of the age of the Universe. Because the Hubble constant changes the Hubble time gives an upper limit on the AGE OF THE UNIVERSE. The value of the Hubble constant is determined in several ways including the REDSHIFT of distant galaxies, the use of GRAVITATIONAL LENSES, and the SUNYAEV–ZEL'DOVICH EFFECT.

Hubble's law The principle that the velocity of recession of a distant galaxy (as measured by its REDSHIFT) is proportional to the distance of the galaxy from the Earth. This law was established by Edwin HUBBLE in 1929.

Hubble Space Telescope A telescope used to take pictures from space, thus avoiding the effects of the atmosphere of the Earth. It was launched in 1990. After it was repaired in 1993 it has provided information about a number of issues including the age of the Universe, black holes, and galaxies.

Hubble time *See* Hubble constant

Hückel, Erich Armand Arthur Joseph (1896–1980) German chemist who worked with Peter DEBYE on electrolytes and who explained the properties of aromatic molecules such as benzene. His work with Debye resulted in the DEBYE–HÜCKEL THEORY of electrolytes in 1923. His work on aromatic molecules in the early 1930s introduced an approximate form of molecular orbital theory known as HÜCKEL THEORY.

Hückel theory A theory to explain the properties of aromatic molecules such as benzene in terms of quantum mechanics. In Hückel theory the σ orbitals are treated separately from the π orbitals and the interactions between nonneighbouring atoms are taken to be zero. Using these approximations Hückel found that the pi-electrons are delocalized and spread above and below the carbon plane. Hückel showed that aromatic compounds have $4n + 2$ pi-electrons, where n is an integer (1, 2, 3, etc.), with $n = 1$ for benzene and n = 2 being naphthalene.

Hulse, Russell Alan (1950–) American astronomer who discovered the BINARY PULSAR with Joseph TAYLOR in the mid 1970s. Hulse and Taylor shared the 1993 Nobel Prize for physics for this discovery.

Humphreys series *See* hydrogen spectrum.

Hund, Friedrich (1896–) German physicist who made important contributions to the quantum mechanics of atoms and molecules including the HUND COUPLING CASES, HUND'S RULES, and work on the development of molecular orbital theory.

Hund coupling cases The five idealized ways in which the different types of angular momentum in a molecule (the electron orbital angular momentum, the electron spin angular momentum, and the angular momentum of nuclear rotation) combine in molecular spectra. Many molecules have actual behavior intermediate between Hundt coupling cases.

Hund's rules Two rules for determining the lowest energy level for an electron configuration in a many-electron atom, put forward by Friedrich HUND in 1925 from empirical evidence of atomic spectra. (1) The state with the lowest energy has the maximum MULTIPLICITY consistent with the PAULI EXCLUSION PRINCIPLE; (2) if more than one state has the maximum multiplicity, the lowest energy state is the state with the maximum total electron orbital angular momentum. These rules can be derived from the quantum theory of electrons in atoms.

Huygens, Christiaan (1629–95) Dutch astronomer and physicst who made a number of contributions to physics. Starting in 1678, and fully described in his *Treatise on Light* published in 1690, Huygens developed a theory of light in terms of waves. His theory described reflection and refraction of light, but could not explain polarization because he incorrectly assumed that the light waves were longitudinal. *See also* Huygens' construction.

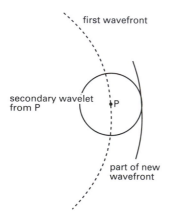

Huygens' construction of wavefronts

Huygens' construction (Huygens' principle) A method for constructing a secondary wavefront from an existing one, first put forward by Christiaan HUYGENS in 1690 in connection with his wave theory of light. Each point on the wavefront is regarded as a source of secondary waves. Characteristic features of waves, such as diffraction, are well described by Huygens' construction.

hybrid orbital *See* orbital.

hydrogen A colorless gaseous element. Hydrogen has some similarities to both the alkali metals (group 1) and the halogens (group 17), but is not normally classified in any particular group of the periodic table. It is the most abundant element in the Universe and the ninth most abundant element in the Earth's crust and atmosphere (by mass). It occurs principally in the form of water and petroleum products; traces of molecular hydrogen are found in some natural gases and in the upper atmosphere.

Natural hydrogen in molecular or combined forms contains about one part in 2000 of deuterium, D, an isotope of hydrogen that contains one proton and one neutron in its nucleus. Although the effect of isotopes on chemical properties is normally small, in the case of hydrogen the difference in mass number leads to a lowering of some reaction rates known as the 'deuterium isotope effect'. *See also* hydrogen molecule.

Symbol: H. Melting pt.: 14.01 K. Boiling pt.: 20.28 K. Density: 0.089 88 kg m^{-3} (0°C). Proton number: 1. Relative atomic mass: 1.0079. Electronic configuration: $1s^1$.

hydrogen bomb *See* nuclear weapons.

hydrogen bond A type of electrostatic attraction between molecules that have hydrogen atoms bound to electronegative atoms such as nitrogen, oxygen, and fluorine. Because these atoms are strongly electronegative they attract the electron of the hydrogen atom, thus making the molecule dipolar. Such dipolar molecules can attract other dipolar molecules by DIPOLE–DIPOLE INTERACTION with the positively charged end of one molecule attracting the negatively charged end of the adjacent molecule. A hydrogen bond is about one tenth as strong as a normal chemical bond, but hydrogen bonding is extremely important. For example, hydrogen bonds are responsible for the fact that water is a liquid at room temperature and for the structure of ice. The two strands of a DNA double helix are held together by hydrogen bonds.

hydrogen molecule A diatomic molecule consisting of two hydrogen atoms held together by a covalent bond. It is the simplest neutral molecule. Hydrogen molecules exist in two forms. In *orthohydrogen* the spins of the two nuclei are aligned parallel. In *parahydrogen* the spins of the nuclei are aligned antiparallel. These two possible alignments of nuclear spin occur in other diatomic molecules, but their effects are more pronounced for hydrogen molecules because of their low mass. At low temperatures parahydrogen strongly dominates but at room temperatures the gas is 75% orthohydrogen. The two forms differ slightly in their physical properties.

hydrogen molecule ion   A diatomic ion (H_2^+) consisting of two protons and a single electron, i.e. a hydrogen molecule that is singly ionized. This is the the only molecular system for which the SCHRÖDINGER EQUATION can be solved exactly.

hydrogen spectrum   The spectrum of atomic hydrogen consists of several series of sharp lines corresponding to transitions between atomic energy levels. In general, the wavelength λ of each line satisfies
$$1/\lambda = R\,(1/n_1^2 - 1/n_2^2)$$
where R is the RYDBERG CONSTANT and n_1 and n_2 are integers, with $n_2 > n_1$. For example, spectral lines in the visible spectrum form the *Balmer series*; i.e. the series in which $n_1 = 2$ and $n_2 = 3, 4, 5, \ldots.$ The Balmer series, found by Johann Jakob BALMER in 1885 on purely empirical grounds, was the first series of the hydrogen spectrum to be found. In the *Lyman series*, discovered by Theodore Lyman in 1906, $n_1 = 1$ and the lines are in the ultraviolet region. The Lyman series corresponds to electrons falling into the lowest energy level. It is the strongest feature of the solar spectrum. There are several series in the infrared region: the *Paschen series* ($n_1 = 3$), the *Brackett series* ($n_1 = 4$), the *Pfund series* ($n_1 = 5$), and the *Humphreys series* ($n_1 = 6$).

A major advance in understanding the hydrogen spectrum, and hence atomic structure, came in 1913 when Niels BOHR put forward his theory of the atom based on quantum theory. Bohr was able to derive the Balmer series (and the other series) from his postulate of quantized energy levels. When quantum mechanics was developed in the mid 1920s the problem of the energy levels of the hydrogen atom was solved exactly by Wolfgang PAULI in 1926 using matrix mechanics and by Erwin SCHRÖDINGER, also in 1926, using wave mechanics.

The FINE STRUCTURE of the hydrogen spectrum is understood in terms of electron spin and relativistic quantum mechanics. There are also some small effects such as the LAMB SHIFT which cannot be understood in terms of the single particle DIRAC EQUATION and require QUANTUM ELECTRODYNAMICS to be fully understood.

hyperbola   *See* conic section.

hypercharge   A number associated with a particle, equal to the sum of the BARYON NUMBER and STRANGENESS of the particle. Hypercharge is a quantity that is conserved in the strong and electromagnetic interactions of particles but not in their weak interactions.

hyperfine structure   Very closely spaced atomic spectral lines caused by energy differences resulting from the interaction between the nucleus of an atom and its electrons.

hypernuclei   Nuclei that contain HYPERONS.

hyperon   Any BARYON that contains strange quarks. Hyperons are heavier than NUCLEONS and have short lifetimes (between about 10^{-10} to 10^{-8} second) before decaying into nucleons.

I

ideal crystal A perfect regular single crystal that contains no impurities or crystal defects.

ideal gas (**perfect gas**) A hypothetical gas made up of atoms that take up no space and do not interact with each other or the walls of the container except by perfectly elastic collisions. The ideal gas is frequently a useful starting point for the analysis of real gases.

imaginary number A number that is a multiple of i, the square root of -1. For example, $\sqrt{-2} = i\sqrt{2}$ is an imaginary number. *See also* Argand diagram; complex number.

index theorem A mathematical theorem that relates the solutions of an equation to TOPOLOGY. Many applications of index theorems occur in gauge theories, quantum gravity, supersymmetry, and superstring theory. The analysis of such entities as INSTANTONS and MAGNETIC MONOPOLES makes use of index theorems.

indium A soft silvery metallic element belonging to group 13 of the periodic table. It is found in minute quantities, primarily in zinc ores and is used in alloys, in several electronic devices, and in electroplating.
Symbol: In. Melting pt.: 155.17°C. Boiling pt.: 2080°C. Relative density: 7.31 (25°C). Proton number: 49. Relative atomic mass: 114.818. Electronic configuration: $[Kr]4d^{10}5s^25p^1$.

induced emission *See* quantum theory of radiation.

induced radioactivity *See* artificial radioactivity.

inelastic collision A collision in which kinetic energy is not conserved. For example, when two billiard balls collide some of the kinetic energy is converted into heat and sound. When two molecules collide some of the kinetic energy is converted into vibrational and rotational energy.

inert gases *See* noble gases.

inertia The resistance of a body to any change in its motion. *See also* Mach's principle; Newton's laws of motion.

inertial frame A frame of reference in which a body that is not acted on by an external force moves in a straight line with a constant velocity (or is at rest). *See also* special relativity theory.

inertial mass *See* mass.

infinity Symbol: ∞ A quantity that has a greater value than any finite assignable value. Minus infinity, $-\infty$, is a quantity that has a value that is less than any assignable value. When the value infinity occurs in the calculation for a physical quantity this indicates that the model used to calculate the quantity is inadequate.

inflation A short period of very rapid expansion of the Universe postulated to occur very soon after the BIG BANG. The hypothesis of inflation in the EARLY UNIVERSE solves a number of cosmological problems. One of these is the *flatness problem*; i.e. the problem that the Universe appears to be almost exactly Euclidean. Another of these problems is the *horizon problem*; i.e. the problem of why the Universe looks the same on opposite horizons in spite of there not being enough time for light to traverse the Universe. The third problem is the *monopole problem*; i.e. the problem of why MAGNETIC MONOPOLES predicted by GRAND UNIFIED THEORIES are not plentiful.

If there was a period of rapid expansion when the size of the Universe was about the same as the PLANCK LENGTH then all these problems disappear. Space–time would become much flatter as a result of such an expansion. Before the inflation the Universe would have been sufficiently small for signals to go the very small distance from one side of it to the other.

In addition to solving these problems inflation provides a solution to the problem of STRUCTURE FORMATION in the Universe. It is postulated that quantum fluctuations in the early Universe were magnified by inflation, producing the irregularities necessary for the formation of structures. This theory made the prediction that there should be very slight variations in the temperature of the COSMIC MICROWAVE BACKGROUND (CMB) as a result of the irregularities in the early Universe. In the early 1990s COBE found precisely the expected variations. This finding was a major success for inflation theory. The theory also predicts that the irregularities in the early Universe produced GRAVITATIONAL RADIATION with distinct features. It is hoped that sufficiently sensitive detectors will detect this radiation within the first few decades of the twenty-first century.

Although the general idea of inflation theory is very successful it has proved very difficult to link it with any particular model of elementary particle theory. Inflation theory was initially proposed by Alan GUTH in 1981 although his initial model was seriously flawed.

information theory The mathematical theory of information. This subject has been used in computing, electrical engineering, and biology. It has been speculated, notably by John WHEELER, that the fundamental principles of physics could be expressed in terms of information theory. The concept of ENTROPY has been expressed using information theory, with an increase in entropy being associated with a decrease in information. *See also* Landauer's principle; quantum information.

infrared radiation Electromagnetic radiation with wavelengths that are longer than visible light but shorter than microwaves or radio waves, i.e. the wavelength range of 0.7 micrometer to 1 millimeter. *See also* electromagnetic spectrum.

infrared slavery *See* quark confinement.

infrared spectroscopy A branch of spectroscopy using infrared radiation. Many materials transparent to visible light are opaque to infrared, including glass. Rock salt, quartz, germanium, or polyethene prisms and lenses are suitable for use with infrared. Infrared radiation is produced by movement of charges on the molecular scale; i.e. by vibrational or rotational motion of molecules. Of particular importance in chemistry is the absorption spectrum of compounds in the infrared region. Certain bonds between pairs of atoms (C–C, C=C, C=O, etc.) have characteristic vibrational frequencies, which correspond to bands in the infrared spectrum. Infrared spectra are thus used in finding the structures of new organic compounds. They are also used to 'fingerprint' and thus identify known compounds. At shorter wavelengths, infrared absorption corresponds to transitions between rotational energy levels, and can be used to find the dimensions of molecules (by their moment of inertia).

inhomogeneous Not homogeneous; i.e. not distributed evenly everywhere.

instanton A solution of the equations of quantum mechanics and quantum field theory that corresponds to TUNNELING. Instanton solutions are particularly important in gauge theories where they have led to an understanding of several nonperturbative features of gauge theories such as the THETA VACUUM.

insulator A substance that is a poor conductor of heat and electricity. *See also* electronic structure of solids.

interaction An effect between two or more bodies or systems which results in some change to one or more of them. See also FUNDAMENTAL INTERACTIONS.

interference The interaction of two or more waves when they pass through the same region. The resultant displacement at any point is the sum of the individual displacements of the waves. If the waves reinforce each other the waves are said to be *in phase* and the interference is said to be *constructive interference*. If the waves cancel each other out the waves are said to be *out of phase* and the interference is said to be *destructive interference*. Interference of light waves was first observed by Thomas YOUNG in 1801, with his DOUBLE-SLIT EXPERIMENT. The characteristic *interference pattern* found in such an experiment is a set of bright fringes, corresponding to constructive interference, and dark fringes, corresponding to destructive interference.

interferometer A device for producing light interference fringes used for accurate measurements. There are many uses of interferometers in physics. A famous example is in the MICHELSON–MORLEY EXPERIMENT. Interferometers are also used in accurate spectroscopic work.

intermediate coupling *See* j–j coupling.

intermolecular forces Forces that occur between molecules. *See* van der Waals' forces.

internal conversion A process in which an atomic nucleus in an excited state decays to the ground state, with the energy released being transferred by electromagnetic coupling to a bound electron of the atom rather than being released as a photon. This electron, known as a *conversion electron*, is ejected from the atom. Its kinetic energy is equal to the energy difference between the nuclear energy levels minus the binding energy of the electron. The ion which results from this process is in an excited state itself and frequently emits an x-ray photon or an Auger electron.

internal energy Symbol: U The total kinetic and potential energy of the particles such as atoms and molecules making up a system. If the temperature of the system is raised the internal energy of the system is increased. The internal energy of a system is also changed by work done on or by the system, with the relation between heat, work, and internal energy being given by the first law of THERMODYNAMICS.

interstellar matter The matter in the space between stars. This matter is not distributed uniformly but are clumped into *interstellar clouds*. There is a lot of hydrogen in interstellar matter but there are also other molecules, including a number of complex organic molecules.

intrinsic semiconductor *See* semiconductor.

invariance The property of remaining unchanged after a transformation. Frequently, the transformation is part of a symmetry GROUP. For example, in special relativity theory there is invariance under LORENTZ TRANSFORMATIONS, with the symmetry group being the LORENTZ GROUP. A quantity that is unchanged by a transformation is called an *invariant*. Conserved quantities such as energy are examples of invariants. The concept is of fundamental importance in theoretical physics.

inverse beta decay A process in which a proton absorbs an electron and becomes a neutron. Inverse beta decay occurs in the formation of NEUTRON STARS.

inverse Compton effect A process in which low-energy photons gain energy when they are scattered by electrons. The inverse Compton effect occurs in the SUNYAEV–ZEL'DOVICH EFFECT.

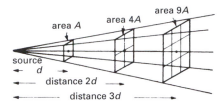

area A
area $4A$
area $9A$
source
d
distance $2d$
distance $3d$

Inverse square law

inverse square law A law in which the effect of a source is inversely proportional to the square of the distance from the source. Newton's law of gravity and Coulomb's law are examples of inverse square laws.

iodine A dark-violet volatile solid element belonging to the halogens (group 17 of the periodic table). It occurs in seawater and is concentrated by various marine organisms in the form of iodides. Significant deposits also occur in the form of iodates.

Symbol: I. Melting pt.: 113.5°C. Boiling pt.: 184°C. Relative density: 4.93 (20°C). Proton number: 53. Relative atomic mass: 126.90447. Electronic configuration: $[Kr]4d^{10}5s^25p^5$.

ion An atom or group of atoms that has either gained at least one electron, making it a *negative ion*, or lost at least one electron, making it a *positive ion*. A negatively charged ion is called an *anion* since it is attracted to the anode during electrolysis. A positively charged ion is called a *cation* since it is attracted to the cathode during electrolysis.

ionic bonding Chemical bonding in which the electrostatic interaction between ions is dominant. For example, ionic bonding occurs in sodium chloride crystals.

ionic crystal A crystal that is held together by ionic bonding.

ionic radius The value given to the radius of an ion in a crystal on the basis of x-ray diffraction. The ionic radius of an ion is found from the interionic distances between the ion and various oppositely charged ions in crystals.

ionization The process of producing ions. There are several ways in which ions may be formed from atoms or molecules. In certain chemical reactions ionization occurs by transfer of electrons; for example, soldium atoms and chlorine atoms react to form sodium chloride, which consists of sodium ions (Na^+) and chloride ions (Cl^-). Certain molecules can ionize in solution; acids, for example, form hydrogen ions as in the reaction:
$$H_2SO_4 \rightarrow 2H^+ + SO_4^{2-}$$
The 'driving force' for ionization in a solution is solvation of the ions by molecules of the solvent. H^+, for example, is solvated as a hydroxonium (hydronium) ion, H_3O^+.

Ions can also be produced by ionizing radiation; i.e. by the impact of particles or photons with sufficient energy to break up molecules or detach electrons from atoms: $A \rightarrow A^+ + e^-$. Negative ions can be formed by capture of electrons by atoms or molecules: $A + e^- \rightarrow A^-$.

ionization chamber A chamber which detects and/or measures IONIZING RADIATION. In such a chamber there is a pair of electrodes with a potential difference between them. When ionizing radiation passes through the chamber electrons and positive ions are produced. This means that there is a small electric current between the electrodes.

ionization potential (**IP**) Symbol: I The energy required to remove an electron from an atom or molecule (or molecule or group) in the gas phase, i.e. the energy required for the process:
$$M \rightarrow M^+ + e^-$$
It gives a measure of the ability of metals to form positive ions. The second ionization potential is the energy required to remove two electrons and form a doubly charged ion:
$$M \rightarrow M^{2+} + e^-$$
Ionization potentials stated in this way are positive; often they are given in electronvolts. *Ionization energy* is the energy

required to ionize one mole of the substance, and is usually stated in kilojoules per mole ($kJ\ mol^{-1}$).

In chemistry, the terms 'second', 'third', etc., ionization potentials are usually used for the formation of doubly, triply, etc., charged ions. However, in spectroscopy and physics, they are often used with a different meaning. The second ionization potential is the energy to remove the second least strongly bound electron in forming a singly charged ion. For lithium ($1s^2 2s^1$) it would refer to removal of a 1s electron to produce an excited ion with the configuration $1s^1 2s^1$.

ionizing radiation Radiation such that the individual particle or quantum has sufficient energy to ionize substances. Electrons with kinetic energy just greater than the ionization potential will cause ionization, but other particles (e.g. molecular positive ions) require higher energies. Gamma-rays and x-rays ionize indirectly by means of the electrons they eject from substances by the photoelectric effect or Compton scattering. Short-wavelength ultraviolet quanta may ionize individual molecules by the photoelectric effect, but the ejected electrons have insufficient kinetic energy to cause further ionization unless an electric field is applied.

At ordinary intensities electromagnetic radiation with quantum energy below the ionization potential cannot cause ionization. But by focusing laser beams it is possible to ionize matter – even when the individual photons have energy less than a hundredth of the ionization potential. Such radiations are not ordinarily classified as ionizing radiations.

Ionizing radiations can be dangerous to health and precautions should be taken with their use.

ionosphere A region in the atmosphere of the Earth that contains ions and free electrons. It exists because of ionization of molecules by ultraviolet radiation from the sun. Radio transmission over long distances is possible because radio waves are reflected by the ionosphere.

iridium A white transition metal that is highly resistant to corrosion. It is used in electrical contacts, in spark plugs, and in jewelry.

Symbol: Ir. Melting pt.: 2410°C. Boiling pt.: 4130°C. Relative density: 22.56 (17°C). Proton number: 77. Relative atomic mass: 192.217. Electronic configuration: $[Xe]4f^{14}5d^7 6s^2$.

iron A transition element occurring in many ores, especially the oxides (haematite and magnetite) and carbonate.

Symbol: Fe. Melting pt.: 1535°C. Boiling pt.: 2750°C. Relative density: 7.874 (20°C). Proton number: 26. Relative atomic mass: 55.845. Electronic configuration: $[Ar]3d^6 4s^2$.

irreducible representation See representation.

irreversible process A process that takes place in only one direction. An irreversible process is associated with an increase in entropy. A process of this type can occur in a system even if the equations describing the bodies in the system such as Newton's laws of motion are invariant under time reversal. This can be understood in terms of statistical mechanics.

Ising model A model for magnetic systems introduced, and solved in the case of one dimension, by Ernest Ising in 1925. Ising showed that in one dimension there is no spontaneous magnetization above absolute zero in the absence of an external magnetic field. The study of phase transitions in the Ising model has been of great importance in the development of the theory of phase transitions. The two-dimensional Ising model was solved exactly by Lars ONSAGER in 1944. In the three-dimensional Ising model it is necessary to use approximation techniques. Frequently these approximation techniques use the RENORMALIZATION GROUP.

isobar One of two or more nuclides that have the same nucleon number but different proton numbers. For example, radium-228, actinium-228, and thorium-228 are isobars with nucleon number 228 and proton numbers of 88, 89, and 90 respectively.

isoelectronic Describing molecules that have the same number of electrons. Diatomic nitrogen (N_2) and carbon monoxide (CO) are examples of isoelectronic molecules. If two molecules are isoelectronic their energy level diagrams are similar.

isomers 1. A pair of nuclei with the same proton and neutron numbers but different quantum states. They have different stability and so act as different nuclides with different half-lives. Such nuclei are sometimes called *nuclear isomers*.
2. Chemical compounds that have the same molecular formulae but have different molecular structures or different arrangements of atoms, or groups of atoms, in space. *Structural isomers* are molecules with different molecular structures. The functional groups can have different positions or be different in this case. *Stereoisomers* have the same molecular formula and functional groups but different arrangements of atoms or functional groups in space. If two molecular structures are different but are mirror images of each other they are said to be *optical isomers* or *enantiomers* (*see also* optical activity). If two molecular structures are *cis–trans isomers* (formerly known as *geometrical isomers*) the isomers have different positions of atoms or functional groups with respect to a central atom or a double bond.

isothermal process A process that takes place at a constant temperature. If necessary, heat is supplied or taken away from the system so as to maintain a constant temperature. The concept of an isothermal process is an idealization but many real processes approximate to it very closely.

isotone One of two or more nuclides that have the same number of neutrons but different numbers of protons. For example, the naturally occurring isotones strontium-88 and yttrium-89 both have 50 neutrons.

isotope One of two or more atoms of the same element that have the same number of protons but different numbers of neutrons in their atomic nucleus. Isotopes of the same element have almost identical chemical properties, since they have the same number of electrons, but slightly different physical properties. Most elements occur naturally as a mixture of isotopes. An unstable isotope is called a *radioactive isotope* or *radioisotope*.

isotope separation The process of separating the different isotopes of an element according to their atomic masses. Most methods depend on small differences in physical properties between isotopes.

On the laboratory scale isotopes of elements can be separated in the mass spectrometer in which the paths of nuclei depend on the ratio of charge to mass. Similar methods have been used on a larger scale. The difference in rates of diffusion of the volatile compounds ^{238}uranium hexafluoride and ^{235}uranium hexafluoride has been used on a large scale to separate the isotopes of uranium. The ultracentrifuge process is also used. $^{235}UF_6$ tends to stay near the axis of the centrifuge and the $^{238}UF_6$ tends to drift to the perimeter. Slight separation occurs at each stage.

Early methods of isolating deuterium used electrolysis of water. Deuterium ions are discharged at a slightly slower rate; thus the residue becomes progressively richer in deuterium during electrolysis. The process presently used is a chemical exchange between water and hydrogen sulfide.

isotopic number (**neutron excess**) The difference between the number of neutrons in a nucleus and the number of protons.

isotopic spin (**isospin; isobaric spin**) A quantum number for hadrons that distinguishes between a set of particles but emphasizes their similarities. Werner HEISENBERG introduced the concept of isotopic spin in 1932. In this picture a proton and a neutron are regarded as different quantum states of the same entity called the nucleon. By analogy to the quantum theory of electron spin, the proton is assigned isotopic spin ½ and the neutron is assigned isotopic spin –½. Later, the concept of isospin was extended to other hadrons such as pions. Isotopic spin is conserved in strong interaction processes but not in processes involving the weak and electromagnetic interactions. The formal analogy between isotopic spin and electric spin means that group theory can be used to analyze isotopic spin. Historically, isotopic spin was the starting point for the development of non-Abelian gauge theories.

isotropic A medium (or space) is isotropic if the value of a quantity does not depend on the direction in the medium.

J

Janssen, Pierre Jules César (1824–1907) French astronomer who, with Norman LOCKYER, discovered the element helium by examining the spectrum of the Sun in 1868.

Jeans, Sir Hames Hopwood (1877–1946) British astrophysicist who made a number of important contributions to physics and cosmology. Much of his early work was concerned with the kinetic theory of gases. He also modified a result of Lord RAYLEIGH to produce the classical RAYLEIGH–JEANS LAW of black-body radiation in 1905.

Jeans instability An instability in a cloud of gas in space when fluctuations in the density of the gas cause clumping of the matter as a result of its mutual gravitational attraction. In 1902, Sir James JEANS worked out a criterion, now known as the *Jeans criterion*, which determines the conditions for this instability to occur in terms of the density and temperature of the gas. This calculation used Newtonian gravity in a static Universe. Extension of this calculation to the case of general relativity theory in an expanding Universe provides the starting point for understanding STRUCTURE FORMATION in the Universe.

Jensen, Johannes Hans Daniel (1907–73) German physicist. *See* Goeppert-Mayer; Maria.

jet A shower of particles associated with quarks and gluons. When electrons and positrons collide at sufficiently high energies there is a quark–antiquark pair created, with the quark and antiquark moving in opposite directions. Because of QUARK CONFINEMENT the quark and anti-quark cannot exist in isolation. As a result quarks come together to form particles such as protons and pions. Consequently, a pair of jets of hadrons moving in opposite directions, corresponding to the original quark–antiquark pair is created. If the energy is greater than about 200 GeV a third jet corresponding to formation of a gluon can be observed. At still higher energies it is possible to have a fourth jet due to both the quark and the antiquark emitting a gluon. These jets were first observed at SPEAR in the 1970s. They are the nearest we come to seeing quarks and gluons directly and have given useful information about quarks. Jets are analyzed theoretically using QUANTUM CHROMODYNAMICS.

JET (**Joint European Torus**) A TOKAMAK experiment located at Culham, England, which is used to investigate NUCLEAR FUSION. The aim is to use controlled nuclear fusion as a source of energy.

j–j coupling A type of coupling of angular moment in many-fermion systems, such as electrons in atoms and nucleons in nuclei. In j–j coupling, in contrast to RUSSELL–SAUNDERS COUPLING, the energies associated with spin–orbit coupling are much greater than the energies associated with electrostatic repulsion. The atomic spectra of atoms with large proton numbers are characterized by j–j coupling. The existence of MAGIC NUMBERS of nuclei can be explained by the nuclear SHELL MODEL in which the multiples are characterized by j–j coupling. In many atoms and nuclei the energies associated with electrostatic repulsion are similar to the energies associated with spin-orbit coupling. This state of affairs is called *intermediate coupling*.

Joliot-Curie, Jean Frédéric (1900–58) and **Joliot-Curie, Irène** (1897–1956) French physicists who discovered ARTIFICIAL RADIOACTIVITY in 1934. Irène was the daughter of Pierre and Marie CURIE. When she married Frédéric Joliot in 1926 they both adopted the name Joliot-Curie. They shared the 1935 Nobel prize for chemistry for their work.

Jordan, Ernest Pascual (1902–80) German physicist who was one of the pioneers of quantum mechanics and quantum field theory. In particular, he developed MATRIX MECHANICS with Max BORN and Werner HEISENBERG in the mid 1920s and subsequently developed the formalism of SECOND QUANTIZATION.

Josephson, Brian David (1940–) British physicist who discovered what are now known as the JOSEPHSON EFFECTS and the JOSEPHSON JUNCTION in 1962. Josephson shared the 1973 Nobel Prize for physics with Leo ESAKI and Ivar GIAVER.

Josephson effects A set of electrical effects associated with a *Josephson junction*, i.e. two pieces of superconducting material separated by a thin layer of insulating material (frequently an oxide layer less than 10^{-8} m thick). It is possible for a *supercurrent*, i.e. an electrical current with no electrical resistance, to flow through the barrier. If the current is increased it is found that at some critical value of the current the superconducting ability of the barrier is lost and only the normal low tunneling through the barrier occurs. If a magnetic field is applied then as the magnetic field is increased the current increases from zero to a maximum value, then drops back to zero again, increases to a lower maximum, drops back to zero again, etc. At some critical value of the magnetic field the superconductivity of the barrier vanishes. If a steady potential difference is applied across the function then a high-speed alternating current flows through the junction, with the frequency of this current depending on the size of the potential difference. The frequency also depends on the ratio e/h, where e is the charge on an electronic and h is the Planck constant. This enables this ratio to be determined very accurately. There are potentially important uses of Josephson junctions in computers.

Josephson junction *See* Josephson effects.

Joule, James Prescott (1818–89) British physicist who is best remembered for his investigations on the conversion of electrical and mechanical work into heat. The results he obtained are important examples of the principle of CONSERVATION OF ENERGY. Joule also worked on the properties of gases.

J/psi particle A meson with a mass of 3097 MeV composed of a charm quark and an anticharm quark. Its independent discovery in 1979 by Burton RICHTER and his colleagues at Stanford and Samuel TING and his colleagues at Brookhaven was a major boost to the QUARK model of hadrons, particularly since its existence had been predicted by Sheldon GLASHOW and his colleagues on theoretical grounds.

Kaluza, Theodor Franz Eduard (1885–1954) German physicist who initiated KALUZA–KLEIN THEORY in a paper written in 1919, and published in 1921.

Kaluza–Klein theory A type of unified field theory in which general relativity theory is written in a higher number of space–time dimensions than four, with the nongravitational interactions coming from the higher dimensions. Kaluza–Klein theories were initiated by Theodor KALUZA in 1919 when he showed that general relativity theory in five space–time dimensions general covers relativity in four space–time dimensions and the Maxwell equations of electromagnetism. In 1926 Oskar KLEIN improved Kaluza's theory by taking quantum theory into account and showing that the fifth dimension could become curled up to a size much smaller than that of an atomic nucleus. To derive the non-Abelian gauge theories used to describe the strong and weak nuclear interactions from a Kaluza–Klein theory it is necessary to have more than five space–time dimensions. When Kaluza–Klein theory is combined with SUPERSYMMETRY eleven space–time dimensional SUPERGRAVITY results. In the early 1980s there was a great deal of interest in deriving the STANDARD MODEL of elementary particles from this type of Kaluza–Klein theory but serious difficulties prevented this from being done successfully. Sometimes the term 'Kaluza–Klein theory' is used to mean any theory involving a higher number of space–time dimensions than four.

Kamiokande A neutrino detector located at the Kamioka mine, in Japan. It has been operational since 1987.

kaon (k-meson) Symbol: K A type of meson that contains a strange quark with either an anti-up quark or an anti-down quark or an anti-strange quark with either an up quark or a down quark. There are positively charged K^+ negatively charged K^- and neutral kaons K^0 and their antiparticles. They are spin-zero particles which all have masses of about 500 MeV. Kaons are unstable particles with several decay modes. Differences in the decay of K^0 and its antiparticle indicate a very small violation of CP symmetry in the weak interactions.

Kapitza, Pyotr Leonidovich (1894–1984) Russian physicist who made important contributions to low-temperature physics, particularly for his work on SUPERFLUIDITY, for which he won the 1978 Nobel Prize for physics.

Kekulé von Stradonitz, Friedrich August (1829–96) German chemist who was one of the pioneers of the concept of structure in chemistry. In particular, he postulated in 1865 that the benzene molecule has a ring structure.

kelvin Symbol: K The SI base unit of thermodynamic temperature. It is defined as the fraction 1/273.16 of the thermodynamic temperature of the triple point of water. Zero kelvin (0 K) is absolute zero. One kelvin is the same as one degree on the Celsius scale of temperature.

Kelvin, Lord (**William Thomson**) (1824–1907) British physicist who made important contributions to thermodynamics and electromagnetism. In 1848 he defined the absolute scale of temperature and the absolute zero temperature. In the mid-nineteenth century he also formulated the second law of thermodynamics and stated the law of conservation of energy. In 1847 he took the first steps towards expressing Faraday's ideas on electric and magnetic fields in a mathematical quantitative way.

Kelvin–Helmholtz timescale   The time of about 20–30 million years that a star such as the Sun could radiate energy by contracting in on itself and converting gravitational energy into heat and light. By the middle of the nineteenth century it was realized that no chemical process could account why the Sun has been shining for millions of years. Hermann HELMHOLTZ and, independently, William Thomson, subsequently Lord KELVIN, suggested that the process of contraction of the Sun as a result of its own weight, which converts gravitational energy into heat and light, would last for about 20–30 million years. However, this timescale is not adequate for the hundreds of millions of years required for the Darwinian evolution of life on Earth. This nineteenth century paradox was not resolved until the twentieth century when it was realized that stars are powered by NUCLEAR FUSION.

Kendall, Henry Way (1926–99)   American physicist. *See* Friedman; Jerome Isaac.

Kerr effect   The occurrence of BIREFRINGENCE in certain substances when they are placed in a strong electric field. The effect was discovered by John Kerr (1824–1907) in 1875.

Kerr–Newman solution   A solution of the Einstein field equation of general relativity theory that corresponds to an electrically charged rotating black hole. This solution was found by Ezra Newman and his colleagues in 1965. It is the most general black hole solution as a consequence of the NO-HAIR THEOREM. Taking the electric charge to be zero gives the KERR SOLUTION, while taking the rotation to be zero gives the REISSNER–NORDSTROM SOLUTION.

Kerr solution   A solution of the Einstein field equation of general relativity theory

that corresponds to an electrically uncharged rotating black hole. This solution was found by Roy Kerr in 1963. It corresponds to the most likely type of black hole to exist in the Universe. In the Kerr solution the SINGULARITY of a black hole is a ring. Taking the rotation to be zero gives the SCHWARZSCHILD SOLUTION.

Ketterle, Wolfgang (1957–)   German physicist who discovered and investigated BOSE–EINSTEIN CONDENSATES in alkali metals. He shared the 2001 Nobel Prize for physics with Eric CORNELL and Carl WIEMAN.

kilogram   Symbol: kg   The SI base unit of mass, equal to the mass of the international prototype of the kilogram, which is a piece of platinum-iridium kept at Sèvres in France.

kinematics   The branch of mechanics that is concerned with the motion of objects but not with the forces responsible for the motion.

kinetic energy   The energy an object has due to its motion. For a body of mass m and velocity v its kinetic energy is mv^2. For a rotating body of moment of inertia I and angular velocity ω its rotational kinetic energy is $I\omega^2$.

kinetic equation   The type of equation such as the BOLTZMANN EQUATION that is used in KINETIC THEORY. Kinetic equations are used to calculate TRANSPORT COEFFICIENTS, such as conductivity. In general, approximation techniques have to be used to solve kinetic equations.

kinetic theory   A theory that explains the physical properties of matter in terms of the motion of its constituent particles. Kinetic theory was initiated by Daniel BERNOULLI in 1738 when he derived Boyle's law from the assumption that the pressure of a gas is due to the impacts of the molecules of the gas with the walls of its container. The theory was developed by Count RUMFORD, James JOULE, and James Clerk MAXWELL. An important aspect of kinetic theory is that the particles do not have the same velocity but have a range of velocities, which is described statistically. This makes it possible to explain why evaporation occurs in a liquid and why it is a cooling process. In a liquid the molecules can move around but because of the attractive forces between the molecules only the fastest moving molecules have sufficient energy to escape from the liquid. This reduces the average kinetic energy of the remaining molecules of the liquid and hence cools the liquid. The properties of ideal gases, in which it is assumed that the molecules take up no volume, only interact with each other and the container via perfectly elastic collisions, and can move freely anywhere within the container, are readily calculated using kinetic theory.

Kirchoff, Gustav Robert (1824–87) German physicist who made a number of important contributions to physics. In 1845 he found laws to describe the currents and electromotive forces in electrical circuits. In 1859 he explained the dark lines found in the solar spectrum by Josef von FRAUNHOFER by postulating that they are due to absorption by substances in the atmosphere of the Sun. In his investigations with Robert BUNSEN he pioneered spectroscopy and discovered cesium and rubidium in this way.

Kirchoff's law A law stating that at a given temperature the rate of emission of electromagnetic energy by an object is equal to the rate at which the object absorbs electromagnetic energy with the same wavelength. KIRCHOFF first stated this law in 1859 and gave a proof of it in 1861.

Klein, Christian Felix (1849–1925) German mathematician who made wide-ranging contributions to mathematics, particularly geometry, and its physical applications. He is best known for the ERLANGEN PROGRAMME of 1872.

Klein, Oskar Benjamin (1894–1977) Swedish physicist who made a number of important contributions to theoretical physics including the incorporation of

quantum theory into what is now called KALUZA–KLEIN THEORY in 1926. In this work of 1926 he derived the KLEIN–GORDON EQUATION.

Klein–Gordon equation The equation for spin-zero particles in relativistic quantum mechanics. This equation was independently discovered by a number of authors, including Oskar KLEIN, Walter Gordon, Vladimir FOCK, and Erwin SCHRÖDINGER. The Klein–Gordon equation was originally thought to be physically irrelevant since it is not viable as an equation in single-particle relativistic quantum mechanics. However, it is viable as an equation in relativistic quantum field theory and describes spin-zero particles such as mesons and predicts the existence of antiparticles of these particles.

Klitzing, Klaus von (1943–) German physicist who discovered the QUANTUM HALL EFFECT. He was awarded the 1985 Nobel Prize for physics.

k-meson *See* kaon.

KM mechanism *See* Kobayashi–Maskawa mechanism.

knot theory The branch of mathematics used to classify and describe knots and entanglements. Knot theory was started in the nineteenth century by Lord KELVIN and his colleague Peter Tait in an attempt to explain the structures of atoms. Certain quantum field theories can be related to knot theory. As a result, knot theory can be used in gauge theories and quantum gravity. Knot theory has also been used in the study of polymers (including biologically important molecules such as DNA) and some models used in the statistical mechanics of phase transitions.

Kobayashi–Maskawa mechanism (KM mechanism) A mechanism for incorporating CP VIOLATION into the STANDARD MODEL of elementary particles. This mechanism, proposed by Maloto Kobayashi and Toshihide Maskawa in 1973, requires the existence of three generations of elementary particles. At present, it is not clear whether this mechanism is a correct explanation of CP violation in the weak interactions.

Kohn, Walter (1923–) Austrian-born American physicist who developed DENSITY FUNCTIONAL THEORY, starting in the mid 1960s. Kohn shared the 1998 Nobel Prize for chemistry with John POPLE for his work.

Kramers, Hendrik Anthony (1894–1952) Dutch physicist who made many important contributions to the development of quantum mechanics, quantum field theory, and statistical mechanics.

Kramers theorem The theorem proved by Hendrick KRAMERS in 1930 that the energy levels of a system, such as an atom, that contains an odd number of spin-½ particles, such as electrons, is at least doubly degenerate in the absence of an external magnetic field. This type of degeneracy is known as *Kramers degeneracy*. It arises because of *time reversal invariance*. Kramers theorem holds in the presence of crystal fields and spin–orbit coupling but does not apply in the presence of an external magnetic field.

krypton A colorless odorless monatomic element of the rare-gas group, known to form unstable compounds with fluorine. It occurs in minute quantities (0.001% by volume) in air. Krypton is used in fluorescent lights.

Symbol: Kr. Melting pt.: $-156.55°$C. Boiling pt.: $-152.3°$C. Density: 3.749 (0°C) kg m^{-3}. Proton number: 36. Relative atomic mass: 83.80. Electronic configuration: $[Ar]3d^{10}4s^24p^6$.

Kusch, Polykarp (1911–93) American physicist who performed very accurate determinations of the MAGNETIC MOMENT of the electron in the years following World War II. This work was very significant because if clearly showed that there was a deviation from the expectation of the single-particle DIRAC EQUATION. This realization provided a motivation for developing QUANTUM ELECTRODYNAMICS (QED) to explain the results of these very accurate experiments. Kusch and Willis LAMB shared the 1955 Nobel Prize for physics for their experimental work relating to QED.

L

labeling The process of replacing an atom in a compound with an isotope of the same element to trace its path through a biological, chemical, or physical system. The compound that is altered in this way is called a *labeled compound*. Frequently the isotope is a *radioisotope* and its path is followed using a GEIGER COUNTER. If the isotope is stable it is detected by a MASS SPECTROMETER. The atom used is called a *label*. When the isotope is a radioisotope labeling is frequently called *radioactive tracing*.

Lagrange, Joseph Louis (1736–1813) Italian–French mathematician and physicist who made important contributions to mechanics, particularly celestial mechanics. He invented the LAGRANGIAN formulation of classical mechanics. In 1788 his great book *Analytical Mechanics*, in which he discussed mechanics entirely in terms of calculus rather than Euclidean geometry, was published.

Lagrangian Symbol: L A function that characterizes the motion of an object, equal to the kinetic energy minus the potential energy at any point in its trajectory. The Lagrangian was used to describe classical mechanics but quantum mechanics can also be expressed in terms of a Lagrangian, as can field theory at both the classical and quantum level. *See also* principle of least action.

Lamb, Willis Eugene (1913–) American physicist who made very precise measurements of the spectrum of the hydrogen atom, leading to the discovery of the LAMB SHIFT. He shared the 1955 Nobel Prize for physics with Polykarp KUSCH.

lambda An electrically neutral particle that is the lightest HYPERON. It consists of one up quark, one down quark, and one strange quark. The lambda particle can be regarded as a neutron in which a down quark has been replaced by a strange quark. By a process analogous to the beta decay of a neutron, the lambda decays via weak interactions to a proton, an electron, and an antineutrino, with the strangeness of –1 of the lambda disappearing due to the strange quark becoming an up quark. The mass of the lambda is 111.56 MeV and its lifetime is 2.6×10^{-10} second.

lambda point The temperature at which liquid helium I becomes the superfluid helium II. *See* helium.

Lamb shift A small difference between two energy levels of the hydrogen atom. According to the single-particle DIRAC EQUATION the two energy levels should be degenerate. The shift occurs because of the quantum mechanical interaction between the electron of the atom and the electromagnetic field. The Lamb shift was discovered by Willis LAMB soon after the end of World War II. It was explained using RENORMALIZATION in quantum electrodynamics.

Landau, Lev Davidovich (1908–68) Azerbaijani physicist who made important contributions to many branches of theoretical physics, including the theory of phase transitions, superconductivity, superfluids, Fermi liquids, ferromagnetism, weak interactions, quantum electrodynamics, and neutron stars. He was awarded the 1962 Nobel Prize for physics, mainly for his work on superfluidity.

Landau ghost A difficulty associated with the possible inconsistency of RENORMALIZATION in quantum electrodynamics (QED) and other quantum field theories that do not have ASYMPTOTIC FREEDOM. In the mid-1950s Lev LANDAU and his colleagues showed that at very high energies there is an inconsistency in the relation between the bare and renormalized electric charge. This difficulty, known as the *Landau ghost* (or sometimes the *Moscow zero*) was demonstrated by Landau using an approximation technique. The existence of the Landau ghost persists when the approximation technique is taken further than the original calculations but it has not been proved in a general way. For QED the difficulty of the Landau ghost is theoretical since the electromagnetic interactions are thought to be unified with the weak and strong interactions at energies which are much lower than the energy at which the difficulty of the Landau ghost appears. The opposite behavior to the Landau ghost at high energies is ASYMPTOTIC FREEDOM, as occurs in QUANTUM CHROMODYNAMICS.

Landau–Ginzburg theory An empirical theory of superconductivity put forward by Lev LANDAU and Vitaly Ginzburg in 1950. It was subsequently realized that the Landau–Ginzburg theory can be derived from the more fundamental BARDEN–COOPER–SCHRIEFFER (BCS) THEORY of superconductivity.

Landau levels The set of energy levels of free electrons in a uniform magnetic field. They are named after Lev LANDAU who analyzed the quantum mechanics of this problem in 1930. These energy levels have discrete values proportional to e/m, where e and m are respectively the charge and mass of an electron.

Landauer's principle A principle that states that when information is erased heat is generated. It was put forward by Rolf Landauer in the 1960s after observing the heat generated by computers. Landauer's principle establishes a connection between information theory and thermodynamics.

Landé's rule A rule for atomic spectra describing multiplets in which SPIN–ORBIT COUPLING is a small perturbation to RUSSELL–SAUNDERS COUPLING. It states that in a given multiplet the energy differences between two successive J levels, where J is the total resultant angular momentum of the coupled electrons, are proportional to the large of the two values of J. The rule was stated by the German-born American physicist Alfred Landé in 1923 on empirical grounds. It can be derived by applying the quantum theory of angular momentum (or equivalently, the theory of the rotation group) to electrons in atoms. The Landé interval rule is not satisfied if SPIN–SPIN INTERACTION between electrons is taken into account.

Landé splitting factor *See* magnetic moment.

Langevin, Paul (1872–1946) French physicist who explained paramagnetism in terms of the theory of electrons. This enabled him to describe how paramagnetism varies as a function of absolute temperature. Langevin also worked on the theory of Brownian motion.

Langmuir, Irving (1881–1957) American chemist who extended the ideas of Gilbert Newton LEWIS on the electron structure of atoms in 1919. He also worked extensively on the properties of surfaces and adsorption, for which he was awarded the 1932 Nobel Prize for chemistry.

lanthanide contraction *See* lanthanides.

lanthanides (lanthanoids; lanthanons; rare-earth elements) A series of elements in the periodic table that is usually taken to range from cerium to lutetium. This corresponds to an increase in proton number from 58 to 71, following lanthanum, and an increase in the number of $4f$ electrons. However, this increase is not smooth since the $4f$ and $5d$ energy levels are close. Lanthanum is sometimes included amongst the lanthanides because it is chemically similar. The 4f electrons do not shield the outer electrons from the nucleus very effectively and the increasing nuclear charge as the proton number increases causes a decrease of the atomic and ionic radii of lanthanides. This decrease of radius with increase of proton number is called the *lanthanide contraction*.

lanthanoids *See* lanthanides.

lanthanum A soft ductile malleable silvery metallic element that is the first member of the lanthanoid series. It is found associated with other lanthanoids in many minerals, including monazite and bastnaesite. Lanthanum is used in several alloys (especially for lighter flints), as a catalyst, and in making optical glass.

Symbol: La. Melting pt.: 921°C. Boiling pt.: 3457°C. Relative density: 6.145 (25°C). Proton number: 57. Relative atomic mass: 138.9055. Electronic configuration: [Xe]5d^{1}6s^2.

Laplace, Marquis Pierre Simon de (1749–1827) French astronomer, mathematician, and physicist who made important contributions to celestial mechanics. In 1796 he considered the possibility of black holes and put forward the suggestion, known as the *nebular hypothesis*, that the Solar System was formed from a rotating mass of gas. In the course of his work on celestial mechanics he developed mathematical tools such as the LAPLACE EQUATION. Laplace also established probability theory on a rigorous basis.

Laplace equation The partial differential equation
$$\partial^2 u/\partial x^2 + \partial^2 u/\partial y^2 + \partial^2 u/\partial z^2 = 0,$$
where u is some potential such as the electric potential or the gravitational potential. This equation can also be written in the form
$$\nabla^2 u = 0,$$
where ∇^2 is known as the *Laplace operator* or the *Laplacian*. The Laplace equation was formulated by Pierre Simon LAPLACE in his work on celestial mechanics but has many other applications in theoretical physics.

Laplacian determinism The idea that if the motion of a mechanical system is known completely at a given time then its future motion could be determined exactly. The dream of Laplacian determinism is fatally undermined both by CHAOS THEORY and by QUANTUM MECHANICS. Chaos theory makes Laplacian determinism impossible because the sensitivity to initial conditions means the initial state of the system has to be known exactly. Both in practice and in principle this is impossible. In quantum mechanics the Heisenberg UNCERTAINTY PRINCIPLE makes it impossible to know both the exact position and the momentum of a particle at the same time. The motion can only be discussed in terms of probability theory rather than the cast-iron certainty of Laplacian determinism.

Large Hadron Collider (LHC) A particle accelerator at CERN, which it is hoped will operate at energies up to 14 TeV. This accelerator may find the Higgs boson and evidence for supersymmetry.

Larmor, Sir Joseph (1857–1942) Irish–British physicist who made important contributions to classical electrodynamics. In particular, he predicted the LARMOR PRECESSION, derived a nonrelativistic expression for the energy radiated by an accelerated electron, and expressed classical electrodynamics in Lagrangian form.

Larmor precession A precession of the motion of a charged particle, such as an electron, in a magnetic field. The motion of an electron orbiting around the nucleus of an atom in a magnetic field can be thought of in terms of the Larmor precession. The existence of this precession was deduced by Sir Joseph LARMOR in 1897.

laser [*l*ight *a*mplification by *s*timulated *e*mission of *r*adiation] A device used to produce a beam of electromagnetic radiation in which the waves are coherent (in phase) and monochromatic (with the same wavelength). Lasers have been built to produce electromagnetic radiation in the infrared, visible, ultraviolet, and x-ray regions of the electromagnetic spectrum. The laser action is produced in a volume of a suitable material, which can be solid, liquid, or gas.

The operation of a laser makes use of the quantum theory of radiation. In a laser the atoms, molecules or ions are 'pumped' to an excited state. This can be done in several ways, including an intense source of light, discharges in gases, and chemical reactions. The aim of the pumping is to obtain *population inversion*, i.e. a situation where a higher energy state E_2 is more populated than a lower energy state E_1. The system is then stimulated to emit photons of energy $(E_2 - E_1)$ by another photon of the same energy, a process known as *stimulated emission*. There are two reflecting surfaces at each end of the laser, with one end being totally reflecting and the other partially reflecting. The repeated reflection between these surfaces results in an intense beam of coherent monochromatic radiation.

Lasers have many applications in science, technology, communications, medicine, and engineering.

laser cooling The process of cooling atoms using lasers. When a moving atom collides with a photon, the frequency of light the atom experiences depends on the velocity of the atom because of the DOPPLER EFFECT. Since atoms have a distribution of velocities this means that the frequency of light is only the right frequency to be absorbed by some atoms, with other atoms experiencing light with different frequencies which cannot be absorbed by the atoms. An atom which has absorbed a photon recoils in the opposite direction. When the atom in the excited state emits a photon by spontaneous emission the photon can go in any direction, with the atom moving in the opposite direction. Thus, the average net change in velocity because of spontaneous emission is zero. For a beam of atoms colliding with a laser beam the net effect of this process is to slow the atoms down. Typically, to cool atoms six laser beams in three oppositely directed pairs are used, with this arrangement being called an *optical molasses*. This type of cooling can cool atoms to what is called the *Doppler limit*, which is about 240 millionths of a degree above absolute zero for sodium atoms. Lasers can also be used to obtain even lower temperatures (*see* Sisyphus effect).

latent heat The heat absorbed or released when a system changes its phase at a constant temperature, from a solid to a liquid at the melting temperature or from a liquid to a gas at the boiling point, etc. For example, when a liquid boils the latent heat is the energy needed to overcome the attractive forces between molecules in the liquid and to do work against atmospheric pressure as the gas expands.

lattice A regular arrangement of points in space. A lattice is used to describe the positons of the particles (atoms, ions, or molecules) in a crystalline solid. The lattice structure can be examined by x-ray diffraction techniques.

lattice energy The energy released when ions of opposite charge are brought together from infinity to form one mole of a given crystal. The lattice energy is a measure of the stability of a solid ionic substance, with respect to ions in the gas.

lattice field theory A formulation of field theory in which a lattice is used as a REGULARIZATION, with the continuum limit being taken at the end of the calculation. This type of regularization violates Lorentz invariance but enables techniques from statistical mechanics to be used.

lattice gauge theory A formulation of gauge theories as theories on a discrete space–time lattice rather than on continuous space–time, with the continuum limit being taken at the end of the calculation. Numerical and computational calculations of gauge theories can be performed using lattice gauge theory. Lattice gauge theories are particularly useful in theories such as quantum chromodynamics in which there is strong coupling and many important features of the theory such as QUARK CONFINEMENT are NONPERTURBATIVE.

lattice vibrations The vibrations of the atoms, molecules, or ions of a crystal about their equilibrium positions. There is still a residual vibration even at absolute zero temperature due to ZERO-POINT ENERGY. As the temperature increases the amplitudes in the lattice increase. This explains the increase in the electrical resistance of metals with temperature since the conduction electrons of the metals are scattered more as the amplitudes of the lattice vibrations increase.

Laughlin, Robert B. (1950–) American physicist who provided a theoretical explanation of the fractional QUANTUM HALL EFFECT in the early 1980s. Laughlin shared the 1998 Nobel Prize for physics with Horst STÖRMER and Daniel TSUI.

Lavoisier, Antoine Laurent (1743–94) French chemist who is widely regarded as one of the founders of modern chemistry. His two most important contributions to chemistry were demonstrating that mass is conserved in chemical reactions and that burning is the combination of a substance with oxygen from the air.

Lawrence, Ernest Orlando (1901–58) American physicist who invented the CYCLOTRON in 1931. Lawrence was awarded the 1939 Nobel Prize for physics for this invention.

lawrencium A radioactive transuranic element of the actinoid series, not found naturally on Earth. Several very short-lived isotopes have been synthesized by bombarding ^{252}Cf with boron nuclei or ^{249}Bk with ^{18}O nuclei.

Symbol: Lr. Proton number: 103. Most stable isotope: ^{262}Lr (half-life 261 minutes). Electronic configuration: [Rn]$5f^{14}6d^{1}7s^{2}$.

lead A dense, dull, gray, soft metallic element; the fifth member of group 14 of the periodic table and the end product of radioactive decay series. Lead is used in accumulators (lead–acid), alloys, radiation shielding, and water and sound proofing. It is also used in the petrochemical, paint, and glass industries.

Symbol: Pb. Melting pt.: 327.5°C. Boiling pt.: 1830°C. Relative density: 11.35 (20°C). Proton number: 82. Relative atomic mass: 207.2. Electronic configuration: [Xe]$4f^{14}5d^{10}6s^{2}6p^{6}$.

lead equivalent A measure of the absorbing power of a radiation screen in terms of the thickness of a lead screen that would give the same protection as the screen being considered. The thickness of the lead screen is expressed in millimeters.

least action *See* principle of least action.

Lederman, Leon Max (1922–) American physicist who, together with Melvin SCHWARTZ and Jack STEINBERGER, showed that the muon neutrino and the electron neutrino are distinct particles. Lederman also led the team that discovered the UPSILON particle. The experimental work on the muon neutrino was carried out at the Brookhaven National Laboratory in the early 1960s. The upsilon particle was discovered at Fermilab in 1977. Lederman, Schwartz, and Steinberger shared the 1988 Nobel Prize for physics for their demonstration of the existence of the muon neutrino.

Lee, David Morriss (1931–) American physicist who, together with Douglas OSHEROFF and Robert RICHARDSON, discovered and investigated the superfluidity of helium-3 in the early 1970s. Lee, Osheroff, and Richardson shared the 1996 Nobel Prize for physics for this work.

Lee, Tsung Dao (1926–) Chinese-born American physicist who, together with Chen Ning YANG, postulated that parity is violated in weak interactions. Other collaborations of Lee and Yang resulted in the prediction that weak interactions are mediated by VECTOR BOSONS, the suggestion that the electron neutrino and the muon neutrino are distinct particles, and important results in the theory of phase transitions. Lee and Yang won the 1957 Nobel Prize for physics for their suggestion of parity violation in weak interactions, just one year after they made their suggestion.

LEED (low-energy electron diffraction) A type of electron diffraction using a beam of low-energy electrons to study solid surfaces and absorption.

Leibniz, Gottfried Wilhelm (1646–1716) German scholar who made many important contributions to mathematics and physics. He invented the calculus independently of Newton, which led to an extremely acrimonious priority dispute. He also found himself in disagreement with Newton over the concept of action at a distance in Newtonian gravity and the concepts of absolute space and absolute time in Newtonian mechanics.

Lemaître, Abbé George Édouard (1894–1966) Belgian astrophysicist, cosmologist, and priest who developed an early version of the BIG-BANG THEORY. In 1927 he discovered, independently of Aleksandr FRIEDMANN, solutions of the Einstein field equation of general relativity theory corresponding to an expanding Universe. This work made little impact until Sir Arthur EDDINGTON arranged for it to be translated into English in 1931. Lemaître extended his result that the Universe is expanding by postulating that the Universe originated in an extremely compact dense state, which he called the *primal atom*, with the big bang being triggered by an event akin to the radioactive decay of an atomic nucleus.

Lenard, Philipp Eduard Anton (1862–1947) German physicist who discovered the features of the PHOTOELECTRIC EFFECT, which Albert EINSTEIN explained by postulating that light has particle-like as well as wavelike aspects. Lenard also investigated cathode rays (electrons). He was awarded the 1905 Nobel Prize for physics.

Lenz's law The principle that direction of an induced e.m.f. always opposes the change producing it. This law was discovered by Heinrich Lenz in 1834 and is an ex-

ample of the principle of energy conservation.

LEP (Large Electron–Positron collider) A circular accelerator 27 km in diameter at CERN. It started operating in 1989 and can reach energies of greater than 100 GeV. This facility has enabled Z bosons to be studied in great detail.

lepton *See* elementary particles.

lepton number A number that is used to indicate which particles are leptons and which are not. Each lepton has a lepton number of +1. Each antilepton has a lepton number of −1. Particles that are not leptons have a lepton number of zero. Lepton number has never been violated in any known experiment. For example, in the beta decay of a neutron: $n \rightarrow p^+ + e^- + \nu_e$, the neutrino produced must be an antineutrino since a neutron and a proton both have lepton numbers of zero and an electron has a lepton number of +1. In GRAND UNIFIED THEORIES it is possible for a lepton number not to be conserved in certain processes. Although such processes have never been observed it is thought that lepton number (and BARYON NUMBER) is violated in processes that occurred in the EARLY UNIVERSE (*see* matter–antimatter asymmetry).

Leucippus (*c.* 500–450 BC) Ancient Greek philosoopher who is frequently credited as the inventor of atomic theory.

Lewis, Gilbert Newton (1875–1946) American chemist who put forward the idea of the *covalent bond* consisting of two atoms sharing a pair of electrons. Lewis used his ideas about electrons in molecules to give more general definitions of an acid and a base than previous definitions. A *Lewis acid* is an atom or compound that can accept an electron pair while a *Lewis base* is an atom or compound that can donate an electron pair. Lewis also made important contributions to the application of thermodynamics to chemistry.

LHC   *See* Large Hadron Collider.

Libby, Willard Frank (1908–80) American chemist who developed the technique of dating with radioisotopes, particularly carbon-14, in the years following World War II. Libby won the 1960 Nobel Prize for chemistry for this work.

Lie group   A type of continuous GROUP that has important applications in elementary particle physics and other branches of physics. The simplest example of a Lie group is the rotation group. Lie groups are named after Sophus Lie who originated the concept in the second half of the nineteenth century. Important contributions to the development of Lie groups were made by Élie CARTAN and Hermann WEYL. The complete classification of Lie groups by Cartan in 1894 indicated that Lie groups can be divided into several series in which the elements of the Lie groups are $N \times N$ matrices and several groups known as *exceptional groups*. The types of Lie group with $N \times N$ matrices of interest to elementary particle physics are denoted $SU(N)$ and $SO(N)$. Lie groups were used in atomic and nuclear spectroscopy in the 1950s. In the early 1960s the group $SU(3)$ was used to systematize the pattern of hadrons known at that time. In the 1970s it became clear that all the nongravitational interactions are described by GAUGE THEORIES involving Lie groups.

lifetime   The average time a system stays in unstable state. The system may be an excited atom, a radioactive nucleus, a free radical, etc.

Lifshitz, Evgeny Mikhailovich (1915–85)   Russian physicist who made important contribution to many branches of theoretical physics, including the theory of intermolecular forces, ferromagnetism, phase transitions, and cosmology.

ligand field theory   An extension of CRYSTAL FIELD THEORY that describes the properties of compounds of transition-metal or rare-earth ions in which covalent bonds between the surrounding molecules (known as *ligands*) and the central ion is taken into account. This can be done by using either molecular-orbital theory or valence bond theory. The splitting pattern of energy levels of the central ion by ligands is determined by group theory in ligand field theory, as in crystal field theory. The optical, spectroscopic, and magnetic properties of compounds are explained very successfully using the theory.

light   A form of electromagnetic radiation that the human eye can detect. The wavelengths of visible light range from about 400 nm to 700 nm, with the precise wavelengths that can be detected depending on the individual. Human eyes have become sensitive to this narrow range of the electromagnetic spectrum because it is produced prolifically by the Sun and is not absorbed by the atmosphere of the Earth.

Light has featured very prominently in the history of physics. Examples of this include the analogies between mechanics and optics, the connection between electricity, magnetism, and light, the importance of the speed of light in special relativity theory, LIGHT-BENDING in general relativity theory, and the dual particle–wave nature of light in quantum theory.

light bending   The deflection of light by a massive body. A quantitative description of this phenomenon is one of the crucial predictions of general relativity theory. This prediction was confirmed by the expeditions to observe the solar eclipse of 1919. *See also* gravitational lens.

light-cone   A pictorial representation of space and time in special relativity theory. In a diagram, called a *Minkowski diagram* after Hermann Minkowski, time is represented by a vertical line and all the spatial dimensions are represented by a horizontal line. The way in which a flash of light spreads from a single point is represented by two straight lines, each of which is at 45 degrees to the horizontal line. If these two lines are rotated around the vertical line then a conical surface known as the *light-cone* is formed. The cone above the horizontal line is called the *future light-cone* and the cone below the horizontal line is called the *past light-cone*. If a signal is moving at less than the speed of light then its path is inside a light-cone. The only events we can have knowledge about are within the past light-cone and the only events which can be influenced are within the future light-cone.

light-emitting diode (LED) A device that uses a semiconductor diode to convert electrical energy into light.

lightest superpartner (LSP) The lightest particle predicted to exist by SUPERSYM-METRY. The LSP must be a stable particle since there is no lighter supersymmetric particle it can decay to. The PHOTINO is a promising candidate to be the LSP. It is not possible to predict the mass of the LSP since the way in which supersymmetry is broken is not known. The LSP may contribute to the DARK MATTER in the Universe.

light scattering by light A specifically quantum mechanical process that is predicted to occur in quantum electrodynamics (QED). There is some evidence that this process has been observed in experiments involving lasers, but this is not yet conclusive. Calculations predicting that light scattering by light should occur were made by Werner HEISENBERG and others in the 1930s, with more complete calculations being performed after RENORMALIZATION in QED was developed in the 1940s.

linac *See* linear accelerator.

Linde, Andrei Dmitrivitch (1948–) Russian-born physicist who made major contributions to INFLATION theory. In the early part of his career he showed that phase transitions occurred in the EARLY UNIVERSE, with the gauge symmetry in the WEINBERG–SALAM MODEL becoming a broken symmetry at this phase transition which occurs when the Universe has cooled to some critical temperature. *See also* multiverse.

linear accelerator (linac) A device for accelerating charged particles in a straight

line. It consists of a series of cylindrical electrodes of increasing length, alternate ones of which are connected together and between which an alternating voltage is applied. A charged particle is accelerated to enter the first cylinder. By the time it emerges the voltages of the cylinders have reversed, and it is now repelled from the first cylinder and attracted towards the second, so accelerating and gaining energy each time it crosses a gap between cylinders. The cylinders have to be made progressively longer as the particles speed up so that the particles arrive at the gaps at the correct point in the voltage cycle.

A more advanced type of linear accelerator for electrons and protons uses a traveling radio-frequency electromagnetic wave in a waveguide. The particles are carried by the electric component of the wave. *See also* tandem generator.

linear combination of atomic orbitals *See* orbital.

linear equation An equation relating two quantities x and y such that a graph of the relation is a straight line. This definition can be extended to more than two quantities.

lines of force Imaginary lines that are used to represent a field. They allow both the direction and the strength of a field to be indicated. The field strength is indicated by the closeness of the lines, with the closer the lines the greater the field strength. The direction of a field at a point is given by the tangent to the line of force at that point. Lines of force are mainly used to describe electric and magnetic fields but can be extended to other fields. The concept of lines of force originated in work by Michael FARADAY on electric and magnetic fields, which was expressed mathematically by James Clerk MAXWELL.

line spectrum A spectrum composed of a number of discrete lines corresponding to single wavelengths of emitted or absorbed radiation. Line spectra are produced by atoms or simple (monatomic) ions in gases. Each line corresponds to a change in electron orbit, with emission or absorption of radiation. *See also* spectrum.

line-width The width of a spectral line. Quantum mechanics means that any spectral line is not perfectly sharp because there is an uncertainty in the energy of a level. In practice, a spectral line is also broadened by the DOPPLER EFFECT and by collisions between atoms or molecules. *See also* monochromatic.

liquid A state of matter intermediate between a solid and a gas. In a liquid the atoms, molecules, or ions can move about but the attractive interactions between them mean that they cannot occupy all their container. A liquid takes up the shape of its container. It is more difficult to construct a theory of liquids than a theory of gases or a theory of solids. Nevertheless, a great deal of progress has been made using STATISTICAL MECHANICS. Experimentally, it has been found that there is order over short distances in liquids.

liquid-drop model A model of the atomic nucleus in which the nucleons are regarded as being analogous to the molecules in a liquid drop, with the shape of the drop being maintained by a force that is a nuclear analog of surface tension. The model was originally proposed by George GAMOW in 1928. After the discovery of the neutron in 1932 the model was developed by Niels BOHR and his colleagues in the 1930s. In particular, Bohr and John WHEELER used the liquid-drop model to give a quantitative description of NUCLEAR FISSION.

lithium A light silvery moderately reactive metal; the first member of the alkali metals (group 1 of the periodic table). It occurs in a number of complex silicates, such as spodumene, lepidolite, and petalite, and a mixed phosphate, tryphilite. It is a rare element accounting for 0.0065% of the Earth's crust.

Symbol: Li. Melting pt.: 180.54°C. Boiling pt.: 1347°C. Relative density: 0.534 (20°C). Proton number: 3. Relative atomic mass: 6.941. Electronic configuration: [He]2s^1.

locality The principle that interactions between different points in space cannot take place faster than the speed of light in a vacuum, in accord with special relativity theory. The Aspect experiment clearly showed that locality is violated by quantum mechanics, i.e. quantum mechanics is characterized by *nonlocality*. *See* Bell's theorem.

localization *See* electronic structure of solids.

local symmetry A symmetry in which the effect of a symmetry transformation varies from point to point in space–time. The symmetries associated with gauge groups are local symmetries. If the symmetry transformations of a symmetry involving a Lie group have the same effect at all points in space–time then the symmetry is said to be a *global symmetry* (sometimes called a *rigid symmetry*).

Lockyer, Sir Joseph Norman (1836–1920) British astronomer who discovered helium in the Sun using spectroscopy in 1868 (nearly 30 years before its discovery on the Earth).

London, Fritz Wolfgang (1900–54) German-born American physicist who, together with Walter HEITLER, explained the covalent bond in the hydrogen molecule in terms of quantum mechanics in 1927. He subsequently made important contributions to the theory of SUPERFLUIDITY and, together with his brother Heinz London, SUPERCONDUCTIVITY.

London formula *See* van der Waals forces.

lone pair A pair of electrons with opposite spin in an orbital of an atom in which the pair does not form a chemical bond with another atom. For example, in the ammonia molecule (NH_3) the nitrogen atom has five electrons. Three of these electrons are paired up with electrons from the

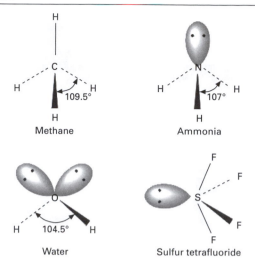

Methane

Ammonia

Water

Sulfur tetrafluoride

Lone pairs in molecules

hydrogen atom, thereby giving the nitrogen atom a share in the stable arrangement of eight electrons. The remaining two electrons form a lone pair. Similarly, there are two lone pairs on the oxygen atom of a water molecule. Lone pairs occupy space in the same way that atom–atom bonds do, and are important in determining the shapes of molecules.

longitudinal wave *See* wave.

loop quantum gravity *See* quantum gravity.

loop variables Quantities such as lines of force in electric and magnetic fields and their analogs for other fields. It is frequently convenient to formulate field theories in terms of loop variables.

Lorentz, Hendrik Antoon (1853–1928) Dutch physicist who made important contributions to electromagnetism, particularly the theory of electrons, which he developed in the 1890s. One result of these investigations was the discovery of the LORENTZ FORCE in 1895. In attempting to understand the negative result of the MICHELSON–MORLEY EXPERIMENT he was led to the LORENTZ–FITZGERALD CONTRACTION and subsequently to the LORENTZ

TRANSFORMATIONS. Lorentz also contributed to the early development of general relativity theory. He shared the 1902 Nobel Prize for physics with Pieter ZEEMAN for his explanation of the (normal) ZEEMAN EFFECT.

Lorentz–Dirac equation An equation put forward by Paul DIRAC in 1938, building on earlier work by Henrik Antoon LORENTZ, which describes the motion of a point charge electron in classical electrodynamics. Although this equation is consistent with special relativity theory and enables the infinite SELF-ENERGY of the electron to be removed it suffers from a number of serious difficulties. For example, it is necessary to know the initial acceleration of the particles in addition to their initial positions and velocities. Also, most initial accelerations lead to *runaway solutions*, i.e. solutions in which the electrons accelerate to speeds that approach the speed of light very rapidly.

Lorentz–Fitzgerald contraction The contraction of the length of a moving body in the direction of its motion which was proposed independently in 1892 by Hendrik Antoon LORENTZ and George Fitzgerald. Lorentz and Fitzgerald showed that the negative result of the MICHELSON–

MORLEY EXPERIMENT could be explained by postulating that the length of an object contracts in the direction of its motion through the ETHER by the factor $\sqrt{(1 - v^2/c^2)}$, where v is its velocity in the ether and c is the speed of light in a vacuum. This result emerges as a result of special relativity theory, but with a different physical interpretation. In special relativity theory it is not necessary to postulate the existence of the ether, the velocity v refers to the relative velocity of the body in one frame of reference with respect to another frame, and the contraction is a mathematical result that depends on the relative velocity of the observer rather than being a real physical contraction of the body.

Lorentz force   The force F on an electric charge Q moving with a velocity v in a magnetic field B. The force is given by $F = BQv \sin \theta$, where θ is the angle between B and v.

Lorentz group   The set of rotations in the four-dimensional space x, y, z, τ, where x, y, and z are space coordinates and $\tau = i c t$, where t is the absolute time and c is the speed of light in a vacuum. Such rotations are called *proper Lorentz transformations*. Combining the Lorentz group with translations in space and time gives the *proper Poincaré group*. In relativistic quantum mechanics analysis of the REPRESENTATIONS of the Lorentz group is used to classify elementary particles.

Lorentzian geometry   A name sometimes given to the geometry used in general relativity theory, i.e. a modification of RIEMANNIAN GEOMETRY in which the space is locally MINKOWSKI SPACE–TIME rather than being Euclidean space locally.

Lorentz–Lorenz equation   An equation that relates the POLARIZABILITY α of a molecule and the REFRACTIVE INDEX n. The Lorentz–Lorenz equation can be written in the form $\alpha = (3/4 \pi N) [(n^2 - 1)/(n^2 + 1)]$, where N is the number of molecules per unit volume. This equation was derived independently by Hendrik Antoon LORENTZ and Ludwig Valentin Lorenz in 1880 using

electrostatics. *See also* Clausius–Mossotti equation.

Lorentz transformation   A set of transformations used in relativistic mechanics in place of the GALILEAN TRANSFORMATIONS used in Newtonian mechanics. The Lorentz transformations from a frame of reference with origin O and coordinates (x, y, z) to a frame moving relative to it with origin at O′ and coordinates (x', y', z') is given by $x' = \beta(x - vt)$, $y' = y$, $z = z'$, $t' = \beta(t - vx/c^2)$, where v is the relative velocity of the two frames, c is the speed of light in a vacuum, v is constant in the xx' direction, O and O′ coincide at $t = t' = 0$ and $\beta = 1/\sqrt{(1 - v^2/c^2)}$.

Lorentz transformations were discovered by Hendrik Antoon LORENTZ in 1904 and occur as a consequence of special relativity theory.

Loschmidt constant   The number of molecules of an ideal gas under standard temperature and pressure conditions (STP) in a cubic meter. This number is one million times the LOSCHMIDT NUMBER.

Loschmidt number   the number of molecules of an ideal gas under standard conditions in a cubic centimeter. This number is named after Johann Joseph Loschmidt who made an estimate of the number of molecules of air in a cubic centimeter in 1865. A modern value for this number is 2.686763×10^{19}.

Lucretius (**Titus Lucretius Carus**) (*c.* 95–55 BC)   Roman philosopher who expounded the ideas of EPICURUS on atoms and the void in a poem entitled *De rerum natura*.

luminescence   The emission of electromagnetic radiation from a substance due to the molecules in the substance being in excited states. If the luminescence lasts a significant time (frequently defined to be 10 nanoseconds) after the cause of the molecule being in an excited state (such as bombardment by photons or electrons) has been removed it is called *phosphorescence*. If the luminescence does not last a significant time the luminescence is called *fluorescence*.

lutetium A silvery element of the lanthanoid series of metals. It occurs in association with other lanthanoids. Lutetium is a very rare element and has few uses.

Symbol: Lu. Melting pt.: 1663°C. Boiling pt.: 3395°C. Relative density: 9.84 (25°C). Proton number: 71. Relative atomic mass: 174.967. Electronic configuration: $[Xe]4f^{14}5d^{1}6s^{2}$.

Lyman series *See* hydrogen spectrum.

M

Mach, Ernest (1838–1916) Austrian physicist whose views on the foundations of mechanics influenced Albert EINSTEIN when he was developing general relativity theory. Mach rejected the Newtonian concepts of absolute space and absolute time and postulated that all motion is relational. Nevertheless, Mach rejected both special and general relativity theory. His view that science should only deal with directly observable quantities also led him to reject the concept of atoms. Mach is best remembered for the *Mach number* which is the ratio of the speed of a body to the speed of sound in a given medium. For example, Mach 2 is twice the speed of sound. *See also* Mach's principle.

MACHOs (**Massive Compact Halo Objects**) Dark matter in the Universe that is made of baryons. The name is a deliberate contrast with WIMPS in which the dark matter is nonbaryonic. Dark matter in the form of MACHOs is thought to exist as BROWN DWARFS. In the early 1990s evidence emerged that the halo of our Galaxy contains MACHOs.

Mach's principle The principle that inertia is not an intrinsic property of a body but results from the interaction of the body with all the other matter in the Universe. This idea, which was stated by Ernest MACH in the 1870s, came from his view that all motion is relational. The general validity of Mach's principle is still a controversial topic of great interest, with its supporters claiming that the cosmological solutions of the Einstein field equation of physical relevance are in accord with Mach's principle.

macroscopic A size-scale that is very much larger than the sizes of atoms and molecules. The motion of macroscopic objects is usually described by classical physics, although some macroscopic motion, notably SUPERFLUIDITY, needs to be explained in terms of quantum mechanics. In addition, many macroscopic properties of matter, such as its electrical conductivity, require quantum mechanics for their explanation.

magic numbers Numbers of protons or neutrons that occur in atomic nuclei that are particularly stable. This stability is indicated by much higher cosmic abundances than for other nuclei that are similar in nucleon number and by high binding energies. The magic numbers for both protons and neutrons are 2, 8, 20, 28, 50, 82, and 126. The existence of magic numbers in nuclei is explained by the SHELL MODEL of the nucleus.

magnesium A light metallic element; the second member of group 2 of the periodic table (the alkaline earths). The element accounts for 2.09% of the Earth's crust and is eighth in order of abundance. It occurs in a wide variety of minerals.

Symbol: Mg. Melting pt.: 649°C. Boiling pt.: 1090°C. Relative density: 1.738 (20°C). Proton number: 12. Relative atomic mass: 24.3050. Electronic configuration: $[Ne]3s^2$.

magnetic bottle A magnetic field used for containment of the plasma in nuclear fusion experiments. Since a plasma temperature can exceed 100 million degrees Celsius it cannot be allowed to touch any material container since it would immediately vaporize it. At this high temperature the electrons have been removed from the atoms and all the matter exists in the form of charged particles. Magnetic fields are used to confine these charged particles in a region away from the boundaries of the container.

magnetic constant   *See* permeability.

magnetic dipole moment   *See* magnetic moment.

magnetic field   A field associated with a magnetic body or a current-carrying conductor. A magnetic field can be represented by LINES OF FORCE. In a magnetic field a moving electric charge is subject to a force and a magnetic dipole experiences a torque.

The strength and direction of a magnetic field can be characterized by its MAGNETIC FLUX DENSITY (sometimes called *magnetic induction*), denoted B. This can be defined from the LORENTZ FORCE F on a particle with a charge Q moving perpendicular to the field with velocity v. B is given by $B = F/Qv$. The SI unit of magnetic flux density is the tesla. The magnetic field can also be characterized by the *magnetic field strength*, denoted H, which is related to B by $B = \mu H$, where μ is the PERMEABILITY of the medium. The SI unit of magnetic field strength is the ampere per meter. Both B and H are vector quantities.

magnetic field strength   *See* magnetic field.

magnetic flux   Symbol: ϕ A measure of a magnetic field passing through an area. For a magnetic field of magnetic flux density B passing through an area A perpendicular to B the magnetic flux ϕ is the magnetic flux density multiplied by the area, i.e. $\phi = BA$. The SI unit of magnetic flux is the weber.

magnetic flux density (magnetic induction)   Symbol: B The flux per unit perpendicular area of a magnetic field; it is sometimes thought of as the number of lines of force per unit area. It is defined by the effect on a current-carrying conductor in the field. For a field B:

$$B = F/Qv$$

where F is the force on a charge Q moving perpendicular to the field with velocity v.

The unit of magnetic flux density is the tesla (T). It is equivalent to the weber per square meter (Wb m^{-2}) since

$$B = \Phi A$$

where Φ is magnetic flux and A is area. In SI the relationship between magnetic flux density and magnetic field strength is:

$$B = \mu_r \mu_0 H$$

where μ_r is the relative permeability of the medium and μ_0 the permeability of free space.

magnetic focusing   The focusing of beams of charged particles using magnetic fields. This technique is used in accelerators, cathode-ray tubes, and electron microscopes.

magnetic mirror   An arrangement used to contain a plasma in a thermonuclear reactor. It consists of an axially symmetric magnetic field with high values of this field at the ends of a containment tube. This has the effect that charged particles are reflected from the ends back to the center.

magnetic moment (**magnetic dipole moment**)   A quantity that gives a measure of the turning force that a magnetic dipole experiences in a magnetic field. If T is the torque observed in a magnetic field with magnetic flux density B then the magnetic moment m is defined by $m = T/B$. If a coil with N turns of area A is carrying a current I then $m = NIA$.

All electrically charged particles such as electrons and protons have nonzero magnetic moments because of the interaction between the moving charges and a magnetic field. Although the neutron is electrically neutral it is made up of electrically charged quarks and so also has a nonzero magnetic moment. Particles can have magnetic moments due to their orbital motion (*orbital magnetic moment*) and their spin (*spin magnetic moment*).

magnetic monopole   The magnetic analog of an isolated electric charge, i.e. an isolated North or South pole. In classical electrodynamics there is always a *magnetic dipole*, i.e. if there is a North pole there must be a South pole and vice versa and never an isolated pole. In 1931 Paul DIRAC postulated that magnetic monopoles could exist in quantum theory, with the existence of a magnetic monopole explaining the discreteness of the electric charge by what is now recognised as a consequence of TOPOLOGY. Very heavy magnetic monopoles are predicted to exist in grand unified theories, Kaluza–Klein theories, and superstring theory. In spite of a great deal of experimental effort no magnetic monopole has been detected. Nevertheless, all theories that explain the discreteness of electric charge involve magnetic monopoles. In addition, magnetic monopoles feature prominently in many attempts to prove the hypothesis of QUARK CONFINEMENT.

magnetic permeability   *See* permeability.

magnetic quantum number   *See* electronic structure of atoms.

magnetic susceptibility   *See* susceptibility.

magnetism   The phenomena associated with magnetic fields. Any motion of electric charge produces a magnetic field. At the microscopic level the orbital motion and spin of electrons produce magnetic fields. This means that atoms can have a net MAGNETIC MOMENT associated with them. The microscopic magnetic properties of a substance are determined by how the electrons interact with an external magnetic field. There are four main types of magnetic behavior.
1. *Diamagnetism* The SUSCEPTIBILITY is negative, i.e. the magnetization is in the opposite direction to that of the applied field. Diamagnetism occurs in all substances but is a weak form of magnetism that may be masked by other stronger forms. It occurs because of changes in the orbital motion of electrons due to the external magnetic field. In accordance with LENZ'S LAW the direction of the change opposes the external

magnetic field, resulting in a negative susceptibility.
2. *Paramagnetism* The susceptibility is positive, i.e. the atoms or molecules have magnetic moments that are capable of being aligned in the direction of the external magnetic field. It occurs because of electron spin and is found in materials containing unpaired electrons.
3. *Ferromagnetism* A much stronger magnetic effect than either diamagnetism or paramagnetism. When it occurs all the magnetic moments in a DOMAIN are aligned in the same direction. This can be explained in terms of EXCHANGE INTERACTIONS between electrons in certain atoms. Above a certain temperature, called the CURIE TEMPERATURE, ferromagnetic materials become paramagnetic. Examples of ferromagnetic materials include iron, nickel, cobalt, and their alloys.
4. *Antiferromagnetism* In this type of magnetism the exchange interactions between electrons lead to an ordered array in which alternate magnetic moments have opposite directions. Above a certain temperature, called the *Néel temperature*, antiferromagnetic materials become paramagnetic. Manganese oxide is an example of a substance that is antiferromagnetic.

A particular type of antiferromagnetism is *ferrimagnetism*, i.e. magnetism in which adjacent magnetic moments have opposite directions but unequal sizes.

magnetochemistry   The branch of chemistry concerned with the magnetic properties of substances and how these properties are related to their chemical properties.

magnetohydrodynamics (MHD)   The study of the interactions between a conducting fluid and a magnetic field. This complex subject is important in understanding plasmas in thermonuclear reactors.

magneton   A unit for measuring the MAGNETIC MOMENTS of electrons, atoms and nuclei. The *Bohr magneton* for an electron, denoted μ_B, is given by $eh/4\pi mc$, where e is the charge of an electron, h is the Planck constant, m is the mass of an electron and c is the speed of light in a vacuum. It has the value 9.274×10^{-24} JT^{-1}. The Bohr magneton was derived as the basic unit of magnetic moments by Niels Bohr in 1913 in terms of his model of orbiting electrons in an atom. For an orbiting electron the orbital magnetic moment is $l\mu_B$, where l is the quantized orbital angular momentum of the electron. The spin magnetic moment of an electron is given by $2s\mu_B$. This means that the GYROMAGNETIC RATIO of the spin motion is twice that of the orbital motion.

magneto-optics The branch of optics concerned with the effects of magnetic fields on light. The FARADAY EFFECT is an example of a magneto-optical effect.

magnon *See* spin wave.

magnox A type of alloy containing magnesium (88%), aluminum (11%), and traces of other elements. It is used for canning the uranium rods in some types of nuclear reactors.

Maiman, Theodore Harold (1927–) American physicist who constructed the first working LASER in 1960.

Main-Smith–Stoner rule The maximum number of electrons in a subshell with orbital quantum number l is $2(2l + 1)$. This rule was stated independently by J. D. Main-Smith and Edmund Stoner in 1924. It is a consequence of the PAULI EXCLUSION PRINCIPLE.

Maldacena conjecture *See* holographic hypothesis.

manganese A transition metal occurring naturally as oxides. Nodules found on the ocean floor are about 25% manganese. Its main use is in alloy steels.
Symbol: Mn. Melting pt.: 1244°C. Boiling pt.: 1962°C. Relative density: 7.44 (20%C). Proton number: 25. Relative atomic mass: 54.93805. Electronic configuration: [Ar]3d^{5}4s^2.

Manhattan project The name given to the project to build the first atomic bomb, in the United States in World War II, under the leadership of Robert OPPENHEIMER. Other notable physicists who took part in this project included Hans BETHE, Enrico FERMI, and Richard FEYNMAN.

manifold A space that is Euclidean locally but not overall. For example, a circle and the surface of a sphere are manifolds.

many-body problem The problem of obtaining exact solutions for systems in which there are interactions between more than two bodies. Both in classical mechanics and quantum mechanics a two-body problem can be solved exactly but a problem with more than two bodies cannot, in general, be solved exactly. In general relativity theory even the two body problem cannot, in general, be solved exactly. This means that to understand many-body problems it is necessary to use APPROXIMATION TECHNIQUES. Sometimes it is possible to obtain qualitative information about a system such as whether it exhibits chaotic behavior. COLLECTIVE EXCITATIONS and QUASIPARTICLES are useful concepts in the quantum theory of many-body systems. The methods of STATISTICAL MECHANICS can be used if there are a great many bodies interacting, as in the molecules of a gas.

many-worlds interpretation An interpretation of quantum mechanics in which it is postulated that whenever there is a choice at the quantum level, such as the choice of which hole an electron will go through in the DOUBLE-SLIT EXPERIMENT, the Universe divides into as many parts as there are choices. For example in this interpretation the SCHRÖDINGER CAT experiment would result in a division of the Universe into two worlds, with one of these worlds having a dead cat and the other world having a live cat. The many-worlds interpretation was proposed by Hugh EVERETT in 1957. It is a very controversial idea but has a number of enthusiastic supporters, particularly among people who work in QUANTUM COSMOLOGY. David Deutsch has proposed that it might be able to test the many-worlds interpretation using a QUAN-TUM COMPUTER.

maser (*m*icrowave *a*mplification by *s*timulated *e*mission of *r*adiation) The analog of a LASER for microwaves. The maser was invented in the early 1950s by Charles TOWNES and, independently, by Nikolai BASOV and Alexander PROKHOROV. A type of maser which uses ammonia molecules is the basis for the AMMONIA CLOCK. Another type of maser is the *solid state maser*, in which paramagnetic atoms or molecules are used, with there being two energy levels, corresponding to whether the spins are parallel to a magnetic field or not.

mass A measure of the amount of matter in a body. Mass can be defined in two ways. The *inertial mass* of a body is a measure of the tendency of a body to resist changes in its motion. The *gravitational mass* of a body is defined in terms of its gravitational attraction for other masses.

Experimentally, it has been shown by the experiments of Roland Eötvös in the nineteenth century, and subsequent experiments in the twentieth century, that the gravitational mass and the inertial mass of a body are equal to an extremely high degree of accuracy. In Newtonian mechanics this has to be accepted as an empirical fact but it is an automatic consequence of general relativity theory, in accord with the EQUIVALENCE PRINCIPLE, that these two masses should be equal.

In special relativity theory the mass of a body can be regarded as a form of energy by the famous equation of Albert EINSTEIN $E = mc^2$, where E is the energy, m is the mass, and c is the speed of light in a vacuum. It is also a consequence of special relativity theory that the mass of a body depends on its velocity. This mass increase is very small for ordinary speeds but a body such as an electron, for example, that is moving at 99 per cent of the speed of light has seven times the mass it has when it is at rest.

mass decrement The difference between the rest mass of a radioactive nu-

cleus and the sum of all the rest masses of the decay products.

mass defect The difference between the rest mass of a nucleus and the sum of the rest masses of all the nucleons in it when they are free. The mass defect is the mass equivalent of the binding energy according to Einstein's equation $E = mc^2$.

mass-energy equation The equation $E = mc^2$, where E is the total energy (rest mass energy + kinetic energy + potential energy) of a mass m, c being the speed of light. The equation is a consequence of Einstein's special theory of relativity; mass is a form of energy and energy also has mass. Conversion of rest-mass energy into kinetic energy is the source of power in radioactive substances and the basis of nuclear-power generation.

massive Having mass. A massive body or particle is a body or particle with a nonzero mass.

mass number *See* nucleon number.

mass spectrometer An instrument for producing ions in a gas and analyzing them according to their charge/mass ratio. The earliest experiments by Francis Aston used a stream of positive ions from a discharge tube, which were deflected by parallel electric and magnetic fields at right angles to the beam. Each type of ion formed a parabolic trace on a photographic plate (a *mass spectrograph*).

In modern instruments, the ions are produced by ionizing the gas with electrons from an electron gun. The positive ions are accelerated out of this ion source into a high-vacuum region. Here, the stream of ions is deflected and focused by a combination of electric and magnetic fields, which can be varied so that different types of ion fall on a detector. In this way, the ions can be analyzed according to their mass, giving a *mass spectrum* of the material. Mass spectrometers are used for accurate measurements of relative atomic masses, for analysis of isotope abundance, and for chemical identification of compounds and mixtures.

mass spectroscopy *See* mass spectrometer.

matrix A rectangular array of numbers (or other mathematical expressions such as polynomials), with the array usually being enclosed in large parentheses. There are specific rules that determine how matrices are added or multiplied. In particular, the multiplication of two matrices A and B is noncommutative, i.e., in general, $AB \neq BA$. There are many applications of matrices in the physical sciences. *See* matrix mechanics.

matrix mechanics The formulation of quantum mechanics in terms of matrices. Matrix mechanics was the first consistent formulation of quantum mechanics, introduced by Werner HEISENBERG in 1925 and developed by Heisenberg, Max BORN, and Pascual JORDAN in 1925–26. In 1926 Erwin SCHRÖDINGER showed that the matrix mechanics and the WAVE MECHANICS formulations of quantum mechanics are equivalent.

matter Matter can be defined as anything that has mass. Since mass depends on relative motion in special relativity theory the concept of *substance* is defined, with the number of particles in a system being a measure of substance.

matter–antimatter asymmetry The asymmetry between matter and antimatter in the Universe; i.e. the fact that all the matter in the Universe appears to be in the form of matter rather than an equal mixture of matter and antimatter, as might be expected from relativistic quantum mechanics. The problem of explaining this asymmetry is a major challenge for cosmology. SAKHAROV'S CONDITIONS provide a general explanation for how this asymmetry can arise. These conditions were implemented in the context of GRAND UNIFIED THEORIES in the late 1970s to provide an explanation of matter–antimatter asymme-

try but it is not known whether this explanation is correct.

matter field A field in which the quantum of the field is a particle with a nonzero mass. Examples of matter fields are electron fields, positron fields, and quark fields.

Maupertuis, Pierre-Louis Moreau de (1698–1759) French mathematician who was one of the first people to formulate the PRINCIPLE OF LEAST ACTION. Maupertuis was led to this formulation of Newtonian mechanics in 1744 by analogy with FERMAT'S PRINCIPLE OF LEAST TIME.

Maxwell, James Clerk (1831–79) Scottish physicist who produced a unified theory of electric and magnetic forces and was one of the founders of the kinetic theory of gases. Maxwell expressed Faraday's ideas of lines of force in quantitative terms in 1856–57. His work, which unified electric and magnetic forces, was published in 1865 and is summarized by MAXWELL'S EQUATIONS. Maxwell predicted the existence of electromagnetic waves from his theory. Since his calculations showed that electromagnetic waves propagate at the speed of light he correctly inferred that light is a form of electromagnetic wave. His work on electricity and magnetism was summarized in his *Treatise on Electricity and Magnetism*, which was published in 1873.

Maxwell was one of the first people to use statistical methods to analyze the behavior of a very large number of molecules in the kinetic theory of gases. This led him to a statistical interpretation of thermodynamics.

Maxwell–Boltzmann distribution The laws for the distribution of speeds or kinetic energies among the molecules of a gas in equilibrium. The number of molecules per unit range of speed at speed v is given by

$$N_v = Av^2 \exp(-mv^2/2kT)$$

where m is the mass of a molecule, T is the kelvin temperature, k is the Boltzmann constant, and A is a constant. The number

of molecules per unit range of kinetic energy at kinetic energy E is given by
$$N_E = \sqrt{(BE)}\exp(-E/kT)$$
where B is a constant.

This distribution is very closely obeyed by gases at ordinary pressures. For large concentrations of particles more exact laws are needed, particularly for the valence electrons in a solid.

See Bose–Einstein statistics; Fermi–Dirac statistics; Maxwell–Boltzmann statistics.

Maxwell–Boltzmann statistics The statistical rules for a large number of particles that obey classical physics. These laws were discovered in the second half of the nineteenth century by James Clerk MAXWELL and Ludwig BOLTZMANN. At low temperatures QUANTUM STATISTICS, either in the form of BOSE–EINSTEIN STATISTICS or FERMI–DIRAC STATISTICS, are obeyed rather than the classical Maxwell–Boltzmann statistics.

Maxwell's corkscrew rule A convenient way to express the relationship between the directions of an electric current and the lines of force of the magnetic field caused by that current. If a corkscrew is being driven in the direction of the current, then the direction of rotation of the corkscrew is the direction of the lines of force.

Maxwell's demon An imaginary being in a thought experiment devised by James Clerk MAXWELL in 1867 (the name was coined by Lord KELVIN). Maxwell envisaged a container of gas, with a partition dividing the gas into two volumes, which each have the same temperature. The demon is able to operate the partition in such a way that when a faster-than-average molecule approaches the partition from one direction, he lets it through and when he sees a slower-than-average molecule approach the partition from the opposite side he lets it through. Repetition of this process many times would lead to one side of the container containing most of the fast molecules and the other side containing most of the slow molecules. Since the side of the gas with the fast molecules would be hotter than the side with the slow molecules this process appears to violate the second law of thermodynamics.

Explaining this paradox has taken a long time and a lot of controversy. In general, it was realized that entropy has to be generated by searching for molecules of the right type and opening the partition. Soon after the importance of measurement in quantum mechanics had been emphasized by Niels BOHR, it was suggested by Leo Szilard that the act of measurement of the molecules would raise the entropy. However, in the 1970s Charles Bennett showed that the source of entropy increase is the erasure of information. As well as solving the problem of Maxwell's demon this work of Bennett was very important in the theory of computing.

Maxwell's equations A set of four differential equations which summarizes classical electrodynamics. Maxwell's equations lead to the prediction of ELECTROMAGNETIC WAVES which move at the speed of light.

Mayer, Julius Robert von (1814–78) German physician and physicist who formulated the principle of the conservation of energy in 1842, independently of Hermann von HELMHOLTZ and James JOULE.

mean-field theory An approximate theory of phase transitions. It is analogous to HARTREE-FOCK THEORY for atoms, with both techniques involving 'averaging out' over interactions in a many-body system. Mean-field theory can be improved upon by using the RENORMALIZATION GROUP.

mean free path Symbol: λ 1. The average distance traveled by the particles of a fluid between collisions. It is given by
$$\lambda = 1/\pi r^2 n$$
where r is the particle radius and n the density of particles.
2. The average distance traveled by electrons between collisions with the lattice in conduction.

mean free time The average time between the collisions of the particles in a system.

mechanics The branch of physics that studies the interaction between matter and the forces acting on it. *Statics* deals with bodies at rest, with the forces being in equilibrium. The study of bodies in motion is divided into *kinematics*, which is the study of the motion of bodies without consideration of the forces causing the motion, while *dynamics* is concerned with the forces and the motions. The branch of mechanics concerned with fluids is called *fluid mechanics*. *See also* classical mechanics; general relativity theory; quantum mechanics; special relativity theory.

megaton A unit used to express the explosive power of nuclear weapons. One megaton is an explosive power equivalent to that of one million tons of TNT.

Meissner effect The exclusion of magnetic flux from a superconductor. The effect was discovered by Walther Meissner in 1933. When a superconductor is in an external magnetic field it is perfectly diamagnetic. A sufficiently large external magnetic field is able to penetrate a superconducting material, thereby destroying its superconductivity.

Meitner, Lise (1878–1968) Austrian-born physicist who, together with Otto HAHN, was one of the first people to investigate NUCLEAR FISSION. It is generally regarded as a serious injustice that she did not share the Nobel Prize for chemistry awarded to Hahn in 1944.

meitnerium A radioactive metallic element not found naturally on Earth. Only a few atoms of the element have ever been detected; it can be made by bombarding a bismuth target with iron nuclei. The isotope ^{266}Mt has a half-life of about 3.4×10^{-3}s.

Symbol: Mt. Proton number: 109. Electronic configuration: [Rn]$5f^{14}6d^77s^2$.

membrane *See* supermembrane.

Mendeleev, Dmitri Ivanovich (1834–1907) Russian chemist who put forward the idea of the PERIODIC TABLE in a paper published in 1869 entitled *On the Relation of the Properties to the Atomic Weights of Elements*. He predicted the existence of new elements and their properties on the basis of gaps in his table.

mendelevium A radioactive transuranic element of the actinoid series, not found naturally on Earth. Several short-lived isotopes have been synthesized.

Symbol: Md. Proton number: 101. Most stable isotope: ^{258}Md (half-life 57 minutes). Electronic configuration: [Rn]$5f^{13}7s^2$.

mercury A transition metal occurring naturally as cinnabar (mercury(II) sulfide); small drops of metallic mercury also occur in cinnabar and in some volcanic rocks. The vapor is very poisonous. Mercury is used in thermometers, special amalgams for dentistry, scientific apparatus, and in mercury cells. Mercury compounds are used as fungicides, timber preservatives, and detonators.

Symbol: Hg. Melting pt.: –38.87°C. Boiling pt.: 356.58°C. Relative density: 13.546 (20°C). Proton number: 80. Relative atomic mass: 200.59. *See* amalgam, zinc group. Electronic configuration: [Xe]$4f^{14}5d^{10}6s^2$.

meson A type of particle forming a subset of the HADRONS. Each meson consists of a quark and an antiquark held together by the strong interaction. Mesons can be electrically neutral or positively or negatively charged. All have integer spins and hence are bosons. KAONS and PIONS are examples of mesons. Mesons were originally postulated to exist (as elementary particles) by Hideki YUKAWA in 1935 as entities that mediate the strong interactions between NUCLEONS. The name meson was chosen because Yukawa calculated that the particle he predicted would have a mass intermediate between an electron and a nucleon. This is true of light mesons such as the pion but is not true of heavy mesons such as the J/PSI PARTICLE.

mesoscopic Intermediate between MACRO-SCOPIC and MICROSCOPIC. Devices used in NANOTECHNOLOGY are examples of meso-scopic systems.

metal Any of a class of chemical elements with certain characteristic properties. In everyday usage, metals are elements (and alloys) such as iron, aluminum, and copper, which are lustrous malleable solids – usually good conductors of heat and electricity. Note that this is not a strict definition – some metals are poor conductors, mercury is a liquid, etc.

In chemistry, metals are distinguished by their chemical properties, and there are two main groups. Reactive metals, such as the alkali metals and alkaline-earth metals, are electropositive elements. They are high in the electromotive series and tend to form compounds by losing electrons to give positive ions. They have basic oxides and hydroxides. This typical metallic behavior decreases across the periodic table and increases down a group in the table.

The other type of metals are the transition elements, which are less reactive, have variable valences, and tend to form complexes. In the solid and liquid states metals have metallic bonds, formed by positive ions with free electrons.

metalloid   *See* semi-metal.

metastable state   A state in which a small perturbation can cause the system to go to a state with lower energy. For example, supercooled water is a metastable state. It is liquid below 0°C (at standard pressure), but a grain of dust or ice introduced into it causes rapid freezing. Certain excited atoms or molecules can also exist in metastable states in which they have a relatively long lifetime before reverting to the ground state.

metric   A mathematical description of the size of the distances between points in space and space–time. In the case of Euclidean geometry these distances are given by the theorem of Pythagoras (and its higher-dimensional generalizations). The use of a metric ensures that the value of the distance between two points is the same, no matter how the coordinates of the position are specified. The metrics used to characterize space–times in general relativity theory are more complicated than those used in Euclidean geometry.

Meyer, Julius Lothar   (1830–1895) German chemist who, independently of Dmitri MENDELEEV, pointed out the periodic pattern of the chemical elements, as shown in the PERIODIC TABLE. Meyer plotted a graph of atomic volume against atomic number (proton number), which clearly demonstrated periodic properties of the elements. The work of Meyer on this topic was published in 1870, one year later than Mendeleev. *See illustration overleaf.*

MHD   *See* magnetohydrodynamics.

Michell, John   (1724–93)   English geologist and astronomer who was the first person to suggest the idea of black holes. In 1784 he put forward the idea that there might be some bodies with an escape velocity so high that even light could not escape from them on the basis of Newtonian gravity. He called such bodies 'dark stars' and calculated some of their properties. Michell also designed an experiment to determine the strength of the gravitational constant, although he did not live long enough to perform it. Henry Cavendish subsequently used the method suggested by Michell. Michell also pointed out from observations that there must be many binary star systems in the Universe.

Michelson, Albert Abraham   (1852–1931)   American physicist who made many precise determinations of the speed of light and who carried out the experiment with Edward Morley which famously found no evidence for the existence of the ether. Michelson won the Nobel Prize for physics in 1907 for this experimental work.

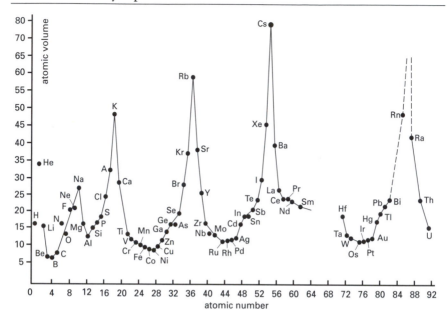

Lothar Meyer's curve for the elements

Michelson–Morley experiment An experiment designed to detect the existence of the ether by measuring the speed of light traveling in the same direction as the motion of the Earth and at right angles to that path. This experiment was performed with an interferometer, firstly by Albert Michelson himself in 1881 and in 1887 by Michelson in collaboration with Edward Morley. The Michelson–Morley experiment, and all subsequent experiments, found no evidence for the existence of the ether. The Michelson–Morley experiment does not appear to have influenced Albert Einstein directly in his development of special relativity theory.

microscope A device which produces a large image of a small object. The resolution of the object depends on the wavelength used. Since the de Broglie wavelength of electrons is very small an electron microscope can produce visible images of much smaller objects than optical microscopes.

microscopic Describing systems at the atomic and subatomic scales. The properties of microscopic systems are generally quite different from those of macroscopic systems and are described by quantum mechanics. *See also* macroscopic; mesoscopic.

microwave *See* electromagnetic spectrum.

microwave spectroscopy A type of SPECTROSCOPY used to determine molecular structure, which makes use of the fact that the energy differences in the rotational energy levels of molecules correspond to the microwave region of the ELECTROMAGNETIC SPECTRUM. Microwave spectroscopy can be used to determine bond lengths, bond angles, and dipole moments of molecules.

Millikan, Robert Andrews (1868–1953) American physicist who measured the charge on the electron accurately, investigated the PHOTOELECTRIC EFFECT, and studied COSMIC RAYS. In a series of experiments, collectively known as the *Millikan* *oil drop experiment*, carried out between 1909 and 1912 he determined the charge of an electron. He did so by producing a fine spray of oil droplets between two horizontal electrically charged plates. The air was then ionized using x-rays, enabling the droplets to pick up electric charges. By measuring the rate at which the droplets fell in an electric field, Millikan found that the electric charge was always an integer multiple of a basic unit of charge, i.e. the charge of an electron. Millikan won the 1923 Nobel Prize for physics for his determination of the charge of the electron.

Millikan oil drop experiment *See* Millikan.

Minkowski, Hermann (1864–1909) Russian-born German mathematician who showed that special relativity theory implies a unified structure of space and time now known as MINKOWSKI SPACE–TIME. Minkowski also made important contributions to several areas of mathematics, notably number theory.

Minkowski space–time The geometrical structure that unifies space and time in special relativity theory. This structure was proposed by Hermann MINKOWSKI in 1908. *See also* world line.

mirror nuclei A pair of nuclei that have the same NUCLEON NUMBER but opposite PROTON NUMBER and NEUTRON NUMBER. An example of a pair of mirror nuclei is given by helium-3, which has two protons and one neutron, and tritium, which is the isotope of hydrogen with one proton and two neutrons. The binding energies of mirror nuclei (taking the electrostatic repulsion between protons into account) provides evidence of the CHARGE INDEPENDENCE of the strong interaction.

mobility edge *See* electronic structure of solids.

mode The pattern of motion in a vibrating body such as a string or a molecule. In a vibrating string the number of nodes in the string characterizes the different modes of vibration. In a molecule the modes of vibration are the different types of molecular vibrations possible.

model A description of a physical system that is simplified but attempts to capture the physical essence of the system. Some models, such as the ISING MODEL of magnetic phase transitions in the two-dimensional case, can be solved exactly. Many models still require approximation techniques to be solved, in spite of being simplifications of the real physical system.

moderator A substance used to slow down rapidly moving neutrons in a NUCLEAR REACTOR. The use of a moderator makes neutrons more likely to cause fission in atoms such as uranium-235, rather than being absorbed in atoms such as uranium-238. Light elements such as deuterium, carbon, and beryllium are used as moderators since neutrons can impart some of their kinetic energy to them without being captured. *See also* thermal neutron.

mole Symbol: mol   The SI base unit of amount of substance, defined as the amount of substance that contains as many elementary entities as there are atoms in 0.012 kilogram of carbon-12. The elementary entities may be atoms, molecules, ions, electrons, photons, etc., and they must be specified. One mole contains $6.022\,52 \times 10^{23}$ entities One mole of an element with relative atomic mass A has a mass of A grams (this was formerly called one *gram-atom*). One mole of a compound with relative molecular mass M has a mass of M grams (this was formerly called one *gram-molecule*).

molecular beam A beam of atoms, ions, or molecules moving in the same direction with few collisions occurring between particles in the beam. Molecular beams are used in spectroscopy and in studies of chemical reactions and surfaces.

molecular orbital *See* orbital.

molecular spectra *See* spectra.

molecular spectroscopy *See* spectroscopy.

molecular weight *See* relative molecular mass.

molecule A number of atoms held together by chemical bonds. A molecule is the smallest amount of a chemical compound that can exist. The chemical bonds in a molecule can be largely covalent, as in diatomic hydrogen, or have a large amount of ionic character, as in hydrogen chloride. Purely ionic compounds such as sodium chloride do not exist as isolated molecules.

molybdenum A transition element that occurs naturally in molybdenite (MoS_2) and wulfenite ($PbMoO_4$). It is used in alloy steels, lamp bulbs, and catalysts. Molybdenum sulfide (MoS_2) is used in lubricants to enhance viscosity.

Symbol: Mo. Melting pt.: 2620°C. Boiling pt.: 4610°C. Relative density: 10.22 (20°C). Proton number: 42. Relative atomic mass: 95.94. Electronic configuration: [Kr]$4d^5 5s^1$.

moment of inertia Symbol: I The analog of mass for rotational motion. The moment of inertia of a body rotating about an axis is given by $I = \Sigma r^2 dm$, which is the sum over all elements of mass dm of the product of dm and the square of r, where r is the distance between the element and the axis of rotation.

momentum The linear momentum p of a body is the product of its mass m and its velocity v. Thus, $p = mv$. *See also* angular momentum.

monochromatic radiation Electromagnetic radiation of an extremely narrow range of wavelengths. (The word means 'of one color'.) It is impossible to produce completely monochromatic radiation, although the output of some lasers is not far off. The 'lines' in line spectra produced

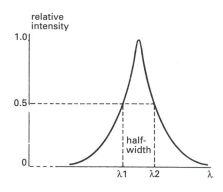

Monochromatic radiation

even in the most ideal circumstances have some width in wavelength terms. The *half-width* is the measure used. It is the range of wavelengths defined in the figure, and contains almost 90% of the energy emitted. The half-width of a sharp line in an optical spectrum is typically 10^{-6} to 10^{-7} of the wavelength. Using lasers, half-widths of the order 10^{-12} λ can be obtained. Low-quantum-energy gamma rays emitted by atoms bound in crystals may have values of the order 10^{-13} λ.

Simple quantum theory leads one to expect perfectly sharp lines in a line spectrum – the energies of the levels concerned appear to be exactly defined, so that λ = $hc/\Delta E$. However, because of the uncertainty principle, no energy level or transition *can* be defined exactly; this means that any line is naturally broadened rather than being sharp. A second broadening influence is the Doppler effect, which is relevant as the radiating particles are always in motion. Thirdly, collisions between emitting particles will broaden the emitted line.

Compare polychromatic radiation.

monopole problem *See* inflation.

Morley, Edward Williams (1838–1923) American chemist and physicist who collaborated with Albert MICHELSON on what became known as the MICHELSON–MORLEY EXPERIMENT.

Moseley, Henry Gwyn Jeffreys (1887–1915) British physicist who developed x-ray spectroscopy. This work enabled him to be the first person to see that the PROTON NUMBER is equal to the charge of the nucleus of an atom. This work greatly clarified the periodic table of the chemical elements and its relation to the structure of atoms. His life was cut short when he was killed by a sniper at Gallipoli in 1915 in World War I. *See also* Moseley's law.

Moseley's law   A law that relates the frequencies of the lines in x-ray spectra of atoms to the proton number of the element. It states that, for a set of elements, the graph of the relation between the square roots of the frequencies of corresponding lines against proton number is a straight line. This law was stated in papers by Henry MOSELEY which were published in 1913 and 1914.

Mössbauer, Rudolf Ludwig (1929–) German physicist who discovered the MÖSSBAUER EFFECT in 1957. He was awarded the 1961 Nobel Prize for physics for this discovery.

Mössbauer effect   The emission (or absorption) of a gamma ray by a nucleus embedded in a solid without recoil. This effect occurs if the nucleus of interest is strongly bound in the structure of the crystal so that the recoil energy is shared by the whole of the crystal lattice. The Mössbauer effect means that gamma rays with very precisely defined wavelengths can be absorbed or emitted.

The Mössbauer effect is made use of in *Mössbauer spectroscopy*. In this technique a gamma source is placed on a moving platform, with a sample near it and a detector that measures gamma rays scattered by the sample. The source is moved toward the sample at a varying speed, meaning that the frequency of the gamma ray is continuously changed because of the DOPPLER EFFECT. If there is a sharp decrease in the signal corresponding to a particular frequency this shows that resonance absorption in the sample nuclei has occurred. This enables nuclear energy levels to be studied.

Mössbauer effect spectroscopy is also used in solid state physics and chemistry.

The precision of the Mössbauer effect enables it to be used to measure the difference in the wavelength of gamma rays due to the difference in the gravitational field between the top and the bottom of a tall building. This enables a very accurate test of general relativity theory to be performed. The Mössbauer effect has also been used in accurate tests of certain predictions of special relativity theory.

Mott, Sir Nevill Francis (1905–96) British physicist who made important contributions to the theory of atomic collisions and the theory of solids. His early work was on the quantum theory of scattering and atomic collisions. For most of his career he worked on the theory of solids including properties of metals, the photographic process, metal–insulator transitions and disordered systems. He shared the 1977 Nobel Prize for physics with Philip ANDERSON and John VAN VLECK.

Mottelson, Benjamin Roy (1926–) American-born Danish physicist who collaborated with Aage BOHR on the theory of nuclear structure. Mottelson, Bohr, and James RAINWATER shared the 1975 Nobel Prize for physics for their work on the theory of nuclear structure.

M-theory   *See* superstring theory.

Müller, Karl Alex (1927–)   Swiss physicist who discovered HIGH-TEMPERATURE SUPERCONDUCTIVITY with Georg BEDNORZ, with whom he shared the 1987 Nobel Prize for physics.

Mulliken, Robert Sanderson (1896–1986)   American chemist and physicist who was one of the major developers of MOLECULAR ORBITAL theory. He was awarded the 1966 Nobel Prize for chemistry.

multicenter bond   A bond between three or more atoms in which there is only one pair of electrons. Multicentered bonds occur in boranes (boron hydrides). These compounds are electron-deficient compounds.

multiple bond A chemical bond between two atoms in which more than one pair of electrons is involved. *See also* chemical bond.

multiplet A set of closely spaced spectral lines in which the splitting from a single line into the set, i.e. the existence of FINE STRUCTURE, is due to the interaction between the orbital angular momentum and the spin angular momentum of electrons. The term is also used to refer to the closely spaced energy levels corresponding to such spectral lines.

multiplication factor Symbol: k The ratio of the rate of production of neutrons in a nuclear chain reaction to the rate of loss of neutrons by absorption or leakage. If $k = 1$, the reaction is said to be *critical*. If $k < 1$ it is *subcritical* and if $k < 1$ it is *supercritical*.

multiplicity The number $2S + 1$, which is the number of ways of vectorially coupling the resultant orbital angular momentum L of an atom with the resultant spin angular momentum S in an atom in which the energy levels are characterized by RUSSELL–SAUNDERS COUPLING. This number is the number of closely spaced energy levels in a MULTIPLET. It is indicated by a left superscript to the value of L.

multipole An array of positive and negative electric charges. A single charge is a *monopole*. A pair of charges is a *dipole*. A set of four charges is a *quadrupole*. A set of eight charges is an *octupole*, etc. Each of these poles with $2n$ charges has a moment associated with it that is the sum over all the charges of the product of the charge Q and r^n, where r is the distance between the charge and the centroid of charge of the system.

multipole interaction Interactions between sets of charged bodies. The potential energy of multipole interactions falls off with increasing rapidity on going from monopole to dipole to quadrupole, etc.

multiverse A speculative concept that the Universe we inhabit is not unique but is merely one of a very large number of Universes, which can have completely different physical laws and possibly even a different number of space–time dimensions. The multiverse concept emerged in the last few years of the 20th century due to the combination of the observation that 'fine tuning' of the basic physical laws is necessary for life to exist in the Universe and the realization that the collapse of stars to black holes and the emergence of 'baby Universes' sprouting into existence in inflation theory may allow other Universes to come into existence. There is no experimental or observational support for the multiverse concept.

mu-meson *See* muon.

muon Symbol: μ An electrically charged LEPTON that is the analog of the electron in the second GENERATION of elementary particles. It has the same charge and spin as an electron but has a mass that is about 207 times as large. The muon is an unstable particle. It decays into an electron, a muon neutrino, and an electron antineutrino, with a half-life of 2.19709 microseconds.

muon catalyzed fusion *See* nuclear fusion.

muon neutrino Symbol: ν_μ The type of neutrino that is associated with the MUON. The muon neutrino was discovered in the early 1960s by Leon LEDERMAN, Melvin SCHWARTZ, and Jack STEINBERGER in experiments that clearly showed that it is a distinct particle from the electron neutrino. The muon neutrino has almost certainly a nonzero mass but at present, it is not known what its value is. It seems very likely that it is less than 1 eV.

N

Nambu, Yoichiro (1921–) Japanese-born American physicist. In 1965 Nambu together with Moo-Young Han solved the serious problem that the quark model of baryons appeared to violate the PAULI EXCLUSION PRINCIPLE by postulating that quarks have an additional quantum number, which is called color. This was a major step in the development of quantum chromodynamics (QCD). In 1970 Nambu was one of the first people to introduce STRING THEORY into elementary particle physics. 1974 Nambu was one of the first people to postulate and give a mechanism for QUARK CONFINEMENT.

nanotechnology The branch of technology concerned with devices, particularly electronic devices, a few nanometers in size, i.e. at the scale of atoms and molecules. Richard FEYNMAN suggested the possibility of nanotechnology in 1959.

natural abundance The ABUNDANCE of a nuclide as it occurs naturally.

natural units A system of units in which the fundamental constants of Nature, such as the Planck constant and the speed of light in a vacuum, are defined to have the value 1. This set of units is mainly used in particle physics. In this system Einstein's equation $E = mc^2$ for the equivalence of energy E and mass m is written $E = m$. This means that the masses of particles are stated in terms of electronvolts (i.e. a unit of energy). It is possible to express masses in terms of grams by putting c^2 back into the calculation.

Néel, Louis Eugène Félix (1904–2000) French physicist who made important contributions to the study of magnetism in solids. In particular, he suggested the existence of antiferromagnetism and ferrimagnetism (*see* magnetism). He won the 1970 Nobel Prize for physics for his work on magnetism.

Ne'eman, Yuval (1925–) Israeli physicist who developed the EIGHT-FOLD WAY in 1961, independently of Murray GELL-MANN.

neodymium A toxic silvery element belonging to the lanthanoid series of metals. It occurs in association with other lanthanoids. Neodymium is used in various alloys, as a catalyst, in compound form in carbon-arc searchlights, etc., and in the glass industry.

Symbol: Nd. Melting pt.: 1021°C. Boiling pt.: 3068°C. Relative density: 7.0 (20°C). Proton number: 60. Relative atomic mass: 144.24. Electronic configuration: $[Xe]4f^46s^2$.

neon An inert colorless odorless monatomic element of the rare-gas group. Neon forms no compounds. It occurs in minute quantities (0.0018% by volume) in air and is obtained from liquid air. It is used in neon signs and lights, electrical equipment, and gas lasers.

Symbol: Ne. Melting pt.: –248.67°C. Boiling pt.: –246.05°C. Density: 0.9 kg m^{-3} (0°C). Proton number: 10. Relative atomic mass: 20.18. Electronic configuration: $[He]2s^22p^6$.

neptunium A toxic radioactive silvery element of the actinoid series of metals. Neptunium was the first transuranic element to be synthesized (1940). Found on Earth only in minute quantities in uranium ores, it is obtained as a by-product from uranium fuel elements.

Symbol: Np. Melting pt.: 640°C. Boiling pt.: 3902°C. Relative density: 20.25 (20°C). Proton number: 93. Most stable isotope: ^{237}Np (half-life 2.14×10^6 years). Electronic configuration: $[Rn]5f^16d^17s^2$.

neptunium series *See* radioactive series.

neutral current A type of weak interaction in which the interaction is mediated by a Z BOSON, i.e. by an electrically neutral boson. Evidence for neutral currents emerged in 1973. *See* Weinberg–Salam model.

neutrino Symbol: ν A type of elementary particle that occurs in three distinct forms (one for each GENERATION of elementary particles): the *electron neutrino*, the *muon neutrino*, and the *tau neutrino*. Neutrinos are electrically neutral, spin-½ leptons with very low masses and move at nearly the speed of light. Each form of neutrino has its own ANTIPARTICLE. Neutrinos take part in weak interactions.

Neutrinos were predicted to exist by Wolfgang PAULI in 1930 in order to account for the fact that there appeared to be some energy missing in the process of beta decay. The neutrino was identified tentatively in 1953 and conclusively in 1956. For many years it was thought that neutrinos are massless. However, a number of experiments and observations, including the SOLAR NEUTRINO EXPERIMENT, indicate that neutrinos do have mass. This is in accord with the theoretical expectations of many GRAND UNIFIED THEORIES. At present the masses of neutrinos have not been established experimentally nor calculated, but it is unlikely that the mass of an electron neutrino is greater than 1 eV. Neutrinos may well contribute to the DARK MATTER in the Universe. If neutrinos are not massless then *neutrino oscillations*, i.e. transformations from one type of neutrino into another, can occur. There is some experimental evidence that muon neutrinos can change into electron neutrinos.

neutron An electrically neutral spin-½ BARYON that is one of the constituents of the atomic nucleus. Inside the nucleus it is stable but outside the nucleus decays into a proton, an electron, and an antineutrino, with a half-life of about 15 minutes. The rest mass of a neutron, which is slightly heavier than that of a proton, is 1.67495×10^{-27} kg. The neutron was discovered by Sir James CHADWICK in 1932. All atomic nuclei except that of the normal hydrogen atom contain neutrons. The neutron consists of two down quarks and one up quark.

neutron beam A beam of neutrons.

neutron bomb *See* nuclear weapons.

neutron diffraction The diffraction of neutrons by matter. The occurrence of this phenomenon demonstrates the quantum-mechanical wave nature of the neutron. The average kinetic energy of thermal neutrons is about 0.025 eV, which gives them a DE BROGLIE WAVELENGTH of about 0.1 nanometer. This means that neutron diffraction can be used in structure determination in a way that complements X-RAY DIFFRACTION. This is the case because neutrons interact with both the nuclear magnetic moments and the electronic magnetic moments, and this is particularly useful for locating light elements such as hydrogen. Protons scatter neutrons strongly whereas, since the hydrogen atom only has one electron, it scatters electrons weakly, thus making it difficult to detect the position of hydrogen atoms by x-ray diffraction.

neutron interferometer A type of interferometer for analyzing the quantum-mechanical wave nature of neutrons by producing interference patterns. Experiments with neutron interferometers have shown the SPINOR nature of fermions such as neutrons.

neutron number Symbol: N The number of neutrons in an atomic nucleus. It is equal to the NUCLEON NUMBER minus the PROTON NUMBER.

neutron scattering   *See* neutron diffraction.

neutron star   A very dense small star that is supported against gravitational collapse by the DEGENERACY PRESSURE of neutrons. It is thought that neutron stars are the end products of stars with masses between the CHANDRASEKHAR LIMIT and the OPPENHEIMER–VOLKOFF LIMIT, i.e. between about 1.4 and 2–3 times the mass of the Sun. For stars that are heavier than a WHITE DWARF star but not sufficiently heavy to form a black hole when their nuclear fuel is exhausted the gravity of the star causes the protons to combine with electrons to form neutrons in INVERSE BETA DECAY. A typical neutron star is about ten kilometers across, and is as dense as an atomic nucleus. The existence of such stars was predicted by Lev LANDAU in 1932, very soon after the discovery of the neutron. Soon afterwards Fritz ZWICKY suggested that neutron stars might be formed in supernova explosions. The idea of neutron stars was not taken seriously by most astronomers until the discovery of PULSARS in the 1960s and the interpretation very soon after their discovery that they are rapidly rotating neutron stars.

Since the discovery of pulsars, a great deal of theoretical and observational work has been done on neutron stars. It is thought that a neutron star has a solid crust made of elements such as iron. Underneath the crust there is a region of 'normal' (as opposed to superfluid) matter made up of neutrons packed together very closely. Towards the center of a neutron star it is thought that the neutron matter becomes superfluid. It has been speculated that the central core of neutron stars consists of quarks. *See also* pulsar.

Newlands, John Alexander Reina (1837–1898)   British chemist who found periodic regularities in the chemical elements. In the mid 1860s he put forward the *law of octaves*, which states that if the elements were listed in order of their atomic weights then the chemical properties of the eighth element starting from a given element are very similar to the first.

These ideas were strongly rejected when they were put forward. After MENDELEEV's work Newlands' paper on chemical periodicity, which had been rejected in the mid 1860s, was eventually published in 1884.

newton   Symbol: N   The SI unit of force, equal to the force needed to accelerate one kilogram by one meter second^{-2}.

$$1 \text{ N} = 1 \text{ kg m s}^{-2}$$

Newton, Sir Isaac (1642–1727)   English physicist and mathematician who was one of the greatest scientists and mathematicians of all time. He discovered the law describing gravity and the laws governing motion which bear his name. He also made important contributions to optics and was one of the inventors of the calculus. Between 1668 and 1701 he was Lucasian Professor of Mathematics at Cambridge University.

He stated his law of gravity and his three laws of motion in a monumental book entitled *Philosophiae Naturalis Principia Mathematica* (*The Mathematical Principles of Natural Philosophy*, generally known as the *Principia*), the first edition of which was published in 1687, although he had obtained many of the main results considerably earlier. In the *Principia* Newton was able to show that his laws were able to explain a great many phenomena including the fall of objects to the ground, the motion of the planets, including the Earth, the motion of the Moon, and the tides.

Newton's work on optics was published in a treatise entitled *Opticks* which was published in 1704 although, as with *Principia*, much of the work had been done earlier. His experiments with prisms showed that white light is a mixture of the colors of the visible spectrum (the rainbow). He also put forward the idea that light is a stream of small particles or corpuscles rather than being a wave, although he held this view with less dogmatism than many of his followers. He also invented a type of reflecting telescope.

Newton invented the calculus, independently of Gottfried Wilheld LEIBNIZ, but his work on this topic entitled *Methodis Fluxionum* (Method of Fluxions) was published posthumously in 1736, in spite of having been written between 1670 and 1671.

Newtonian gravity The description of gravity in terms of NEWTON'S LAW OF GRAVITY. This is an accurate description of gravity for the dynamics of stars, planets, and their satellites, except for very massive or dense bodies such as NEUTRON STARS or BLACK HOLES. For bodies with strong gravitational fields GENERAL RELATIVITY THEORY gives a more accurate description than Newtonian gravity. Newtonian gravity is a special limiting case of general relativity theory.

Newtonian mechanics The type of mechanics that is described by NEWTON'S LAWS OF MOTION. Newtonian mechanics describes the motion of bodies that are moving at speeds much less than the speed of light and are much larger than atoms (i.e. macroscopic systems). The mechanics of bodies moving at speeds at or near the speed of light is described by SPECIAL RELATIVITY THEORY. The appropriate mechanics for bodies at the atomic and subatomic scale is QUANTUM MECHANICS. *See also* classical mechanics.

Newton's constant Symbol: G The constant of proportionality which appears in NEWTON'S LAW OF GRAVITY. *See* gravitational constant.

Newton's law of gravity The law stating that the gravitational force F between two bodies of mass m_1 and m_2 is given by $F = Gm_1m_2/r^2$, where r is the distance between the two bodies and G is Newton's constant (the GRAVITATIONAL CONSTANT). This law was stated by Sir Isaac NEWTON in *Principia* in 1687.

Newton's laws of motion The three fundamental laws for the motion of bodies. (1) Every body continues in a state of rest or uniform motion in a straight line unless it is acted on by an external force. (2) When a force is applied to a body the momentum of the body is changed, with the rate of change of momentum being pro-

portional to the force and in the same direction that the force is applied. If the mass m of the body is constant then the force F is given by $F = ma$, where a is the acceleration produced. (3) If the body exerts a force on another body then the second body exerts an equal and opposite force (called the *reaction force*) on the first body. This third law is sometimes stated in the form, 'to every action there is an equal and opposite reaction'.

nickel A transition metal that occurs naturally as the sulfide and silicate. Nickel is used as a catalyst in the hydrogenation of alkenes, e.g. margarine manufacture, and in coinage alloys.
Symbol: Ni. Melting pt.: 1453°C. Boiling pt.: 2732°C. Relative density: 8.902 (25°C). Proton number: 28. Relative atomic mass: 58.6934. Electronic configuration: [Ar]$3d^84s^2$.

niobium A soft silvery transition element used in welding, special steels, and nuclear reactor work.
Symbol: Nb. Melting pt.: 2468°C. Boiling pt.: 4742°C. Relative density: 8.570 (20°C). Proton number: 41. Relative atomic mass: 92.90638. Electronic configuration: [Kr]$4d^45s^1$.

Nishijima, Kazuhiko (1926–) Japanese physicist who put forward the concept of STRANGENESS in 1953, independently of Murray GELL-MANN.

nitrogen The first element of group 15 of the periodic table; a very electronegative element existing in the uncombined state as gaseous diatomic N_2 molecules. Nitrogen accounts for about 78% of the atmosphere (by volume) and it also occurs as sodium nitrate in various mineral deposits. It is separated for industrial use by the fractionation of liquid air. Nitrogen has two isotopes: ^{14}N, the common isotope, and ^{15}N (natural abundance 0.366%), which is used as a marker in mass spectrometric studies.
Symbol: N. Melting pt.: –209.86°C. Boiling pt.: –195.8°C. Density: 1.2506 kg m^{-3} (0°C). Proton number: 7. Relative

atomic mass: 14.00674. Electronic configuration: [He]$2s^2 2p^3$.

nobelium A radioactive transuranic element of the actinoid series, not found naturally on Earth. Several very short-lived isotopes have been produced.

Symbol: No. Proton number: 102. Most stable isotope: ^{259}No (half-life 58 minutes). Electronic configuration: [Rn]$5f^{14} 7s^2$.

noble gases (inert gases; rare gases; group 18 elements) A group of gaseous elements which make up group 18 of the periodic table. It consists of helium, neon, argon, krypton, xenon, and radon. The molecules of these gases are monatomic, i.e. they exist as isolated atoms. All these gases are chemically unreactive and are not readily ionized. These features arise because the atoms have closed-shell configurations. For many years it was widely thought that the closed-shell configurations meant that no compounds of the noble gases could exist. However, in 1962 it was found that compounds of xenon can be formed with fluorine.

no-boundary condition A proposal put forward by James Hartle and Stephen HAWKING in 1983 in which EUCLIDEAN QUANTUM GRAVITY is used to get round the problem of singularities at the BIG BANG by postulating that there are no singularities but that space–time is finite and closed with no boundaries, and hence no beginning or end. There is no direct evidence for this proposal but Hawking and his colleagues have successfully described some important features of cosmology using it.

no-cloning theorem *See* quantum teleportation.

node A point in a stationary wave in which the disturbance has a minimum or zero value. The distance between two nodes is equal to half the wavelength of the wave.

Noether, Emmy (1882–1935) German mathematician who stated and proved what is known as NOETHER'S THEOREM. Most of her work was in abstract algebra.

Noether's theorem The result that every continuous symmetry of a system is associated with a conservation law. For example, if a system is invariant under rotations about a point then angular momentum is conserved. Noether's theorem was stated and proved by Emmy NOETHER in 1918. However, some conservation laws are associated with TOPOLOGY rather than symmetry, particularly for SOLITONS.

no-hair theorem The result of general relativity theory that a BLACK HOLE is characterized uniquely by its mass, electric charge, and angular momentum. This result was established by a number of authors in the late 1960s.

non-Abelian gauge theory *See* gauge theory.

non-Abelian group *See* group.

noncommutative geometry A type of geometry in which the basic objects are noncommuting entities such as matrices rather than points. Noncommutative geometry has been developed since the 1980s, mainly by Alain CONNES. It has many physical applications including phase-space in quantum mechanics, the quantum Hall effect, regularization, renormalization, and superstring theory. At present there is no experimental evidence for the suggested applications of noncommutative geometry to the theory of elementary particles.

nonequilibrium statistical mechanics *See* statistical mechanics.

nonequilibrium thermodynamics *See* thermodynamics.

non-Euclidean geometry Any type of geometry that does not satisfy all the axioms of EUCLIDEAN GEOMETRY. There are many types of non-Euclidean geometry, RIEMANNIAN GEOMETRY being the main one of interest in physics.

noninertial frame A frame of reference that is not an INERTIAL FRAME.

nonlinear equation An equation in which the relation between two quantities x and y is not linear. For example, $y = x^2$ is a nonlinear equation.

nonlinear optics The branch of optics concerned with the optical properties of matter subjected to intense external electromagnetic fields. This is the case when the external field is not small compared to the internal fields of the atoms and molecules composing the matter. The subject of nonlinear optics has come into being largely as a result of the invention of the LASER.

nonlocality The term used to describe the behavior of a quantum mechanical system in which ENTANGLEMENT occurs, so that what happens in one part of the system instantaneously affects another part of the system, even if the parts are a long way away from each other. Experiments such as the Aspect experiment have shown that nonlocality is a real feature of quantum mechanical systems. *See* Bell's inequality.

nonmetal A material which is not a metal. Nonmetals are poor conductors of electricity and heat.

nonperturbative Not able to be described using PERTURBATION THEORY.

nonrenormalizable *See* renormalization.

nonrenormalization theorems Theorems concerning certain calculations in quantum field theory, which state that the correct result of the calculation is given by the first term in perturbation theory since there are no corrections due to the higher-order terms in the perturbation series. Examples of nonrenormalization theorems occur in certain field theories in which there is SUPERSUMMETRY and in calculations of chiral anomalies.

nova (*plural* **novae**) A star that becomes thousands of times brighter over a period of a few days. The name arises because in ancient times it appeared that new stars suddenly came into existence. After the development of photographic techniques it became apparent that faint stars precede novae. Each year there are usually between 10 and 15 nova events in the Milky Way.

It is thought that novae occur in binary star systems in which one of the stars is a white dwarf and the other is a red giant. Matter flows from the red giant to the white dwarf. When enough matter has built up on the white dwarf the pressure leads to nuclear fusion reactions. Such explosive reactions blast matter into space and dramatically increase the brightness. The matter ejected in this way is a source of heavy elements. A SUPERNOVA releases about a million times more energy than a nova.

Novikov, Igor Dmitrievich (1935–) Russian physicist who has made important contributions to cosmology and astrophysics. In 1962 Novikov and Andrei Doroshkevich predicted the existence of the cosmic microwave background (CMB). Novikov also contributed to the theory of black holes and suggested the possibility of mini black holes.

nuclear abundance *See* abundance.

nuclear energy Energy that is released in the processes of nuclear fission, nuclear fusion, and, more generally, NUCLEAR REACTIONS. In such reactions the energy released is equivalent to the mass lost in accord with Einstein's equation $E = mc^2$. NUCLEAR POWER and NUCLEAR WEAPONS are applications of nuclear energy. Far more energy is released in nuclear energy than in chemical reactions for the same amount of starting mass. For example, mass for mass, uranium fission gives about 2.5×10^6 times as much energy as the combustion of carbon. Similarly, mass for mass, the fusion of deuterium to give helium gives about 400 times as much energy as the fission of uranium. *See also* nuclear power; nuclear weapons.

nuclear fission A type of nuclear reaction in which a heavy atomic nucleus (such as that of uranium) splits into two or more smaller nuclei (called *fission products*). Fission can occur spontaneously or be induced by irradiation, usually by neutrons. When fission occurs a quantity of energy is released equal to the difference between the rest masses of the fission products and that of the original nucleus. Fission reactions generally result in the emission of neutrons. There are different final products possible in a fission reaction. The number of neutrons produced in a fission reaction also varies but is usually either 2 or 3. The neutrons given off in fission reactions can produce fission in nearby nuclei, thereby causing a CHAIN REACTION. The energy from nuclear fission, which is very much greater per unit mass than the energy in chemical reactions, is used for the production of nuclear power and in nuclear weapons. An example of a nuclear fission process is the reaction:

$$^{235}_{92}U + n \rightarrow {}^{143}_{56}Ba + {}^{90}_{36}Kr + 3n.$$

Fission can be described theoretically in terms of the LIQUID DROP MODEL of nuclear structure. This was done in 1939 by Lise MEITNER and Otto FRISCH and, in more detail, by Niels BOHR and John WHEELER. Although the liquid drop model gives a nice physical picture of nuclear fission it does not give a complete quantitative description of fission, particularly as regards the asymmetry of mass of the fission products. Much more advanced theories based on more complicated models of nuclear structure have been constructed. For example, in the late 1960s theories of nuclear fission that combine the liquid drop model with the SHELL MODEL were initiated. Advances in computing power have enabled calculations to be performed for these theories, which give accurate results for the mass distribution of fission products.

nuclear fuel Material that will sustain a nuclear fission chain reaction and can be used as a source of nuclear energy. The FISSILE MATERIAL used in nuclear fuel can be uranium-233, uranium-235, plutonium-239, and plutonium-241. The isotope uranium-235 occurs naturally as 1 part in 140 of naturally occurring uranium. The other isotopes used as fissile material have to be made artificially from FERTILE MATERIAL. The isotope uranium-233 is produced when thorium-232 captures a neutron and plutonium 239 is produced when uranium-238 captures a neutron.

nuclear fusion A type of nuclear reaction in which light atomic nuclei come together to produce a heavier nucleus, with the release of a large amount of energy. Since nuclei are all positively charged there is a large repulsive force between two close nuclei, which means that the nuclei must have high kinetic energies to overcome this electrostatic repulsion. If the temperature is about 10^8 K there are sufficiently many nuclei with enough energy for fusion to occur. Because of the high temperatures involved this process is called a *thermonuclear reaction*. The fusion occurs at these temperatures because enough light nuclei, particularly protons, have sufficiently high energy to tunnel into nuclei. Fusion reactions occur in stars (*see* nucleosynthesis). Because of the large amount of energy given off in nuclear fusion such reactions are self-sustaining. The hydrogen bomb (*see* nuclear weapons) produces energy by nuclear fusion.

It is hoped that controlled nuclear fusion can be used as a source of power, using a THERMONUCLEAR REACTOR. However, serious difficulties in maintaining stability in the plasma state in which nuclear fusion occurs have so-far prevented fusion from being a practical source of power.

It is possible for nuclear fusion to occur at lower temperatures than are needed to overcome the electrostatic repulsion between nuclei using muons. This process is called *muon catalyzed fusion* (MCF). In MCF the electrons of deuterium atoms are replaced by muons. Since a muon is about 207 times heavier than an electron a 'muonic atom' of deuterium is much smaller than a normal deuterium atom and so can approach another deuterium atom more closely, thus allowing nuclear fusion to occur. The process is described as 'catalyzed' since the muon is released, enabling it to form another muonic atom, and hence to continue the process. However, since a muon has a short lifetime the number of fusion reactions it can catalyze is limited.

nuclear isomerism A condition in which two or more atomic nuclei have the same proton and neutron numbers but are in different quantum states. This means that they have different stabilities in general, and hence have different half-lives. For example, there are two isomers of protoactinium-234 that arise from the beta decay of thorium-234. These two isomers have half-lives of about 70 seconds and about 6.7 hours respectively, with both of these isomers undergoing further beta decay to uranium-234. It is possible for about 1 in 700 nuclei of the less stable isomer (i.e. the one with the shorter half-life) to emit a gamma ray, and hence make a transition to the more stable isomer.

nuclear magnetic resonance (NMR) The absorption of electromagnetic radiation of specific frequencies by certain atomic nuclei in an external magnetic field. This phenomenon occurs if the nucleus has a net spin that is nonzero. Such nonzero spin nuclei behave as small magnets. When the nuclei are in external magnetic fields this means that the magnetic moment vector of the nucleus precesses about the direction of the magnetic field, with quantum mechanics only allowing certain orientations. These different orientations have slightly different energies. Nuclear magnetic resonance is the absorption of electromagnetic radiation when the energy of the photon is equal to the energy difference between two nuclear spin states. Since the magnetic moment of a nucleus is much smaller than the magnetic moment of an electron the energy level splitting is much smaller for NMR than for the analogous case of ELECTRON SPIN RESONANCE (ESR). This means that the frequencies required for NMR are smaller than for ESR. The frequencies required for NMR are short-wavelength radio frequencies. NMR was discovered in 1946 by Felix BLOCH and his colleagues and independently by Edward Purcell and his colleagues.

It is possible to use NMR in chemical analysis and in the determination of the structure of molecules – a technique known as *NMR spectroscopy*. The electrons surrounding a nucleus shield the nucleus from the external magnetic field to a certain extent. This causes a slight shift, known as the *chemical shift*, in the energy levels of the nuclear states and hence in the resonance frequency for a fixed magnetic field (or a slightly changed magnetic field for a fixed frequency). In *continuous wave (CW) NMR* a sample is subjected to a strong external magnetic field which is varied slightly, with the electromagnetic radiation having a fixed frequency. In *Fourier transform (FT) NMR* the magnetic field is fixed and the frequencies varied. The most commonly studied nucleus is the hydrogen nucleus, which consists of one proton.

A standard example of the application of NMR to the determination of molecular structure is the ethanol molecule CH_3CH_2OH. Since there are three different electron environments for the hydrogen nuclei there are three peaks in the NMR spectrum, with the areas below the peaks being in the ratio 3:2:1 corresponding to the number of hydrogen atoms in each different electronic environment. When instruments with high resolution are used there is fine structure in the peaks known as *spin–spin splitting* caused by interactions between nuclear spins.

In medicine NMR is used to obtain images inside the body of a person by converting the NMR signals from the nuclei in molecules inside the body into images. This technique, known as *magnetic resonance imaging*, is not invasive and is not harmful to health, unlike frequent exposure to x-rays.

nuclear magneton Symbol: μ_N A unit of magnetic moment used in nuclear physics, in which the mass of the proton replaces the mass of the electron in the formula for the BOHR MAGNETON. Thus, $\mu_N = eh/4\pi Mc$, where e is the charge of an electron, h is the Planck constant, M is the mass of the proton and c is the speed of light in a vacuum. Since the mass of a proton is about 1840 times the mass of an electron the value of the nuclear magneton is about 1840 times smaller than the value of the Bohr magneton. The value of the magnetic moment of the proton is +2.793 nuclear magnetons. The value of the neutron, which might be expected to be zero since the neutron is electrically neutral is −1.913 nuclear magnetons. These facts clearly indicate that protons and neutrons are not structureless pointlike particles.

nuclear medicine   A broad term which is taken to mean all the applications of nuclear physics to medicine. This includes the use of gamma rays to destroy cancer cells and the use of radioactive isotopes as tracers in the body.

nuclear physics   The study of the properties, structure, and reactions of atomic nuclei. Applications of nuclear physics include the production of nuclear power, the manufacture of nuclear weapons, and the production of radioactive isotopes, which are used in medicine and in industry.

nuclear power   The generation of power from the heat of nuclear reactions. Up to the present time all commercial nuclear power stations have used NUCLEAR FISSION. Power based on fission has the disadvantage that care must be taken in the handling of radioactive material and the disposal of radioactive waste. It is hoped that power based on NUCLEAR FUSION will be a practical possibility in the future, but serious difficulties remain in spite of a vast amount of research in this area. *See also* nuclear reactor.

nuclear reaction   A reaction involving atomic nuclei. This includes radioactive decay and processes in which nuclei are bombarded with other nuclei or with particles. The energies involved in nuclear reactions are generally much larger than for chemical reactions. Nuclear reactions are represented, like chemical reactions, with the reactants on the left, an arrow, and the products on the right. For example, the reaction involved in the discovery by Ernest RUTHERFORD in 1919 that nuclei can be broken up by bombardment by alpha particles can be written:

$$^{14}_{7}\text{N} + ^{4}_{2}\text{He} \rightarrow ^{17}_{8}\text{O} + ^{1}_{1}\text{H}.$$

The theory of nuclear reactions combines the various theoretical models that have been used to analyze nuclear structure with the quantum theory of scattering.

An important concept in the theory of nuclear reactions is the *compound nucleus*, introduced and developed by Niels BOHR in the second half of the 1930s in terms of the LIQUID DROP MODEL. In this picture, a target nucleus is struck by an incident particle, with a new nucleus, called the compound nucleus, being formed. The proton number and mass number of the compound nucleus is the sum of the proton numbers and mass numbers respectively of the original particles. For example, in the nuclear reaction discussed above one can consider the reaction to proceed by an intermediate compound nucleus (denoted by square brackets):

$$^{14}_{7}\text{N} + ^{4}_{2}\text{He} \rightarrow [^{18}_{9}\text{F}] \rightarrow ^{17}_{8}\text{O} + ^{1}_{1}\text{H}.$$

A compound nucleus does not 'remember' how it was formed since the energy from the incident particle is shared by all the particles of the compound nucleus. The lifetimes of compound nuclei are about 10^{-16} s, much longer than the 10^{-21} s required for a moderately energetic nuclear particle to pass through a nucleus. Although a given compound nucleus can, in general, decay in several different ways a particular decay mode is usually favored by a compound nucleus in a specific excited state. If a high-energy particle causes a large number of particles to be ejected from the compound nucleus it forms then the nuclear reaction is called a *spallation*.

There are many nuclear reactions that cannot be described by the compound-nucleus model. This is the case when the projectile nucleus colliding with the target nucleus interacts with all or part of the target nucleus, with emission taking place almost immediately. This type of reaction, in which there is no relatively long-lived intermediate compound nucleus, is called a *direct reaction*. Direct reactions occur on the short time-scales of about 10^{-21} s. This is the time it takes for the projectile to travel past the nucleus rather than the 10^{-16} s for which a compound nucleus exists.

Most direct reactions involve transfer either from or to the projectile nucleus as it passes the target nucleus. A direct reaction involving transfer is called a *stripping reaction* or a *pick-up reaction*, according to whether the projectile nucleus loses or gains nucleons in the reaction. An example of a stripping reaction is what is known as a *(d,p) reaction*. In this reaction a deuteron loses a neutron when it collides with a target nucleus, with the proton from the deuteron continuing on its path. Transfer reactions are used to produce new isotopes or nuclei that are very difficult or impossible to produce by other methods.

nuclear reactor A device in which nuclear reactions take place for the purpose of producing energy. Reactors are also used to produce new nuclides or to conduct research in nuclear physics.

So far, *fission reactors*, i.e. reactors that depend on controlled, sustained fission of uranium or plutonium have been the only practical nuclear reactors. *Fusion reactors*, i.e. reactors based on the fusion of hydrogen, are still in the experimental stage.

In fission reactors the reaction used is the fission of uranium-235. When a nucleus of this isotope is struck by a neutron

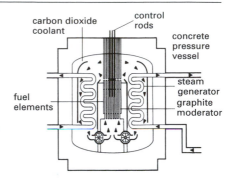

Advanced gas-cooled reactor

it is converted into uranium-236, which splits into two equal fragments with the production of two or three neutrons. The basis of the operation is to have a controlled chain reaction in the *core* of the reactor to produce heat, which is used to raise steam to drive a turbine.

There are two ways of producing a chain reaction. One is to use natural uranium, which contains about 0.7% of uranium-235, with the remainder being uranium-238. This uranium-238 absorbs the fast neutrons produced by fission and prevents the chain reaction propagating. To maintain a chain reaction the neutrons are slowed down to thermal energies using a MODERATOR, so that they are not absorbed by uranium-238. Reactors of this type are called *thermal reactors*. The other method of producing a sustainable chain reaction is to *enrich* the natural uranium, either with extra uranium-235 or with plutonium-239, which is also fissile. There are then enough neutron collisions to propagate the chain reaction despite absorption by uranium-238. Reactors of this type do not have a moderator. They are called *fast reactors* (because they use fast neutrons).

It is possible to use certain fast reactors as *breeder reactors* or *converter reactors*, with a breeder reactor being a reactor that produces more of the same fissile material (such as plutonium-239). This is produced by placing a *blanket* of natural uranium around the reactor core. It is also possible to construct thermal breeder reactors. This is done using uranium-233 as a fuel, with more of this fissile material being gener-

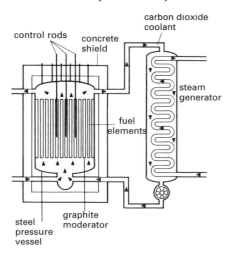

Magnox reactor

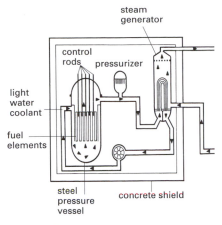

Pressurized-water reactor

ated by capturing neutrons in nonfissile thorium-232.

Energy is carried from the core by a *coolant*. The earliest thermal reactors were *Magnox reactors*, which use carbon dioxide as a coolant and the fuel is natural uranium metal clad in a magnesium alloy. *Advanced gas-cooled reactors* (*AGRs*) use enriched uranium oxide pellets clad in steel. Most modern thermal reactors are *pressurized-water reactors* (*PWRs*), which use the uranium oxide fuel and use water as coolant and moderator. In *fast reactors* the core becomes very hot and liquid sodium is the usual coolant.

The rate at which a reaction occurs is controlled by *control rods*, i.e. rods of neu-

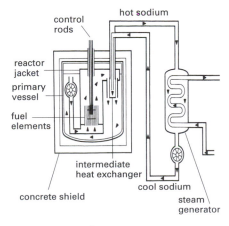

Fast reactor

tron absorbing material such as boron or cadmium, being moved into or out of the core. *See also* thermonuclear reactor.

nuclear structure The structure of atomic nuclei in terms of its constituent protons and neutrons. It is not possible to work out nuclear structure starting from the strong interactions between nucleons, which, in turn are vestiges of the strong interactions of quarks and gluons within nucleons. Nevertheless, a large degree of understanding of nuclear structure has been obtained by expressing the interactions between nucleons in terms of potentials described by quantum mechanics. This is still a many-body problems of considerable complexity.

Because of this complexity a great deal of progress in understanding nuclear structure has come from the study of *nuclear models*, i.e. simplified models of nuclei used to explain various aspects of their behavior. For example, in the SHELL MODEL the nucleons are regarded as almost independent. On the other hand, in the LIQUID DROP MODEL all the nucleons in a nucleus are regarded as acting collectively in a way that is analogous to the way that molecules act in a liquid. Since these models appear to be very different physically this raises the problems of establishing under which circumstances the different models are applicable and attempting to find a unified understanding of the different models. To a large extent these problems have been solved using concepts from the quantum theory of many-body systems, notably the interactions between QUASIPARTICLES and COLLECTIVE EXCITATIONS. The work of Aage BOHR, Ben MOTTELSON, and Leo RAINWATER in the 1950s and subsequently was very important in establishing this unified understanding. An analogy that is frequently used to describe this unified picture is of a swarm of bees, with the swarm moving slowly through the air, but inside the swarm each bee moves rapidly. This analogy correctly emphasizes that nuclear structure is not a static rigid structure like that of a solid.

nuclear waste *See* radioactive waste.

nuclear weapons Devices in which an explosion is produced by nuclear energy. In a *fission bomb* (sometimes called an *atomic bomb*) two subcritical masses (*see* critical mass) of a fissile material (such as uranium-235 or plutonium-239) are brought together to produce one supercritical mass. The resulting chain reaction causes a nuclear explosion, which is typically in the *kiloton* range, i.e. an explosive power equivalent to one thousand tons of TNT, with temperatures of about 10^8K being reached. In a *fusion bomb* (sometimes called a *hydrogen bomb* or a *thermonuclear weapon*) the explosion is caused by the fusion of light elements such as *deuterium*. A fission reaction is used to obtain sufficiently high temperatures to start a self-sustaining fusion reaction. The explosion produced by a fusion bomb is typically in the *megaton* range, i.e. an explosive power equivalent to one million tons of TNT. A particular type of fission-fusion bomb is the *neutron bomb*, in which most of the energy is released in the form of high-energy neutrons. These neutrons destroy people but there is less blast and consequently less destruction of buildings.

nucleon A constituent of atomic nuclei, i.e. a proton or a neutron.

nucleonics The technological aspect of nuclear physics, such as the design of nuclear reactors and of devices to produce and detect radiation and techniques for the disposal of RADIOACTIVE WASTE.

nucleon number (**mass number**) Symbol: *A* The number of nucleons in an atomic nucleus.

nucleosynthesis The way in which the nuclei of heavier elements have been built from the nuclei of hydrogen. There is no single process that can explain the origin of all the chemical elements.

In *primordial nucleosynthesis*, which was complete by a time of 3 minutes and 46 seconds after the BIG BANG, most of the helium that exists was formed by the fusion of protons and neutrons. Small amounts of other very light elements, such as deuterium and lithium, were also formed. About 25 per cent of the mass of the matter in the Universe that exists in the form of baryons does so as nuclei of helium. It is not possible to account for this number of nuclei by processes that occur in stars. On the other hand, only the very light elements were formed in primordial nucleosynthesis. All but the lightest elements have been subsequently formed from hydrogen and helium nuclei inside stars, this process being called *stellar nucleosynthesis*.

This process starts by the addition of helium-4 nuclei to existing nuclei. The first step is that two helium-4 nuclei are fused together to form beryllium-8. When this picture of stellar nucleophysics was first proposed this caused a major difficulty because beryllium-8 is very unstable, existing for only a small fraction of a second before reverting back to two helium nuclei. Fred HOYLE solved this problem in the early 1950s by suggesting the *triple alpha process*, in which he predicted that there is a 'resonant' excited state of carbon-12 with exactly the right energy for beryllium-8 to fuse with helium-4. Following Hoyle's suggestion, William FOWLER and his colleagues found this excited state of carbon-12 at an energy very near to the value of 7.65 MeV which Hoyle had predicted. Once the beryllium 'barrier' is overcome the fusion of further alpha particles to create heavier nuclei is straightforward. Other elements are formed from these elements by radioactive decay, mostly of electrons but sometimes positrons.

This process halts when iron-56 and related elements such as nickel-56 and cobalt-56 are produced. These elements have the most stable nuclei. This is frequently represented by iron-56 being at the bottom of a 'valley of stability', with all lighter elements to the left of it and all heavier elements to the right of it. In other words, iron-56 is the element with the least energy per nucleon. This means that energy is released both when there is fusion between two very light elements and when fission of a very heavy element occurs. This in turn means that a large amount of energy must be put in for elements heavier than iron to be produced.

Elements heavier than iron can be made if there are high-energy neutrons that can interact with nuclei. This is the case inside stars since there are always high-energy neutrons around as products of various nuclear reactions. In the *s-process* (*slow process*) a nucleus captures a neutron and is likely to decay before it captures another neutron. Elements up to bismuth-209 are thought to be formed inside stars in this way. However, the s-process cannot go beyond bismuth-209 because when bismuth-209 captures a neutron it very rapidly undergoes alpha decay.

Elements heavier than bismuth-209, and also many isotopes of nuclei between iron-56 and bismuth-209, are produced by the *r-process* (*rapid process*) in which a nucleus captures several neutrons before decay occurs. The r-process requires a very high concentration of high-energy neutrons. Such a high concentration is only available in a SUPERNOVA explosion.

Although the r-process and the s-process explain the origin of most heavy nuclei they cannot explain the origin of all of them. Another process known to occur in supernova explosions is the *p-process* (*proton-capture process*) in which protons are captured to give proton-rich nuclei.

The combination of the big bang for the origin of the lightest elements and reactions in stars for all other elements explains the observations of the abundances of elements in the Universe very well in a quantitative way. It is possible to understand the nuclear reactions that are postulated to occur in element formation by studying them in accelerators on the Earth.

Although the theory of nucleosynthesis based on processes just after the big bang and reactions in stars is very successful there are a few elements for which the abundances cannot be explained. This is particularly true of light elements such as beryllium and boron. It is thought that these elements originate in spallation reactions in interstellar space involving cosmic rays.

nucleus (of an atom)   The compact central core of an atom. An atomic nucleus contains most of the mass of an atom and all of its positive charge. The constituents of an atomic nucleus are electrically neutral neutrons and positively charged protons. Thus, the positive charge of a nucleus is equal to its number of protons. In a neutral atom this number is balanced by an equal number of electrons which move around it. The diameter of a nucleus is between 10 000 and 100 000 times smaller than the diameter of an atom, with an atom having a diameter of about 10^{-10} meter while the diameters of nuclei vary from 10^{-15} meter to 10^{-14} meter.

The number of protons in the nucleus (proton number) determines the chemical element. The simplest nucleus is the hydrogen nucleus, which consists of one proton. All other nuclei contain at least one neutron.

The constituents of nuclei are held together by the strong interactions. The biggest naturally occurring nucleus is uranium-238, which has 92 protons and 146 neutrons. Not all combinations of protons and protons form stable nuclei. Some undergo spontaneous radioactive decay. For example, all nuclei with more than 92 protons are radioactive since the repulsive force due to the accumulation of positive charges outweighs the strong interaction. Nuclei that are very neutron rich are also unstable. This is the case because there are 'stacks' of energy levels for both protons and neutrons in nuclei as protons and neutrons are distinguishable. Since protons and neutrons are both fermions they are subject to the Pauli exclusion principle. The proton energy levels start off higher than the neutron energy levels because additional energy must be supplied to the protons in a nucleus to allow them to overcome their electrostatic repulsion, with this 'energy gap' between proton energy levels and neutron energy levels increasing as the number of protons increases. For a stable nucleus the tops of the two energy level stacks have about the same energy. This explains why stable nuclei have more neutrons than protons, but that it is energetically unfavorable for the nucleus to be too neutron-rich. It also explains why a neutron is stable inside a nucleus but unstable outside it. If a neutron decayed inside a nucleus the proton left behind would have to occupy one of the proton energy levels. Since all the lower proton energy levels in a nucleus are full the Pauli exclusion principle dictates that the proton has to fill the energy level at the top of the stack. This is energetically unfavorable and so does not occur.

A nucleus is indicated by the symbol for the element, with the nucleon number as a left superscript and the proton number as a left subscript. For example, uranium-238 is written as $^{238}_{92}U$.

nuclide A type of atomic nucleus with a given number of protons and neutrons. Thus, a nuclide refers to a specific isotope of a given element. For example, $^{23}_{11}Na$ which has 11 protons and 12 neutrons, $^{24}_{11}Na$ which has 11 protons and 13 neutrons and $^{24}_{12}Mg$ which has 12 protons and 12 neutrons are all different nuclides.

octet A set of eight electrons in the outer shell of an atom (or ion). This is a particularly stable configuration, as seen in the NOBLE GASES neon and argon. The relation between octets of electrons in atoms and their stability was proposed by Gilbert Newton LEWIS and was subsequently justified by the application of quantum mechanics to the ELECTRONIC STRUCTURE OF ATOMS.

odd-A nucleus An atomic nucleus that contains an odd number of nucleons.

odd–even nucleus An atomic nucleus that contains an odd number of protons and an even number of neutrons.

odd–odd nucleus An atomic nucleus that contains an odd number of protons and an odd number of neutrons. There are very few odd–odd nuclei known. The only four stable ones are $^2_1$H, $^6_3$Li, $^{10}_5$B, and $^{14}_7$N. Odd–odd nuclei are much less common than even–odd nuclei or odd–even nuclei, with even–even nuclei being much more common than any other type of nucleus. This pattern occurs because pairing between individual pairs of protons and individual pairs of neutrons is energetically favorable. *See also* pairing energy.

Oersted, Hans Christian (1777–1851) Danish physicist who discovered that an electric current produces a magnetic field. Oersted stated this discovery in 1820, stimulating a great deal of interest, with André AMPERE and Michael FARADAY taking his investigations of the links between electricity and magnetism further.

ohm Symbol: Ω The SI unit of electrical resistance, equal to a resistance that passes a current of one ampere when there is an electric potential difference of one volt across it. $1\ \Omega = 1\ \text{V A}^{-1}$. Formerly, it was defined in terms of the resistance of a column of mercury under specified conditions.

Ohm's law The principle that the ratio of the potential difference V between the ends of an electrical conductor to the current I flowing through it, is a constant known as the resistance R. This law was stated by Georg Simon Ohm in 1827. Not all materials obey Ohm's law; materials that do so are called *ohmic conductors*.

Oklo reactor A uranium mine located at Oklo in Gabon in West Africa, which acts as a natural NUCLEAR REACTOR. About two billion years ago a deposit of uranium ore formed a self-sustaining chain reaction. This phenomenon is of great interest to elementary-particle theorists because it makes it possible to determine whether some of the FUNDAMENTAL CONSTANTS, such as the fine structure constant, have actually remained constant over the last two billion years. The abundance of the decay products from two billion years ago would be very sensitive to slight changes in the values of the fundamental constants over this period of time. At present, investigations of the Oklo reactor do not appear to give any support for the idea that changes have occurred in the value of the fundamental constants with time.

old quantum theory A term usually taken to mean the theory of the atom developed by Niels BOHR, Arnold SOMMERFELD, and others between 1913 and 1925. The old quantum theory was not a consistent theory but a series of *ad hoc* rules of quantization for electrons in atoms based on empirical observations from chemistry and atomic spectroscopy. The successes and failures of the old quantum theory paved the way for the development of QUANTUM MECHANICS.

omega minus Symbol: ω⁻ A spin-3/2 baryon with a mass of about 1672 MeV, a strangeness of –3, and an electric charge of –1. The existence of this particle, together with its properties, was predicted by Murray GELL-MANN in the early 1960s as part of the EIGHT-FOLD WAY of classifying baryons. The particle was subsequently detected at Brookhaven National Laboratory in 1964. The prediction and discovery of the omega minus particle were important advances in particle physics and paved the way for the development of the QUARK model. An omega minus particle is thought to be made up of three strange quarks. The average lifetime of the particle is about 10^{-10} s. The antiparticle of the omega minus has a charge of +1 and a strangeness of +3.

Onnes, Heike Kamerlingh (1853–1926) Dutch physicist who made important contributions to the study of matter at low temperatures. In particular, he became the first person to liquefy helium (in 1908) and in 1911 he discovered the phenomenon of SUPERCONDUCTIVITY. Onnes won the 1913 Nobel Prize for physics for his pioneering contributions to low-temperature physics.

Onsager, Lars (1903–76) Norwegian-born American chemist who made various important contributions to theoretical chemistry and theoretical physics. In 1944 he solved the ISING MODEL in two dimensions. This was a major landmark in the theory of phase transitions since it was the first time an exact solution to a nontrivial model had been found. He was awarded the 1968 Nobel Prize for chemistry for his 1931 work on nonequilibrium thermodynamics.

open shell A shell of electrons in an atom or nucleons in a nucleus that is not a complete shell.

operator A mathematical entity that changes one function into another function. For example, the square root operator √ applied to x in $\sqrt{x}$ indicates that the square root of x should be taken. The differential operator d/dx applied to y in dy/dx indicates that the derivative of y with respect to x, should be taken. It is possible to express quantum mechanics using a formalism that depends on operators (*operator formalism*).

Oppenheimer, Julius Robert (1904–67) American physicist who is best remembered for leading the development of the atomic bomb in the MANHATTAN PROJECT. Oppenheimer also made important contributions to quantum mechanics, quantum field theory, and astrophysics. These include the BORN–OPPENHEIMER APPROXIMATION and the OPPENHEIMER–VOLKOFF LIMIT for neutron stars.

Oppenheimer–Volkoff limit A limit for a neutron star that is the analog of the CHANDRASEKHAR LIMIT for a white dwarf. If a NEUTRON STAR has a mass greater than the Oppenheimer–Volkoff limit then it will undergo gravitational collapse to become a BLACK HOLE. The existence of this limit was pointed out by Robert Oppenheimer and George Volkoff in 1939. It is more difficult to calculate the Oppenheimer–Volkoff limit accurately than the Chandrasekhar limit because it is more difficult to obtain an accurate equation of state for matter in a neutron star than for matter in a white dwarf. The best calculations available at present indicate that the limit is between two and three times the mass of the Sun.

optical activity The ability of certain crystals or compounds in solution to rotate the plane of polarization of plane polarized light. Compounds that are optically active have an asymmetric molecular structure such that their molecules can exist in left- and right-handed forms (the forms are mirror images of each other; the property of having such forms is called *chirality*). Particular forms of the compound are classified as *dextrorotatory* (right turning) or *levorotatory* (left turning). Levorotatory compounds rotate the plane of polarization to the left; i.e. anticlockwise as viewed facing the oncoming light. Dextrorotatory compounds rotate the plane to the right (i.e. in the opposite sense). Many naturally occurring substances (e.g. sugars) are optically active.

optical model *See* cloudy crystal ball model.

optical pumping *See* laser.

optical rotary dispersion (ORD) The dependence of the amount of rotation in OPTICAL ACTIVITY on wavelength. Graphs of rotation against wavelength, with their characteristic peaks and troughs, can be used to give information on molecular structure.

optic axis A direction in a birefringent crystal in which light travels without double refraction. A uniaxial crystal has one such axis and a biaxial crystal has two such axes.

optics The study of light (and, more generally, electromagnetic waves). Optics is divided into *geometrical optics*, in which the wave nature of light is not considered and light is analyzed in terms of RAYS, and *physical optics*, which is concerned with the wave nature of light. Thus, geometrical optics deals with phenomena such as reflection and refraction while physical optics is concerned with phenomena such as interference, diffraction, and polarization. It was shown by Sir William HAMILTON that geometrical optics and classical mechanics can be formulated in very similar ways. This analogy between optics and mechanics was completed by Erwin SCHRÖDINGER when he showed that quantum mechanics is to classical mechanics as physical optics is to geometrical optics. Both classical mechanics and geometrical optics are derived as limiting theories from quantum mechanics and physical optics respectively as the wavelength λ tends to 0.

orbit The path along which a body moves relative to some selected point. The elliptical, almost circular, path of a planet

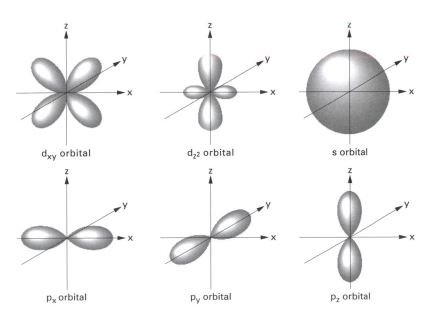

Shapes of atomic orbitals

such as the Earth going round the Sun is a familiar example of an orbit, but an orbit need not be a closed path. In an analogy with the closed orbits of celestial mechanics, Niels Bohr postulated that electrons in an atom move around the nucleus in orbits (*see* Bohr model). However, it subsequently became clear that the concept of a precise orbit is not permitted in quantum mechanics because of the Heisenberg UNCERTAINTY PRINCIPLE. In quantum mechanics the concept of an orbit is replaced by that of an ORBITAL.

optoelectronics   *See* photonics.

orbital   A region in which an electron in an atom or molecule has a high probability of being found. Because of the dual wave–particle nature of the electron and the Heisenberg uncertainty principle it is not accurate to describe an electron as

going round a nucleus in a definite orbit (as in the BOHR MODEL of the atom).

An orbital can be defined as a wave function ψ, which is a solution of the SCHRÖDINGER EQUATION for an electron in the atom or molecule. This is closely related to the idea of an orbital as a region since Born's interpretation of the wave function ψ can be stated in the form that the probability that an electron is in a volume element $d\tau = dxdydz$ is given by $\psi^2d\tau$ – i.e. the probability is equal to the square of the wave function.

In an atom the regions in which electrons can be found are known as *atomic orbitals*. For example, in the case of the hydrogen atom with one proton and one orbiting electron the Bohr model puts the electron in a definite circular orbit with a known radius (the Bohr radius). The quantum mechanical picture is much less definite. The electron can be found anywhere

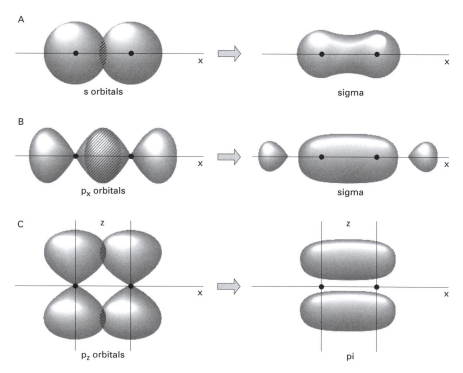

Formation of molecular orbitals

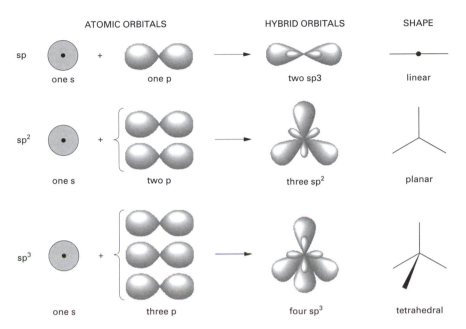

	ATOMIC ORBITALS	HYBRID ORBITALS	SHAPE

sp   one s   one p   two sp3   linear

sp² one s   two p   three sp²   planar

sp³   one s   three p   four sp³   tetrahedral

Formation of hybrid orbitals

from close to the nucleus to an infinite distance away in a spherical region around the nucleus. However, the maximum probability of finding the electron is in a region at a distance equal to the Bohr radius.

In an atom an orbital has three quantum numbers: the principal quantum number, the orbital quantum number, and the magnetic quantum number (*see* electronic structure of atoms). Since an electron state of an atom is characterized by four quantum numbers, with the spin quantum number being the fourth quantum number, each orbital can contain two electrons.

Orbitals are different from the orbits in the Bohr model in that they have characteristic shapes. The shape of an orbital is determined by the value of its orbital quantum number. There is one *s-orbital* for each shell, corresponding to the fact that for $l = 0$ there is only one possible value of the magnetic quantum number m_l. An *s*-orbital is spherically symmetric. There are three *p-orbitals* in a shell (for which the principal quantum number $n \geq 2$). This corresponds to the fact that for $l = 1$ there are three pos-

sible values of m_l. The *p*-orbitals are not spherical. Each of the *p*-orbitals has two lobes, with one lobe on either side of the nucleus and the three *p*-orbitals are at right angles to each other. Each of the two electrons in a *p*-orbital occupies both lobes. There are five *d-orbitals* in a shell (for which $n \geq 3$). The *d*-orbitals have more complicated shapes than either *s*-orbitals or *p*-orbitals.

Chemical bonding can be discussed in terms of orbitals. In a molecule the combination of two atomic orbitals gives rise to two *molecular orbitals*. These two molecular orbitals have different energies, with the orbital with lower energy (called the *bonding orbital*) being associated with a chemical bond holding the atoms together. The orbital with higher energy (called the *antibonding orbital*) is associated with the atoms being pushed apart. In a bonding orbital there is a high probability of two valence electrons being between the nuclei of two atoms, which are held together in this way. More generally, a molecular orbital can be built up from a *linear combination of atomic orbitals* (LCAO).

An important concept in theories of chemical bonding is *hybridization*. A typical example of this occurs in the methane molecule (CH_4). The outer shell of a carbon atom has one *s*-orbital and three *p*-orbitals. In a methane molecule these orbitals can be thought of as combining to form four equivalent sp^3 hybrid orbitals, each of which points towards the corner of a tetrahedron, with the carbon nucleus being at the center of the tetrahedron. These hybrid orbitals overlap with the *s*-orbitals of the hydrogen atoms to give the C–H bonds of the methane molecule.

If one is considering two atoms A and B then if a molecular orbital is made up of two *s*-orbitals or two *p*-orbitals (or hybrid orbitals) that both have lobes along the AB axis the orbital is said to be a *sigma orbital* (σ orbital) and the electrons involved are called *sigma electrons*. A molecular orbital is formed from two *p*-orbitals in which each of the *p*-orbitals has a lobe above and below the nucleus of its atom is called a *pi orbital* (π orbital) and the electrons involved are called *pi electrons*. Sigma and pi orbitals have different shapes, with a sigma orbital being along the axis of AB while a pi orbital has regions above and below AB.

orbital quantum number *See* electronic structure of atoms.

ORD *See* optical rotary dispersion.

order parameter A quantity that characterizes the order of a phase of a system. This phase is frequently the low-temperature phase of a system that undergoes a phase transition at some particular transition temperature. The magnetization of a ferromagnetic system is an example of an order parameter. In a continuous phase transition the order parameter goes to zero continuously as the transition temperature is approached from below. A *disorder parameter* is a quantity that is nonzero above

the transition temperature and zero below it. The concept of an order parameter is closely related to the concept of broken symmetry.

ortho-positronium *See* positronium.

Osheroff, Douglas Dean (1945–) American physicist who did pioneering work on the SUPERFLUIDITY of helium-3 in the early 1970s. Osheroff shared the 1996 Nobel Prize for physics with his colleagues David LEE and Robert RICHARDSON for this work.

osmium A transition metal that is found associated with platinum. Osmium is the most dense of all metals. It is used in catalysts and in alloys for pen nibs, pivots, and electrical contacts.

Symbol: Os. Melting pt.: 3054°C. Boiling pt.: 5027°C. Relative density: 22.59 (20°C). Proton number: 76. Relative atomic mass: 190.23. Electronic configuration: $[Xe]4f^{14}5d^66s^2$.

overtone *See* harmonic.

oxygen A colorless odorless diatomic gas; the first member of group 16 of the periodic table. Oxygen is the most plentiful element in the Earth's crust accounting for over 40% by weight. It is present in the atmosphere (20%) and is a constituent of the majority of minerals and rocks as well as water. Oxygen is an essential element for almost all living things. Elemental oxygen has two forms: the diatomic molecule O_2 and the less stable molecule trioxygen (ozone), O_3, which is formed by passing an electric discharge through oxygen gas.

Symbol: O. Melting pt.: –218.4°C. Boiling pt.: –182.962°C. Density: 1.429 kg m^{-3} (0°C). Proton number: 8. Relative atomic mass: 15.9994. Electronic configuration: $[He]2s^22p^4$.

P

packing fraction The difference between the relative atomic mass of an isotope and its mass number, divided by its mass number.

pair creation The production of a particle–antiparticle pair (such as an electron–positron pair) from energy in accord with Einstein's equation $E = mc^2$. It can happen when a gamma-ray photon passes close to an atomic nucleus. The occurrence of pair creation is one of the successful predictions of relativistic quantum mechanics.

pairing energy The energy associated with the tendency of nucleons of the same type to pair up. The pairing energy is responsible for EVEN–EVEN NUCLEI being much more common than odd-A nuclei and in turn that odd-A nuclei are much more common than odd–odd nuclei. The existence of the pairing energy is an additional assumption in the SHELL MODEL that gives rise to GOEPPERT-MAYER'S RULE. A more precise description of the pairing energy uses techniques from the quantum theory of many-body systems. These techniques are also used to describe pairing in the theory of superconductivity.

palladium A silvery white ductile transition metal occurring in platinum ores in Canada and as the native metal in Brazil. Most palladium is obtained as a by-product in the extraction of copper and zinc. It is used in electrical relays and as a catalyst in hydrogenation processes. Hydrogen will diffuse through a hot palladium barrier.

Symbol: Pd. Melting pt.: 1552°C. Boiling pt.: 3140°C. Relative density: 12.02 (20°C). Proton number: 46. Relative atomic mass: 106.42. Electronic configuration: [Kr]4d^{10}.

parallel spins Adjacent fermions of the same type, frequently electrons, in which the spins, and hence magnetic moments, are aligned in the same direction. A ferromagnetic system (*see* magnetism) is an example of a system with parallel spins. *See* antiparallel spins.

paramagnetism *See* magnetism.

para-positronium *See* positronium.

parent A nuclide that undergoes radioactive decay to another nuclide (called the *daughter*). The term is also used for an ion that breaks up to give a smaller (daughter) ion.

parity Symbol: *P* A property of the wave function of a quantum mechanical system associated with reversing all the spatial coordinates, i.e. replacing x, y, z with $-x$, $-y$, $-z$. A wave function ψ is said to have *even parity* if it satisfies the equation $\psi(x,y,z) = \psi(-x,-y,-z)$ and *odd parity* if it satisfies the equation $\psi(x,y,z) = -\psi(-x,-y,-z)$. One has the general equation: $\psi(x,y,z) = P\psi(-x,-y,-z)$, where P is a quantum number called parity which can only have the values +1 or −1.

Until the mid 1950s it was generally assumed that parity is a conserved quantity in all processes. This assumption is correct for processes involving electromagnetic or strong interactions and imposes SELECTION RULES on such processes. However, in the mid 1950s it was found that there is *parity violation* in the weak interactions. This was discovered by analyzing the decays of particles known as the theta and tau. These two particles appeared to be exactly the same except that the decay of the theta resulted in two PIONS whereas the decay of the tau resulted in three pions. This is a problem because a single pion has odd parity, meaning that a pair of pions has even parity and a triplet of pions has odd parity. This led to the suggestion that the theta and the tau are actually the same particle, now known as the KAON, and that parity is not conserved in the weak interactions. This suggestion, made by Chen Ning YANG and Tsung Dau LEE in 1956, was confirmed by Chien-Shiung WU in 1957 in experiments on beta decay. It appears that there is a fundamental asymmetry in Nature between left and right. It is not known why parity violation occurs in weak interactions but not in strong or electromagnetic interactions.

partial derivative The derivative of a function of two or more variables with respect to one of the variables, while the other variables remain constant. The partial derivative of $z = f(x,y)$ is denoted by $\partial f/\partial x$.

partial differential equation An equation in which partial derivatives appear. All the main laws known in physics can be expressed as partial differential equations. Many other of the main equations of physics, such as the WAVE EQUATION, are partial differential equations.

particle 1. *See* elementary particle.
2. In mechanics an idealization of a real body that is regarded as having mass but no volume.

particle accelerator *See* accelerator.

particle in a box A quantum mechanical system that is sufficiently simple that it can be solved exactly but illustrates important features of quantum mechanics, such as the quantization of energy levels and the existence of ZERO-POINT ENERGY. It consists of a particle of mass m allowed to move between two walls that have the coordinates $x = 0$ and $x = L$, with the potential energy of the particle being taken to be zero between the walls and infinite outside the walls. The time-independent Schrödinger equation can be solved exactly for this problem, with the wave functions ψ_n being given by $\psi_n = (2/L)^{1/2} \sin (nMx/L)$ and the values E_n of the energies being given by $E_n = n^2 h^2/(8mL^2)$, where $n = 1, 2, 3, ...$, and h is the Planck constant. Delocalized electrons in a molecule or a metal can be modeled, in a very approximate way, by particles in a box.

parton *See* parton model.

parton model A model developed by James BJORKEN, Richard FEYNMAN, and others in the late 1960s and early 1970s, which describes high-energy electron scattering experiments of baryons such as protons and neutrons in terms of pointlike particles called *partons*. The partons are now identified as quarks and the successes of the parton model are understood in terms of ASYMPTOTIC FREEDOM in QUANTUM CHROMODYNAMICS.

pascal Symbol: Pa   The SI unit of pressure, equal to a pressure of one newton per square meter ($1\ Pa = 1\ N\ m^{-2}$).

Paschen–Back effect In spectra, a pattern of atomic energy-level splitting that occurs in a strong magnetic field. It is found when the fine-structure splitting is much less than the splitting due to the external magnetic field. This energy-level splitting pattern is very like that of the normal ZEEMAN EFFECT, with the fine-structure splitting due to spin–orbit coupling being a small correction to the main pattern. The effect was discovered by Friedrich Paschen and Ernst Back in 1912.

Paschen series *See* hydrogen spectrum.

path integral A concept, also known as 'sum over histories', which was introduced by Richard FEYNMAN to reformulate quantum mechanics. In this formulation the probability of an event that can occur in a number of ways, such as finding an electron at a particular place, is calculated by adding the sum of the probabilities of all possible ways the event could occur. Feynman was able to show how the WAVE FUNCTION description of quantum mechanics emerges in this approach and why variational principles such as FERMAT'S PRINCIPLE OF LEAST TIME and the PRINCIPLE OF LEAST ACTION can be used in classical physics.

Pauli, Wolfgang (1900–1958) Austrian-born Swiss physicist who made many important contributions to quantum mechanics and quantum field theory, particularly the PAULI EXCLUSION PRINCIPLE, which he put forward in 1925. Among his other contributions were the concept of the neutrino (1930), the spin–statistics theorem (1940), and the CPT theorem (1955). Pauli was awarded the Nobel Prize for physics in 1945 for his exclusion principle.

Pauli exclusion principle A fundamental principle of quantum mechanics that applies to fermions but not to bosons. It states that no two identical fermions in a system, such as electrons in an atom or protons in a nucleus, can have the same set of quantum numbers. Among other things, the Pauli exclusion principle is responsible for the shell structure in atoms and nuclei (*see* shell model). *See also* spin–statistics theorem.

Pauli matrices A set of 2×2 matrices that is used in the description of spin in nonrelativistic quantum mechanics, introduced by Wolfgang PAULI in 1927. Their mathematical structure is very closely related to QUATERNIONS. Pauli matrices play the same role in nonrelativistic quantum mechanics that DIRAC MATRICES play in relativistic quantum mechanics.

Pauli spinors Two-component SPINORS that are used in the description of spin in nonrelativistic quantum mechanics, introduced by Wolfgang PAULI in 1927. Pauli spinors play the same role in nonrelativistic quantum mechanics that DIRAC SPINORS play in relativistic quantum mechanics.

Pauling, Linus Carl (1901–94) American chemist who made fundamental contributions to understanding the theory of chemical bond and molecular structure in the 1930s. In particular, he developed VA-LENCE BOND THEORY. Pauling summarized his work on chemical bonding and molecular structure in his classic book *The Nature of the Chemical Bond*, the first edition of which was published in 1939. He was awarded the 1954 Nobel Prize for chemistry for his work on the chemical bond. Pauling was also a campaigner against the testing of nuclear weapons. His activities in this area led to his award of the 1962 Nobel Peace Prize.

p-block elements The block of elements in the periodic table that consists of groups 13, 14, 15, 16, 17 and 18. The outer electronic configurations of these elements has the form ns^2np^x, where x runs from 1 to 6.

p-brane *See* superstring theory.

Peebles, Phillip James Edwin (1935–) Canadian-born American physicist who has made a number of major contributions to cosmology. In particular, Peebles and Robert Dicke predicted the existence of the COSMIC MICROWAVE BACKGROUND in 1965, just before it was discovered. He also calculated the amount of light elements, such as helium and deuterium, produced in the early Universe. In the 1960s and 1970s he worked extensively on the problem of the formation of large-scale structures such as galaxies in the Universe. In 1979, together with Dicke, he pointed out a number of difficulties with the standard big-bang cosmology such as the FLATNESS PROBLEM which led to the hypothesis of INFLATION. In the 1970s, together with Jeremiah Ostriker, Peebles pointed out that the stability of galaxies requires the existence of DARK MATTER.

Peierls, Sir Rudolf Ernst (1907–95) German-born British theoretical physicist who made important contributions to several branches of theoretical physics. Together with Otto FRISCH, Peierls wrote a secret memorandum in 1940, which demonstrated the possibility of developing nuclear weapons.

Penrose, Sir Roger (1931–) British mathematician and theoretical physicist who is best known for his contributions to general relativity theory, particularly concerning black holes and singularities. In 1965 he showed that under very general conditions there is a singularity at the center of a black hole. In 1970, together with Stephen HAWKING, Penrose extended this analysis to the BIG BANG at the beginning of the Universe. In 1969 Penrose put forward the cosmic censorship conjecture that singularities must be concealed by an event horizon. Also in 1969, Penrose suggested the PENROSE PROCESS for the extraction of energy from a rotating black hole.

Since the 1960s Penrose has been concerned with the problem of how to combine general relativity theory and quantum mechanics. One of his suggestions was that space–time is a discrete structure at the quantum level built out of SPIN NETWORKS. Another of his suggestions, which he and his colleagues have developed much more extensively, is TWISTOR theory in which geometry with complex numbers features prominently.

In 1974 he devised a pattern for tiling a plane, known as *Penrose tiling*, with two types of rhombus that do not have a periodically repeating pattern. A three-dimensional version of these patterns was subsequently found to occur in Nature when QUASICRYSTALS were discovered.

Penrose process A process for extracting energy from a rotating black hole. If an object falls into a region near a rotating black hole and then splits into two, one part of the object falls into the black hole while the other part escapes with more mass–energy than the original particle. This occurs because rotational energy of the black hole is transferred to the object that escapes.

Penzias, Arno Allan (1933–) German-born American physicist who, together with Robert WILSON, discovered the cosmic microwave background in 1965. Penzias and Wilson shared the 1978 Nobel Prize for physics for this discovery.

period Symbol: T The time it takes to complete one cycle of a regularly repeated motion, such as an oscillation. It is the reciprocal of the frequency.

periodic table A table of the chemical elements in order of increasing proton number in which the elements are grouped so that the similarities of groups of elements are apparent. The periodic table was first proposed by Dmitri MENDELEEV in 1869. The periodic table is explained by quantum mechanics. In particular, the SHELL structure of electrons in atoms is dictated by the PAULI EXCLUSION PRINCIPLE. The elements are divided into vertical columns called *groups*. On going down a group, the elements all have the same electronic structure in the outer shell but the number of inner shells increases. It is customary for the alkali metals to be shown on the left of the table and the noble gases on the right. The horizontal rows of the table are called *periods*. Within any period the atoms of the elements have the same number of inner shells but the number of electrons in the outer shell increases on going from left to right. The groups in the table are numbered from 1 to 18. It is possible to divide the periodic table into four *blocks*, with each block characterized by the type of shell being filled: the S-BLOCK, the P-BLOCK, the D-BLOCK, and the F-BLOCK. The d-block includes the TRANSITION METALS. The f-block includes the ACTINIDES and the LANTHANIDES.

Perl, Martin Lewis (1927–) American physicist who discovered the TAU particle in experiments at SLAC in the mid 1970s. Perl won the 1995 Nobel Prize for physics for this discovery.

permeability Symbol: μ A quantity that gives a measure of how a substance changes the strength of a magnetic field in which it is placed. Formally, the magnetic permeability of a substance is defined as the ratio of the magnetic flux density (B) in a substance to the magnetic field strength (H) outside the substance, i.e. $\mu = B/H$. The SI unit of permeability is the henry per meter.

The *permeability of free space*, also known as the *magnetic constant*, denoted μ_0, has the value $4\pi \times 10^{-7}$ Hm^{-1} in SI units. A useful quantity is the *relative permeability*, denoted μ_r, which is the ratio of the permeability of a substance to the permeability of free space, i.e. $\mu_r = \mu/\mu_0$. The relative permeability of a substance is a number (with no units). It is related to the magnetic SUSCEPTIBILITY χ of a substance by $\mu_r = 1 + \chi$.

permeability of free space *See* permeability.

permittivity Symbol: E A quantity that is the analog of PERMEABILITY for electricity. If two electric charges Q_1 and Q_2 are in a dielectric medium a distance r apart the force F between them is given by $F = Q_1Q_2/(4\pi\varepsilon r^2)$, where ε is a constant called the permittivity of the medium. The SI unit of permittivity is the coulomb^2 newton^{-1} meter^{-2} (C^{2}N^{-1}m^{-2}) or farad meter^{-1} (Fm^{-1}).

In the case of free space the constant ε is denoted ε_0, and is called the *permittivity of free space*, also known as the *electric constant*, and has the value 8.854×10^{-12} Fm^{-1}. The *relative permittivity*, also known as the *dielectric constant*, denoted ε_r, is the ratio of the permittivity of a substance to the permittivity of free space, i.e. $\varepsilon_r = \varepsilon/\varepsilon_0$. Since this is a ratio the relative permittivity of a substance is a number (with no units). The relative permittivity is related to the electric SUSCEPTIBILITY χ by $\varepsilon_r = 1 + \chi$.

permittivity of free space *See* permittivity.

permutation group The set of all permutations of a system with identical particles. This group has been used extensively in the quantum theory of many-fermion systems.

Perrin, Jean Baptiste (1870–1942) French physicist who provided important evidence for the existence of atoms using Brownian motion. In 1895 he conducted important research on cathode rays, showing that they are electrically charged particles. This paved the way for the subsequent work of J. J. THOMSON. His most important work was published in 1909 when he was able to demonstrate the physical reality of atoms conclusively in experiments on Brownian motion. Perrin was able to show that his experimental results were in accord with the theoretical analysis of Brownian motion published by Albert EINSTEIN in 1905. This enabled Perrin to give estimates of the size of water molecules and the value of the AVOGADRO CONSTANT. This work convinced even the most sceptical of the existence of atoms.

perturbation A small influence on a physical system that modifies the behavior of the system in a minor way. For example, in celestial mechanics the gravitational interactions between planets can be treated as perturbations of the gravitational interaction between a planet and the Sun. This means that the orbits of the planets are slightly different from the orbits determined only by the interaction between the planet and the Sun. In an analogous way in atoms, interactions between electrons are perturbations of the behavior of electrons determined by the interaction between an electron and the nucleus. Both in classical physics and quantum mechanics, the effects of perturbations are calculated using PERTURBATION THEORY.

perturbation theory A technique used in both classical physics and quantum mechanics to calculate the effect of perturbations on a system. The perturbed system is divided into a part that is exactly solvable and a small perturbation. The effects of the small term are calculated using an infinite series, which in general is an ASYMPTOTIC SERIES. This series is called a *perturbation series*, with each term in the series being a 'correction term' to the solutions of the exactly soluble system.

In celestial mechanics, perturbation theory is used to calculate the modifications to the exactly elliptical interactions between planets. In general relativity theory, perturbation theory is used to calculate the effect of density fluctuations on a homogeneous expanding Universe. In quantum mechanics, perturbation theory is used to calculate the energy levels and properties of atoms, molecules, and solids.

Pfund series *See* hydrogen spectrum.

phase 1. A homogeneous part of a mixture distinguished from other parts by boundaries. A mixture of ice and water has two phases. A mixture of ice crystals and salt crystals also has two phases. A solution of salt in water is a single phase.
2. The stage in a cycle that a wave (or other periodic system) has reached at a particular time (taken from some reference point). Two waves are *in phase* if their maxima and minima coincide.

For a simple wave represented by the equation

$$y = a\sin2\pi(ft - x/\lambda)$$

the phase of the wave is the expression $2\pi(ft - x/\lambda)$. The *phase difference* between two points distances x_1 and x_2 from the origin is $2\pi(x_1 - x_2)/\lambda$.

A more general equation for a progressive wave is

$$y = a\sin2\pi(ft - x/\lambda - \phi)$$

Here, ϕ is the *phase constant* — the phase when t and x are zero. Two waves that are out of phase have different phase constants (they 'start' at different stages at the origin). The phase difference is $|\phi_1 - \phi_2|$. It is equal to $2\pi x/\lambda$, where x is the distance between corresponding points on the two waves. It is the *phase angle* between the two waves; the angle between two rotating vectors (phasors) representing the waves.

phase space A $2n$-dimensional space that can be used to define the state of a system with n degrees of freedom. The coordinates of phase space are $(q_1, q_2, \dots q_n, p_1, p_2, \dots p_n)$, where $q_1, q_2, \dots q_n$ are the coordinates and $p_1, p_2, \dots p_n$ are the corresponding momenta. Each point in phase space represents a state of the system. As time evolves a representative point in phase space traces out a curve. The concept of phase space is useful in CHAOS THEORY.

phase transition A change in a physical system from one phase (state) into another. Examples of phase transitions include

melting, freezing, and boiling. Other examples of phase transitions include the transition from ferromagnetism to paramagnetism at the Curie temperature and the transition from superconductivity to normal metallic conduction. Phase transitions can be brought about by changing variables such as temperature and pressure. It is almost invariably the case that a phase transition caused by the temperature being reduced results in a state with a lower symmetry than the high-temperature phase, i.e. there is BROKEN SYMMETRY in the low-temperature phase.

It is thought that in the EARLY UNIVERSE there was a sequence of phase transitions as the Universe expanded and cooled, resulting in the broken symmetry pattern of elementary particles now found in Nature.

Phase transitions are divided into *first-order transitions* and *second-order transitions*. In a first-order transition there is nonzero *latent heat*, i.e. heat is absorbed or released during the phase transition. In a *second-order transition* the latent heat is zero. *See also* Ising model.

Phillips, William (1948–) American physicist who worked on laser cooling. He was able to cool beams of atoms down to the very low temperature of 17 millikelvins in the late 1980s. Phillips shared the 1997 Nobel Prize for physics with Steven CHU and Claude COHEN-TANNOUDJI.

phonon A quantum of the energy of vibration of a crystal lattice. The theory of lattice vibrations is analogous in many ways to the theory of the electromagnetic field, with phonons being the equivalent of quanta of light, i.e. photons. The concept of phonons is very useful in the theories of thermal and electrical conductivity of solids.

phosphorescence *See* luminescence.

phosphorus A reactive solid nonmetallic element; the second element in group 15 of the periodic table. Phosphorus is widespread throughout the world; economic sources are phosphate rock ($Ca_3(PO_4)_2$) and the apatites, variously occurring as both fluoroapatite ($3Ca_3(PO_4)_2.CaF_2$) and as chloroapatite ($3Ca_3(PO_4)_2CaCl_2$). Guano formed from the skeletal phosphate of fish in sea-bird droppings is also an important source of phosphorus. The largest amounts of phosphorus compounds produced are used as fertilizers, with the detergents industry producing increasingly large tonnages of phosphates. Phosphorus is an essential constituent of living tissue and bones, and it plays a very important part in metabolic processes and muscle action.

Symbol: P. Melting pt.: 44.1°C (white) 410°C (red under pressure). Boiling pt.: 280.5°C. Relative density: 1.82 (white) 2.2 (red) 2.69 (black) (all at 20°C). Proton number: 15. Relative atomic mass: 30.973762. Electronic configuration: [Ne]$3s^23p^3$.

photino *See* supersymmetry.

photochemistry The branch of chemistry concerned with chemical reactions caused by light or ultraviolet radiation.

photodisintegration The ejection of particles from atomic nuclei on absorption of electromagnetic radiation. Most commonly a neutron is ejected.

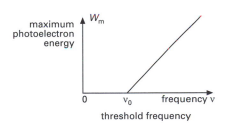

The photoelectric effect

photoelectric effect The ejection of electrons from a substance by electromagnetic radiation (usually ultraviolet radiation). The number of such electrons (known as *photoelectrons*) is proportional to the intensity of radiation at a given frequency. There is a threshold frequency below which electrons are emitted. The value of this frequency depends on the substance. The photoelectric effect was very

important in the development of quantum mechanics since it led Albert EINSTEIN to postulate in 1905 that light has particle-like behavior as well as wavelike behavior.

Einstein was able to explain the experimental results concerning the photoelectric effect by postulating that light exists in the form of particle-like entities called PHOTONS with the energy E of a photon being given by $E = h\nu$, where h is the Planck constant and ν is the frequency of the light. This leads to Einstein's equation that the maximum kinetic energy E_m of the ejected electrons is given by $E_m = h\nu - \phi$, where ϕ is the minimum energy needed to remove an electron. When the emitted electrons are valence electrons from solids ϕ is called the *work function*. The minimum frequency ν_0, known as the *threshold frequency*, at which an electron is ejected is given by $h\nu_0 = \phi$. The effect occurs in liquids and gases as well as in solids. *See also* photoionization.

photoelectron *See* photoelectric effect.

photoelectron spectroscopy A technique for determining the structure of molecules which makes use of the photoelectric effect. A sample of the material (usually in the form of a gas) is irradiated with a beam of monochromatic ultraviolet radiation. Photoelectrons are produced, with their energies being in accord with Einstein's equation for the photoelectric effect. The photoelectrons are then deflected by either electric or magnetic fields to give an energy spectrum. This enables the ionization potentials, and hence the orbital energies, of the molecule to be obtained.

Photoelectron spectromicroscopy (*PESM*) is used to investigate surfaces by bombarding the surface with ionizing radiation and then focusing the ejected electrons.

In *electron spectroscopy for chemical analysis* (*ESCA*) x-rays are used rather than ultraviolet radiation. This means that the ejected electrons are from the inner shells of atoms. The energies of these shells may have slight differences depending on the configuration of the outer shells (i.e. on the chemical bonding).

photoionization The ionization of an atom or molecule by electromagnetic radiation. As with the photoelectric effect it is necessary for the photon to have a minimum frequency determined by $h\nu_0 = I$, where I is the ionization potential.

photomultiplier A device in which electrons originally emitted from a photocathode initiate a cascade of electrons by secondary emission in an electron multiplier. It is much more sensitive as a radiation detector than a single photoelectric cell.

photon A zero-rest-mass spin-one particle that is the quantum of the electromagnetic field. In a vacuum all photons move at the speed of light. The energy E of a photon is given by $E = h\nu$, where h is the Planck constant and ν is the frequency of the electromagnetic radiation. Photons are needed to explain phenomena such as the Compton effect and the photoelectric effect, in which light, and other electromagnetic radiation behaves as if it were composed of particles rather than waves.

photonics The study of devices analogous to electronic devices but using photons instead of electrons. Photonic devices depend on the transmission, modulation, reflection, refraction, amplification, detection, and guidance of light. The study of lasers can be regarded as a branch of photonics. Photonics is closely related to *optoelectronics*, i.e. the branch of electronics concerned with generating, transmitting, modulating, and detecting electromagnetic radiation, particularly light. There are many important applications of photonics in telecommunications.

photonuclear reaction A nuclear reaction that is induced by a photon (gamma ray).

physical optics *See* optics.

pi-bond *See* orbital.

pick-up reaction *See* nuclear reaction.

pilot wave A wave that is postulated to guide a particle such as an electron or a photon in a version of quantum mechanics put forward by Louis DE BROGLIE and extended by David BOHM. The idea of pilot waves was put forward in an attempt to explain the apparent paradoxes in wave-particle duality. The idea is that electrons are particles but their motion is guided by a wave – the pilot wave. There is no experimental evidence for the existence of pilot waves.

pi-meson *See* pion.

pion (pi-meson) Symbol: π A type of spin-zero meson that is involved in mediating the strong interactions between nucleons. There are three types of pion: positively charged (π^+), electrically neutral (π^0), and negatively charged (π^-). The charged pions have a rest mass of about 140 MeV and decay into a muon and a neutrino, with a mean-life of 2.6×10^{-8} s. Neutral pions have a rest mass of about 135 MeV and decay into two photons, with a mean-life of 8×10^{-17} s. The π^+ is made up of an up quark and an anti-down quark. The π^- is made up of a down quark and an anti-up quark. The π^0 is made up of an up quark and an anti-up quark.

Planck, Max Karl Ernst Ludwig (1858–1947) German physicist who initiated the quantum theory in 1900 when he postulated that the problem of BLACK-BODY RADIATION could be solved by assuming that electromagnetic radiation is absorbed and emitted in discrete amounts called *quanta*. Planck was led to this problem by consideration of THERMODYNAMICS, a subject in which he was a leading expert. Planck was also one of the first physicists to appreciate the importance of special relativity theory.

Planck constant Symbol: h A fundamental constant of Nature which relates the energy E of a quantum of electromagnetic radiation (i.e. a PHOTON) to the frequency of the radiation by the equation $E = h\nu$. The presence of the Planck constant in a calculation always indicates that quantum mechanics is involved. The value of the Planck constant is 6.626×10^{-34} Js. The very small size of this constant explains why quantum-mechanical phenomena are not usually directly apparent at the macroscopic level. *See also* Dirac constant.

Planck length The length scale at which classical descriptions of gravity and space–time break down and it is necessary to take quantum mechanics into account. It is the combination of the constants G, $\hbar$, and c which give a unit of length, where G is the gravitational constant, $\hbar$ is the DIRAC CONSTANT, and c is the speed of light in a vacuum. It is given by $(G\hbar/c^3)^{1/2}$ and is about 10^{-35} m, i.e. about 10^{-20} times smaller than the size of a proton. It is widely thought that quantum gravity and space–time is described by discrete structures such as SPACE–TIME FOAM and SPIN NETWORKS at the Planck length.

Planck mass The mass of a particle that has a COMPTON WAVELENGTH equal to the PLANCK LENGTH. The Planck mass is given by $(\hbar c/G)^{1/2}$, where $\hbar$ is the DIRAC CONSTANT, c is the speed of light in a vacuum, and G is the gravitational constant. It has a value of 10^{-5} g, which, by $E = mc^2$, is equivalent to an energy of 10^{19} GeV. It is thought that quantum gravity is important at such energies. Although such energies cannot remotely be attained using particle accelerators it is thought that they did occur in the EARLY UNIVERSE.

Planck's radiation law The law that gives the distribution of energy radiated by a BLACK BODY, i.e. a body that absorbs all the radiation that falls on it. It states that the energy E_λ emitted per unit area per unit time per unit wavelength at the wavelength λ from a black body at the absolute temperature T is given by: $E_\lambda = 2\pi hc^2\lambda^{-5}/[\exp(hc/kt\lambda)-1]$, where h is the Planck constant, c is the speed of light in a vacuum, and k is the Boltzmann constant.

This law was first suggested by Max PLANCK in 1900. Planck was only able to obtain agreement between theory and experiment by postulating that energy exists in discrete amounts called *quanta*. It was historically important because it was the start of the development of quantum mechanics. The necessity of using quantum theory is made clear by the fact that the analysis of the problem of black-body radiation in terms of classical statistical mechanics leads to the *ultraviolet catastrophe* predicted by the RAYLEIGH–JEANS LAW. Planck's radiation law was subsequently derived from quantum statistical mechanics by Satyendra Nath BOSE in 1924.

Planck time The time it takes light to move across a distance equal to the PLANCK LENGTH, i.e. the Planck length divided by c, the speed of light in a vacuum. This is given by $(G\hbar/c^5)^{1/2}$, where G is the gravitational constant and $\hbar$ is the DIRAC CONSTANT. It has a value of about 10^{-43} s. It is necessary to have a theory of quantum gravity to describe the EARLY UNIVERSE for a time up to the Planck time following the big bang.

Planck units A system of units in which length, mass, and time are expressed as multiples of the PLANCK LENGTH, the PLANCK MASS, and the PLANCK TIME. These are natural units to use in theories of quantum gravity.

plasma A state of matter consisting of free electrons and positively charged ions. Plasmas occur in discharge tubes, stars such as the Sun, and in thermonuclear reactors. A plasma is sometimes described as a fourth state of matter. The study of plasmas, known as *plasma physics*, is important both in studying stars and in practical experiments on nuclear fusion.

plasma oscillation An oscillation that occurs in a plasma. There are several types. A simple one, analyzed by Irving Langmuir in 1928, consists of electrons moving to screen the positively charged ions, overshooting, being pulled back, overshooting again, etc. This results in the electric charge density oscillating with a simple harmonic motion, with the frequency of this motion being called the *Langmuir frequency*.

plasmon A collective excitation for the quantized plasma oscillations of the electrons in a metal.

platinum A silvery-white malleable ductile transition metal. It occurs naturally in Australia and Canada, either free or in association with other platinum metals. Platinum is used as a catalyst for ammonia oxidation (to make nitric acid) and in catalytic converters. It is also used in jewelry.
Symbol: Pt. Melting pt.: 1772°C. Boiling pt.: 3830 ± 100°C. Relative density: 21.45 (20°C). Proton number: 78. Relative atomic mass: 195.08. Electronic configuration: $[Xe]4f^{14}5d^96s^1$.

plutonium A radioactive silvery element of the actinoid series of metals. It is a transuranic element found on Earth only in minute quantities in uranium ores but readily obtained, as ^{239}Pu, by neutron bombardment of natural uranium. The readily fissionable ^{239}Pu is a major nuclear fuel and nuclear explosive. Plutonium is highly toxic because of its radioactivity; in the body it accumulates in bone.
Symbol: Pu. Melting pt.: 641°C. Boiling pt.: 3232°C. Relative density: 19.84 (25°C). Proton number: 94. Most stable isotope: ^{244}Pu (half-life 8.2×10^7 years). Electronic configuration: $[Rn]5f^67s^2$.

Poincaré, Jules Henri (1854–1912) French mathematician who made many contributions to several branches of mathematical physics. In the late nineteenth century his work on celestial mechanics, prompted by his investigations on the stability of the Solar System, paved the way for CHAOS THEORY. Independently of Albert EINSTEIN, Poincaré derived many of the results, and some of the concepts, of special relativity theory. He also wrote some influential works on the philosophy of science.

Poincaré group The *proper Poincaré group* is the group formed by the combination of the LORENTZ GROUP and translations in space and time. The *unrestricted Poincaré group* is the group formed by the combination of the proper Poincaré group and inversions in space and time. *See also* Lorentz group.

Poincaré stresses Nonelectrostatic forces that were postulated by Henri Poincaré in 1906 to give stability to a model of the ELECTRON in which the electron has a definite size (i.e. is not a point). It is now thought that the description of any structure of the electron requires quantum theory rather than classical physics.

point group The set of all symmetry operations that can be applied to the arrangement of objects around a point that leaves the pattern unchanged. An example of a point group is the set of all symmetry operations of a molecule.

Poisson, Simeon-Denis (1781–1840) French mathematician and physicist who made important contributions to several branches of physics. In particular, he extended the concept of potential from gravitation to electricity and magnetism. He also contributed to classical mechanics, optics, and the theory of heat conduction and was one of the founders of the theory of elasticity. He is also well known for his work on probability theory.

Poisson equation An equation in POTENTIAL THEORY which relates the electric potential V to the electric charge density ρ. It has the form $\nabla^2 V = -\rho/\varepsilon$, where ∇^2 is the Laplace operator (*see* Laplace equation) and ε is the PERMITTIVITY. There is also a version of the Poisson equation for the gravitational potential. In the case when $\rho = 0$ the Poisson equation reduces to the Laplace equation.

polar coordinates A system of coordinates used in analytic geometry which is used to locate a point P in two or three-dimensional space. In two dimensions there are two polar coordinates: the distance r from the origin to P and the angle θ between the x-axis and the *radius vector* OP. In three dimensions P can either be regarded as lying on the surface of a cylinder, giving *cylindrical polar coordinates*, or on the surface of a sphere, giving *spherical* *polar coordinates.* In two dimensions polar coordinates are denoted (r, θ). Cylindrical polar coordinates are denoted (r,θ,z). Spherical polar coordinates are denoted (r, θ, ϕ).

polarizability Symbol: α A measure of how a molecule responds to an external electric field. The field changes the distribution of electrons, producing a DIPOLE MOMENT. The polarizability α is the constant of proportionality between the induced electric dipole moment μ and the electric field strength E: i.e. $\mu = \alpha E$. In principle, it is possible to calculate the polarizability of a molecule from first principles using quantum mechanics but it is frequently convenient to regard it as an empirical parameter.

polarization   The confinement of the vibrations of a transverse wave (usually a light wave) to one direction. In *plane polarized light* all the electric oscillations are parallel to each other. It is possible to plane polarize light by reflection or by passing it through certain substances such as Polaroid. If light is polarized by one polarizer then it is not transmitted by a second polarizer set to polarize at right angles to the first.

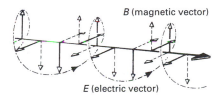

B (magnetic vector)

E (electric vector)

Circular polarization

Another type of polarization gives rise to *circularly polarized light* in which the tip of the electric vector of the wave describes a circular helix about the direction of propagation of the wave with a frequency equal to the frequency of the wave.

A more general type of polarization gives rise to *elliptically polarized light* in which the electric vector rotates about the direction of propagation but, unlike circu-

larly polarized light in which the magnitude of the vector remains constant, the amplitude changes.

Polaroid   A synthetic doubly refracting material that plane-polarizes light that passes through it. It is made from a plastic sheet which is made birefringent by high straining to align the molescules.

polar vector   A vector that changes its sign when the inversion operator is applied to the coordinate system (x,y,z), i.e. $(x,y,z) \rightarrow (-x,-y,-z)$.

polonium   A radioactive metallic element belonging to group 16 of the periodic table. It occurs in very minute quantities in uranium ores. Over 30 radioisotopes are known, nearly all alpha-particle emitters. Polonium is a volatile metal and evaporates with time. It is also strongly radioactive; a quantity of polonium quickly reaches a temperature of a few hundred °C because of the alpha emission. For this reason it has been used as a lightweight heat supply in space satellites.

Symbol: Po. Melting pt.: 254°C. Boiling pt.: 962°C. Relative density: 9.32 (20°C). Proton number: 84; stablest isotope ^{209}Po (half-life 102 years). Electronic configuration: $[Xe]4f^{14}5d^{10}6s^26p^4$.

polyatomic molecule   A molecule with more than two atoms. Simple examples of polyatomic molecules are water (H_2O), ammonia (NH_3), methane (CH_4), and benzene (C_6H_6).

polymer   A substance in which there is a basic unit, called the *monomer*, which is repeated many times. Some polymers occur naturally and some are synthesized in the laboratory. Statistical mechanics is used extensively in the theoretical analysis of the physical properties of polymers. Some polymers are electrically conducting because of SOLITONS.

Pople, John Anthony (1925–   ) British-born theoretical chemist who has been a leading figure in the development of computational methods in the electronic structure of molecules. This work led to Pople sharing the 1998 Nobel Prize for chemistry with Walter KOHN.

p-orbital *See* orbital.

positron Symbol: e⁺ The positively charged particle that is the antiparticle of the electron. Positrons occur naturally in cosmic rays and in certain types of beta decay. When they encounter an electron they are rapidly annihilated. *See also* annihilation; elementary particles; pair creation; positronium.

positronium A short lived bound state of an electron and a positron. There are two forms of positronium. In *orthopositronium* the spins of the electron and positron are parallel. This form decays into three photons, with this process having a lifetime of about 1.5×10^{-7} s. In *parapositronium* the spins of the electron and positron are antiparallel. This decays into two photons, with the shorter lifetime of about 10^{-10} s.

potassium A soft reactive metal; the third member of the alkali metals (group 1 of the periodic table). Potassium accounts for 2.4% of the lithosphere and occurs in large salt deposits such as carnallite ($KCl.MgCl_2.6H_2O$) and arcanite (K_2SO_4). Large amounts also occur in mineral forms that are not of much use for recovery of potassium, e.g., orthoclase, $K_2Al_2Si_6O_{10}$.

Symbol: K. Melting pt.: 63.65°C. Boiling pt.: 774°C. Relative density: 0.862 (20°C). Proton number: 19. Relative atomic mass: 39.0983. Electronic configuration: [Ar]4s¹.

potassium–argon dating A method of radioactive dating used to estimate the age of certain rocks such as micas and feldspar. It depends on the decay of potassium-40 to argon-40 (with a half-life of 1.27×10^{10} years). In calculating the ages of rocks in this way it is assumed that all the argon-40 present in the mineral is due to the decay of potassium-40 and that all the argon-40 produced accumulates within the mineral.

potential barrier A region in space in which there is a higher potential than that for an adjacent region, such that this potential difference acts as a barrier for objects in the lower-potential region. For example, electrons in a metal are in a region of lower electrical potential inside the metal than in the space outside. There is a potential barrier at the surface of the metal. In classical theory, the electrons cannot escape from the metal unless they have sufficient energy to overcome the barrier (*see* photoelectric effect). This would be the case even if the barrier were very narrow. In quantum mechanics there is a finite possibility that particles can pass through a potential barrier if the barrier has finite width (*see* field emission; tuneling).

potential energy The energy that a body has because of its position. For example, an object has more gravitational potential energy at the top of a building than at the bottom. Other examples occur for objects in electric and magnetic fields. The energy stored in a compressed spring is another example of potential energy.

potential energy curve A graph of potential energy against internuclear distance for two atoms. If the two atoms form a chemical bond there is a minimum in the curve corresponding to the equilibrium internuclear distance R_e. The difference between the potential energy at R_e and the value of the potential energy at infinite separation is called the *equilibrium dissociation energy* D_e. When nuclear motion is taken into account the *bond dissociation energy* D_0 is less than D_e because of the ZERO-POINT ENERGY of vibrational motion.

potential energy surface A generalization of a potential energy curve to a surface in which the potential energy is calculated as a function of the distances between atoms for a set of atoms in chosen arrangements. Potential energy surfaces are used to analyze the dynamics of chemical reactions since they can be regarded as maps of the reacting systems. Such surfaces are calculated using quantum mechanics, a possibility first pointed out by Fritz LONDON in

1928 and developed by Henry EYRING in the 1930s.

potential theory   The branch of mathematical physics which is concerned with analyzing gravitational, electric, and magnetic fields in terms of their potentials. The LAPLACE EQUATION and the POISSON EQUATION are the key equations in potential theory.

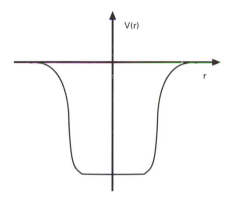

Potential well in a nucleus

potential well   An attractive potential in which bound states of particles can be formed. Examples of potential wells include the Coulomb potential, the harmonic oscillator potential, and the PARTICLE IN A BOX in which the walls have a finite height. Some potential wall problems in quantum mechanics are simple enough to be solved exactly but are still good approximations to real physical systems.

Powell, Cecil Frank (1903–69)   British physicist who studied elementary particles by the size and the paths of the tracks they leave in photographic emulsions. In 1947 he discovered the PION from the tracks of cosmic rays. He was awarded the 1950 Nobel Prize for physics for this discovery.

praseodymium   A soft ductile malleable silvery element of the lanthanoid series of metals. It occurs in association with other lanthanoids. Praseodymium is used in several alloys, as a catalyst, and in enamel and yellow glass for eye protection.

Symbol: Pr. Melting pt.: 931°C. Boiling pt.: 3512°C. Relative density: 6.773 (20°C). Proton number: 59. Relative atomic mass: 140.91. Electronic configuration: $[Xe]4f^36s^2$.

precession   A type of motion that can occur in a rotating body in which the axis of rotation itself rotates around another axis at an angle to it. Precession occurs when a torque is applied to it. For example, in a spinning top in which the axis is not exactly vertical there is a torque on the top because of gravity. Precession can also occur at the subatomic level (*see* Larmor precession).

preon   A hypothetical particle that has been postulated as a fundamental 'building block' for quarks and leptons. There is no experimental evidence for preons and nobody has succeeded in constructing a successful theory of quarks and leptons in terms of preons.

Prigogine, Ilya (1917–)   Russian-born Belgian chemist who extended the application of thermodynamics to systems that are not in equilibrium. There are important applications of his ideas in biology. Prigogine won the 1977 Nobel Prize for chemistry for this work.

primordial nucleosynthesis   *See* nucleosynthesis.

principle of least action   The principle that the ACTION, i.e. the integral with respect to time of the kinetic energy minus the potential energy, in the actual trajectory of a body is always less than for any other trajectory. This idea, which is equivalent to Newton's laws of motion, can be regarded as the fundamental principle of classical mechanics.

principle of superposition   In general, if $\psi_1, \psi_2, \ldots \psi_n$ are individual solutions of a linear equation then any linear combination of them, i.e. any sum of the form $\psi =$

$\Sigma a_i \psi_i$, where the a_i are constants, will also be a solution. In particular, the principle of superposition applies to the SCHRÖDINGER EQUATION and the WAVE EQUATION.

probability The likelihood (chance) that a particular event will occur. If an experiment has n equally likely outcomes and an event E can occur in a ways then the probability that the event E occurs is a/n. For example, if a die is thrown there are three possible ways in which an odd number can occur out of the six possible outcomes. Thus, the probability of an odd number being thrown is $3/6 = 1/2$.

probability amplitude The amplitude of the wave function in the Schrödinger equation. The name arises because of the interpretation of the wave function in terms of probability. This interpretation leads to the wave function being referred to as a *probability wave*.

***p*-process** *See* nucleosynthesis.

Prokhorov, Aleksandr Mikhaylovich (1916–2000) Russian physicist who did fundamental work in microwave spectroscopy with Nikolai BASOV, which paved the way for masers, and subsequently lasers. He shared the 1964 Nobel Prize for physics with Basov and Charles TOWNES.

promethium A radioactive element of the lanthanoid series of metals. It is not found naturally on Earth but can be produced artificially by the fission of uranium. It is used in some miniature batteries.
Symbol: Pm. Melting pt.: 1168°C. Boiling pt.: 2730°C (approx.). Relative density: 7.22 (20°C). Proton number: 61; stablest isotope ^{145}Pm (half-life 18 years). Electronic configuration: [Xe]$4f^5 6s^2$.

proportional counter A counter for ionizing radiation in which the potential difference applied is high enough for multiplication of ions, so that the height of the pulse is proportional to the number of ions produced by the particle, and thus to its energy loss. A counter used in this way is said to be in the *proportional region*.

protactinium A toxic radioactive element of the actinoid series of metals. It occurs in minute quantities in uranium ores as a radioactive decay product of actinium.
Symbol: Pa. Melting pt.: 1840°C. Boiling pt.: 4000°C (approx.). Relative density: 15.4 (calc.). Proton number: 91. Most stable isotope: ^{231}Pa (half-life 32 500 years). Electronic configuration: [Rn]$5f^2 6d^1 7s^2$.

proton A stable, positively charged spin-1/2 elementary particle. The size of its charge is equal to but opposite that of an electron. It has a rest mass of 1.672614×10^{-27} kg, which is about 1836 times bigger than that of the electron. Protons are nucleons that are found in all nuclei. A proton is made up of two up quarks and a down quark.

proton decay A decay process of a proton into a positron and a pion ($p \rightarrow e^+ + \pi^0$) predicted by GRAND UNIFIED THEORIES (GUTS). Such theories predict that a proton has an average lifetime of about 10^{32} years. This means that if one has about 10^{32} protons then one of the protons should decay each year. Careful experiments have conclusively shown that this does not happen, and this rules out the simplest types of GUT. However when GUTs are combined with SUPERSYMMETRY the predicted lifetime of the proton is much longer, and the current experiments are unable to rule out this possibility.

proton number (**atomic number**) Symbol: Z The number of protons in the nucleus of an atom.

proton–proton reaction A series of nuclear fusion reactions that converts hydrogen into helium, with the release of a great deal of energy. This process is thought to be the main source of energy production in stars such as the Sun. The proton–proton reaction was proposed as the source of solar energy by Hans BETHE and Charles Critchfield in 1938.
The reaction starts when two protons fuse to form a deuteron, with the emission of a positron and a neutrino.
$$^1H + {}^1H \rightarrow {}^2H + e^+ + \nu.$$

Then, another proton fuses with the deuteron to make a helium-3 nucleus:

$$^2H + {}^1H \rightarrow {}^3He.$$

In the final step of the process two helium-3 nuclei react to give one helium-4 nucleus, with two protons being released:

$$^3He + {}^3He \rightarrow {}^4He + 2^1H.$$

The last step occurs for about 95 per cent of the helium-3 nuclei. The remainder of the helium-3 nuclei take part in more complicated nuclear reactions that also result in the production of helium-4.

Prout's hypothesis The hypothesis put forward by William Prout in 1815 that the atomic weight of all atoms is an exact multiple of the atomic weight of hydrogen. Subsequent experiments in the nineteenth century showed that there are some exceptions, notably chlorine. This difficulty was resolved by the discovery of isotopes. Nevertheless, the hypothesis was an early insight into the fact that elements have more fundamental 'building blocks'.

pseudo-scalar A scalar quantity that changes its sign when there is a transformation in the coordinate system from (x, y, z) to $(-x,-y,-z)$. The SCALAR PRODUCT of an axial vector and a polar vector gives rise to a pseudo-scalar.

pseudo-vector *See* axial vector.

psi particle *See* J/psi particle.

pulsar A celestial source of electromagnetic radiation that is emitted in regular bursts. Pulsars are confidently thought to be rotating neutron stars with large magnetic fields. This strong magnetic field concentrates electronically charged near the two magnetic poles of the neutron star, meaning that electromagnetic radiation is emitted in two directional beams (like a lighthouse) as the neutron star rotates. Pulsars were first discovered in 1967 by Jocelyn Bell.

Hundreds of pulsars are now known, with most having periods of about one second. Pulsars have strong gravitational fields and is necessary to use general relativity theory to describe their dynamics. They provide information about nuclear physics and astrophysics as well as general relativity theory. *See also* binary pulsar.

Purcell, Edward Mills (1912–) American physicist who discovered nuclear magnetic resonance in 1946, independently of Felix BLOCH. Bloch and Purcell shared the 1952 Nobel Prize for physics for this discovery.

QCD *See* quantum chromodynamics.

QED *See* quantum electrodynamics.

QFD *See* quantum flavordynamics.

quadrupole A set of four electric charges that has no net charge and no net dipole moment.

quadrupole moment *See* multipole.

quadrupole radiation The electromagnetic radiation given off in a transition involving a quadrupole. These transitions are much weaker than those involving dipoles but transitions can occur that are forbidden in dipole-moment transitions.

quantization The process of constructing a quantum theory for a system, starting from the corresponding classical theory. Quantization is achieved using a formalism, such as MATRIX MECHANICS or WAVE MECHANICS. The process of quantization leads to the conclusion that quantities such as the angular momentum and energy of electrons in atoms have a set of discrete values rather than varying continuously.

quantization of space–time *See* quantum gravity.

quantum (*pl.* **quanta**) The minimum amount by which certain quantities, such as the angular momentum or energy of a system, can change. This means that such quantities change by integer multiples of the relevant quantum rather than continuously. A quantum can be regarded as an excitation of a wave or field, thus giving a particle-like behavior to the wave or field. For example, the quantum of the electro-

magnetic field is the PHOTON and the quantum of the gravitational field is the GRAVITON.

quantum biochemistry The application of QUANTUM CHEMISTRY to molecules of biological interest such as amino acids, proteins, and DNA.

quantum chaos The quantum mechanical description of systems that can exhibit chaos in the classical limit (*see* chaos theory). Albert EINSTEIN initiated this subject in 1917 when he showed that the quantization conditions of the BOHR–SOMMERFELD MODEL need to be modified in classical systems that can exhibit chaos. Quantum chaos has been extensively investigated since the 1970s but some basic issues have not been completely clarified. It does not appear to be the case that systems that exhibit chaos in classical mechanics necessarily do so in quantum mechanics.

quantum chemistry The application of quantum mechanics to the electronic structure of atoms and molecules, and hence to explanations of chemical bonds and chemical reactions. Quantum chemistry was initiated in the second half of the 1920s, soon after the discovery of quantum mechanics, and has been developed very extensively since then.

quantum chromodynamics (QCD) The theory that describes the strong interactions of quarks and gluons. Quantum chromodynamics is a non-Abelian gauge theory analogous to QUANTUM ELECTRODYNAMICS (QED), with quarks being analogous to electrons, gluons being analogous to photons, and a property of quarks and gluons called *color charge* being analogous to electric charge. A major difference between QCD and QED is that in QCD the gauge group is SU(3), i.e. a non-Abelian group, whereas in QED the gauge group is U(1), i.e. an Abelian group. This distinction makes QCD much more complicated than QED. In QCD the gauge symmetry is not a BROKEN SYMMETRY.

An important feature of QCD is AS-YMPTOTIC FREEDOM. This feature means that at very high energies (and hence, short distances) the interactions between quarks tend to zero as the distance between them goes to zero. This explains why high-energy scattering experiments in the 1960s indicated the presence of particles, initially called PARTONS and subsequently identified as quarks, moving freely inside nucleons. The asymptotic freedom of QCD means that perturbation theory can be used to describe the interactions of quarks and gluons at high energies. This enables corrections to the PARTON MODEL to be calculated in a systematic way using perturbation theory.

quantum computer A computing machine based on the principles of quantum mechanics, in particular one that uses EN-TANGLEMENT to perform a large number of computations simultaneously. A quantum computer would be much more powerful than classical computers. A major difficulty which has, so far, prevented the construction of a practical quantum computer is that quantum states are extremely delicate entities due to DECOHERENCE.

quantum cosmology The application of quantum mechanics to the origin of the Universe and the EARLY UNIVERSE. Quantum cosmology has the potential to solve problems in cosmology that the big-bang theory, based on general relativity theory, cannot. For example, quantum cosmology should be able to explain why the big bang occurred and avoid the singularities of general relativity theory. Ultimately, quantum cosmology requires a theory of QUANTUM GRAVITY to be complete. Nevertheless, important progress in quantum cosmology has been obtained using the STANDARD MODEL of particle physics, GRAND UNIFIED

THEORIES, INFLATION, and, more speculatively, other models used in particle physics and quantum gravity.

quantum cryptography Cryptography that makes use of the principles of quantum mechanics. In particular, it has been suggested that if a message was sent in the form of quantum states, such as the polarization states of photons, then it would be possible to detect an eavesdropper since the act of eavesdropping alters the quantum states of systems. Quantum cryptography has been investigated in the laboratory but has not yet found practical applications.

quantum dot A system of a small number of electrons confined to a small area in an electronic device. Such a system can be regarded as an artificial two-dimensional atom that is hundreds of times bigger than a real atom. Quantum dots have a great deal of potential both for investigating the fundamentals of quantum mechanics and in the development of new devices.

quantum electrodynamics (QED) The quantum theory of electromagnetic interactions, i.e. the interactions of electrically charged particles such as electrons and positrons with each other and with electromagnetic radiation such as light. In QED the interaction between two charged particles is viewed as being mediated by the exchange of photons.

Calculations of quantities of interest in QED can be performed using PERTURBA-TION THEORY. Although such calculations gave infinities when they were first performed in the late 1920s and early 1930s it was found in the late 1940s that these infinities could be removed by the process of RENORMALIZATION. Quantities of interest could then be calculated very accurately. Richard FEYNMAN, who was one of the major pioneers of renormalization and its application to QED, observed that the accuracy of the calculations is comparable to calculating the distance between New York and Los Angeles to within the width of a human hair. FEYNMAN DIAGRAMS are very useful both in performing the calculations and picturing the processes involved. Quantities such as the ANOMALOUS MAGNETIC MOMENT and the LAMB SHIFT can be calculated in this way. The agreement of these calculations with very precise experiments demonstrates the physical reality of fluctuations and polarization in the vacuum state.

QED is a gauge theory in which the gauge group is the Abelian group U(1). In spite of the spectacular success of the agreement between theory and experiment it has not been established that QED is a mathematically consistent theory because of the possible difficulty of the LANDAU GHOST at extremely high energies.

quantum electronics The branch of electronics concerned with the quantum-mechanical properties of electrons and light.

quantum field theory The quantum theory of a system with an infinite number of degrees of freedom. It is frequently convenient to represent a nonrelativistic system with a large number of particles, such as the ions and electrons in a metal or the nucleons in a nucleus, as a quantum field theory. *Relativistic quantum field theory* combines field theory, quantum mechanics, and special relativity theory. All the FUNDAMENTAL INTERACTIONS can be described by relativistic quantum field theories except for gravity. It is possible to obtain important general results about elementary particles, such as the existence of antiparticles, the SPIN–STATISTICS THEOREM, and the CPT THEOREM, from the basic principles of relativistic quantum field theory. In spite of their success, it has not been established whether any of the quantum field theories used to describe the fundamental interactions are mathematically consistent. At present, it is not known whether QUANTUM GRAVITY can be described successfully using a quantum field theory or whether it is possible to find a quantum field theory that gives a unified description of all the elementary particles and fundamental interactions.

quantum field theory in curved space–time *See* quantum gravity.

quantum flavordynamics (QFD) *See* Weinberg–Salam model.

quantum fluctuation A fluctuation from the average value of a quantity that arises because a quantum mechanical system can never be completely at rest due to the Heisenberg UNCERTAINTY PRINCIPLE (*see also* zero-point energy). Quantum fluctuations in the vacuum state in quantum field theory give rise to observable phenomena, such as a major contribution to the LAMB SHIFT. There is substantial evidence from accurate observations of the cosmic microwave background radiation for the theory that quantum fluctuations in the EARLY UNIVERSE were ultimately responsible for large-scale STRUCTURE FORMATION such as galaxies, in the Universe.

quantum geochemistry The application of the methods of QUANTUM CHEMISTRY to the materials and chemical processes of interest in geology. Theoretical calculations in quantum geochemistry are particularly useful for conditions involving high temperatures and pressures like those inside the Earth, which cannot readily be reproduced experimentally.

quantum gravity A theory that successfully combines general relativity theory and quantum mechanics. At present such a theory has not yet been found. However, some features of a true theory of quantum gravity are apparent. If quantum gravity is regarded as a quantum field theory then the gravitational interactions between particles with nonzero masses are mediated by the exchange of massless spin-two particles called GRAVITONS. Regarding quantum gravity as a quantum field theory leads to the serious difficulty that it gives rise to infinities that cannot be removed by the process of RENORMALIZATION. Nevertheless, significant progress has been made in quantum gravity since the 1980s using *loop quantum gravity*, i.e. a formulation of quantum gravity as a quantum field theory

in which the main variables are LOOP VARI-
ABLES.

Some important features of quantum
gravity emerge from *quantum field theory
in curved space–time* in which the gravitational interactions are treated classically
while other interactions are treated quantum mechanically. For example, quantum
field theory in curved space–time predicts
HAWKING RADIATION.

As an alternative to treating quantum
gravity as a quantum field theory many authors have suggested that quantum gravity
requires discrete structures such as SPIN
NETWORKS and SPACE–TIME FOAM. Since the
gravitational field is closely associated with
space–time in classical general relativity
theory the quantization of the gravitational
field is closely associated with the quantization of space–time. This means that
space–time might well have a discrete
structure. The theory of TWISTORS is another technique which has been used extensively in quantum gravity. It is widely
expected that the problem of SINGULARI-
TIES, which occurs in classical general relativity theory does not occur in quantum
gravity.

An important unresolved problem in
quantum gravity is whether a viable quantum theory of gravity can be found for the
gravitational interactions alone or whether
it requires a theory that unifies the gravitational interactions with all the nongravitational interactions. It is difficult to envisage
how gravity can be unified with the nongravitational interactions in a quantum
field theory (*see* unified field theory). There
is substantial evidence that SUPERSTRING
THEORY, in which the fundamental entities
are one-dimensional extended objects
called *strings*, could give a mathematically
consistent quantum theory of gravity, free
of infinities, which unifies the gravitational
interactions with the nongravitational interactions.

There are many problems concerning
the very early Universe, black holes, and elementary particles that require a theory of
quantum gravity. It has been suggested
that very high-energy photons and cosmic
rays might provide observational evidence
for certain theories of quantum gravity.

quantum Hall effect A quantum mechanical feature of the HALL EFFECT, which
occurs in two-dimensional electronic systems at low temperatures. In 1980 it was
found by Klaus von KLITZING that the resistance associated with the Hall effect does
not increase directly with the strength of
the magnetic field as in the classical Hall effect but is quantized and varies in a series
of steps that are integer multiples of h/e^2,
where h is the Planck constant and e is the
charge of an electron. This is the *integer
Hall effect*. It can be used for precise determination of e and h. In the *fractional quantum Hall effect* the quantized steps are
fractional, as discovered by Horst STÖRMER
and Daniel TSUI in 1982.

It is possible to understand the integer
quantum Hall effect in terms of noninteracting electrons using TOPOLOGY whereas
the fractional quantum Hall effect is understood in terms of the interactions of
electrons.

quantum mechanics The type of mechanics that describes the motion of particles at the atomic and subatomic scale.
There is a DUALITY in quantum mechanics
in which objects such as electrons, which
are regarded simply as particles in classical
mechanics, behave as both particles and
waves in quantum mechanics. Similarly,
light behaves as both a particle and a wave
in quantum mechanics.

There are several mathematically equivalent formulations of quantum mechanics.
The analogy between mechanics and optics
is emphasized in WAVE MECHANICS. Other
formulations include MATRIX MECHANICS,
the PATH INTEGRAL, and OPERATORS in
HILBERT SPACE.

quantum numbers A set of numbers
that characterizes a quantum state of a system. For example, an electron state of an
atom is characterized by a set of four quantum numbers (*see* electronic structure of
atoms). Frequently, quantum numbers are
GROUP THEORY labels associated with the
symmetry of a system. For example, in
atoms and nuclei quantum numbers are associated with the ROTATION GROUP whereas
in molecules they are associated with the

POINT GROUP of the molecule. Some quantum numbers can be associated with TOPOLOGY.

quantum optics The branch of optics that is concerned with specifically quantum-mechanical phenomena. Quantum optics includes the study of LASERS.

quantum plasma A plasma in which the motion of the particles is governed by quantum mechanics. The electrons and ions in a metal make up a quantum plasma.

quantum state The state of a quantum mechanical system such as a nucleus, atom, or molecule, as specified by its set of QUANTUM NUMBERS. Each state is characterized by a certain energy. It is possible for more than one quantum state to have the same energy (the states are then said to be *degenerate*). The quantum state with the lowest energy is called the *ground state*. All other quantum states are called *excited states*.

quantum statistics The statistical description of a system with identical particles described by quantum mechanics. There are two types of quantum statistics: *Bose–Einstein statistics* and *Fermi–Dirac statistics*. Bose–Einstein statistics apply to *bosons*, with any number of particles occupying a given quantum state. Fermi–Dirac statistics apply to *fermions*, with only one particle occupying a particular quantum state. In the classical limit of quantum statistics the MAXWELL–BOLTZMANN STATISTICS of classical statistical mechanics emerge.

quantum teleportation The process of sending a copy of a quantum mechanical system across space; i.e. the system disappears at one place and reappears at another. For quantum teleportation to occur it is necessary to prepare a quantum state involving two entities in which there is ENTANGLEMENT. Quantum teleportation has been achieved in the laboratory with laser beams and with photons and electrons.

quantum theory of radiation The study of the absorption and emission of electromagnetic radiation in terms of quantum mechanics. This theory was initiated by Albert EINSTEIN in 1916–1917. There are three processes to consider. The first is *spontaneous emission* in which an atom emits a photon as an electron in an excited state returns to the ground state. In *absorption* an electron in the ground state absorbs a photon to go to an excited state. In *stimulated emission* (which lies behind the operation of the LASER) an atom emits a photon as an electron in an excited state returns to the ground state when the atom is exposed to electromagnetic radiation of the same wavelength as the photon. A complete quantum theory of radiation requires QUANTUM ELECTRODYNAMICS.

quark A type of elementary particle that is thought to be one of the fundamental building blocks of matter. There are six *flavors* of quark, with there being two flavors in each of the three GENERATIONS of elementary particles. The quarks in the first generation are called the *up quark* and the *down quark*. Those in the second generation are called the *strange quark* and the *charm quark*, while those in the third generation are called the *bottom quark* (sometimes called *beauty quark*) and the *top quark* (sometimes called *truth quark*). All quarks are spin 1/2 particles with an electric charge which is either +2/3 or –1/3. All HADRONS are made up of quarks. Quarks take part in strong interactions, as governed by QUANTUM CHROMODYNAMICS. This means that a quark is associated with a COLOR charge. There are three types of color charge: *red, green,* and *blue.*

Isolated quarks have never been observed. This has given rise to the hypothesis of QUARK CONFINEMENT. Nevertheless, there is still a great deal of experimental evidence in favor of quarks.

In the course of weak interactions of hadrons one flavor of quark turns into another flavor of quark. For example, in the beta decay of a neutron to a proton (with the emission of an electron and an antineutrino) a neutron, which has two down quarks and up quark, is converted into a proton, which has two up quarks and one down quark. Since quarks are electrically charged they take part in electromagnetic interactions and since they have nonzero rest masses they also take part in gravitational interactions.

quark confinement The hypothesis that isolated quarks can never be observed. This hypothesis is a consequence of QUANTUM CHROMODYNAMICS (QCD), which governs the strong interactions of quarks. The feature of QCD known as ASYMPTOTIC FREEDOM means that the greater the distance between quarks the stronger the interaction between them. The quark confinement hypothesis extrapolates from this to postulate that since the strength of the interactions between quarks become stronger with distance it is impossible to isolate a single quark since the energies involved would lead to HADRON production. Although this hypothesis is very plausible, and there is a great deal of evidence in its favor theoretically and from numerical calculations, a conclusive mathematical proof has remained elusive since the mid 1970s when the hypothesis was first put forward. It is thought that at very high temperatures, such as those in the early Universe, the vacuum state of QCD is altered, with there being a phase transition to a phase in which quarks are not confined. The temperature at which this phase transition occurs is called the *deconfinement temperature*.

quark–gluon plasma A state of matter thought to have existed in which isolated hadrons did not exist but in which quarks and gluons formed a 'hot soup'. It is thought that this state ended abut 10^{-5} seconds after the big bang when there was a phase transition in which hadrons formed as the Universe cooled. Experiments at BROOKHAVEN and CERN are attempting to recreate the quark–gluon plasma of the early Universe by accelerating the nuclei of heavy elements to speeds that are very near to the speed of light and then smashing these nuclei into each other in head-on collisions.

quarkonium A general name given to any bound state system consisting of a quark and an antiquark. For example, CHARMONIUM which consists of a charm quark and an anti-charm quark, is a type of quarkonium.

quasar A class of astronomical object thought to exist at the core of a galaxy, in which there is a great deal of violent activity. It is generally believed that quasars are supermassive black holes with a mass of about 100 million times the mass of the Sun, with the energy coming from the black hole swallowing up material from its surrounding galaxy. The name *quasar* comes from 'quasi-stellar object'.

quasicrystal A solid that has order, as manifested in x-ray diffraction, but does not have the periodicity of a crystal lattice. Crystals such as AlMn have the point group symmetry of an icosahedron, which is not allowed in crystals.

quasiparticle A long-lived single particle EXCITATION in a quantum mechanical many-body system in which interactions between an individual particle and the surrounding medium modify the single particle excitation.

quaternion A generalization of a complex number, which can be written in the form $Q_0 + Q$, where Q_0 is a SCALAR and Q is a VECTOR with components Q_1, Q_2, and Q_3. Quaternions are used in problems in classical mechanics and are related to the rotation group, the Lorentz group, and the Pauli matrices. They were invented by Sir William HAMILTON in 1843 and subsequently developed extensively by him.

R

Rabi, Isidor Isaac (1898–1988) Austrian-born American physicist who developed the technique of magnetic resonance to measure the magnetic moments of particles. He was awarded the 1944 Nobel Prize for physics.

rad A former unit of absorbed dose of ionizing radiation, equal to 1/100 gray.

radian Symbol: rad The SI unit of plane angle; 2π radian is one complete revolution (360°).

radiation In general, particles or waves that are emitted from a source and carry energy. Examples of wave radiation are *electromagnetic radiation* (including light and radiant heat) and *gravitational radiation*. In radioactivity the radiation consists of both particles (alpha particles and beta particles) and electromagnetic radiation (gamma rays).

radiation damage Damage to inanimate materials or living organisms as a result of exposure to electromagnetic radiation or energetic particles such as electrons, nucleons, or fission fragments. In inanimate materials this damage can take the form of electronic excitation, ionization, or atomic displacement. The radiation may kill cells or alter their genetic structure. In the case of human beings this can lead to *radiation burns*, *radiation sickness*, and eventually to cancer, particularly leukemia.

radiation pressure The pressure exerted on a surface by radiation, usually electromagnetic radiation. This can be thought of as caused by photons transferring momentum as they hit the surface. For large bodies the effect of radiation pressure is very small but the effect can be significant for smaller bodies. For example, radiation pressure from the Sun is responsible for the tails of comets. Radiation pressure is large near the centers of stars with light masses.

radiation units Units that are used to measure the radioactivity of a radioactive nuclide and the dose of ionizing radiation. Non-SI units such as the *curie* have been used for many years. The SI unit of activity is the *becquerel* (Bq), which is the activity of a radioactive nuclide that decays at an average rate of one spontaneous nuclear transition per second. The curie is equal to 3.7×10^{10} Bq.

The SI unit of absorbed dose is the *gray* (Gy), which is the dose absorbed when the energy per unit mass imparted to matter by ionizing radiation is one joule per kilogram.

radiative corrections Terms in calculations in relativistic quantum field theory using PERTURBATION THEORY that go beyond the first order in the coupling constant of the theory (such as the FINE STRUCTURE CONSTANT in quantum electrodynamics). FEYNMAN DIAGRAMS are convenient for calculating radiative corrections. It is necessary to take radiative corrections into account to obtain agreement between theory and very precise measurements of quantities such as the MAGNETIC MOMENT of the electron and the LAMB SHIFT in QED. Radiative corrections have also been calculated in other theories such as the WEINBERG–SALAM MODEL.

radioactive dating (radiometric dating) Any method for dating the age of materials, such as archeological or geological specimens, that makes use of radioactivity. *See* carbon dating; fission-track dating; potassium–argon dating; rubidium–strontium dating; uranium–lead dating.

radioactive decay *See* decay.

radioactive series A series of radioactive nuclides, each member of which is formed by the decay of the previous member. All radioactive series terminate with a stable nuclide. There are three naturally occurring radioactive series: the *thorium series*, which starts with thorium-232 and terminates with lead-208 after six alpha decays and four beta decays, the *actinium series*, which starts with uranium-235 and terminates with lead-207, and the *uranium series*, which starts with uranium-238 and terminates with lead-206. In addition, a radioactive series not found in Nature is the *neptunium series*, which starts with neptunium-237 and ends with bismuth-209.

radioactive tracing The process of replacing a stable atom in a compound by a radioisotope of the same element and following the path of the radioisotope through a mechanical, chemical, or biological system by tracing the radiation it emits. Radioactive tracing is useful in investigating the mechanisms of chemical reactions. *See also* labeling.

radioactive waste (**nuclear waste**) Any waste material that contains radionuclides. Radioactive waste can come from the mining, extraction, and processing of radioactive ores, the normal operation of nuclear reactors, decommissioned nuclear reactors, the manufacture of nuclear weapons, and from radioactive materials used in hospitals, laboratories, and industry.

Since radioactive waste can be very harmful to all living matter and since it can contain radionuclides that have half-lives of many thousands of years the problem of how to dispose of radioactive waste is very important. Different levels of radioactivity have to be dealt with. *High-level waste* has to be stored until its radioactivity has been reduced to a level at which it can be processed. *Intermediate-level waste* can be contained by burial in deep mines or below the seabed. *Low-level waste* can be stored in steel drums at special sites. Spent nuclear fuel is an example of high-level waste. Reactor components and processing plant sludge are examples of intermediate level waste. Solids or liquids slightly contaminated by radioactive substances are examples of low-level waste.

radioactivity The spontaneous disintegration of certain atomic nuclei with the emission of radiation. There are three types of emission: alpha particles (helium-4 nuclei), beta particles (usually electrons but sometimes positrons), and gamma rays (short-wavelength electromagnetic radiation).

Natural radioactivity is radioactivity that occurs due to the spontaneous disintegration of naturally occurring nuclei whereas ARTIFICIAL RADIOACTIVITY (sometimes called *induced radioactivity*) is radioactivity that occurs as a result of nuclear reactions (e.g. bombardment of nuclei by neutrons or light nuclei).

See also alpha decay; beta decay; gamma decay.

radiography The technique of producing an image of an opaque object on a photographic film or a fluorescent screen using radiation. The radiation can either be short-wavelength electromagnetic waves (such as x-rays or gamma rays) or particles. A photograph produced in this way is called a *radiograph*. Radiography is used in RADIOLOGY and to detect flaws in industrial products.

radioisotope (**radioactive isotope**) An isotope of a chemical element that is radioactive. For example, TRITIUM is a radioisotope of hydrogen. Radioisotopes are used for RADIOACTIVE TRACING and as sources of radiation. In medicine, radioisotopes are used in diagnosis and in treatment.

radiology The study and applications of radioactive materials, x-rays, and other ionizing radiations in medicine, particularly for diagnosis (*diagnostic radiology*) and the treatment of diseases such as cancer (*radiotherapy*).

radiolysis The production of chemical reactions by using ionizing radiation. Different types of ionizing radiation are used, including alpha particles, electrons, neutrons, x-rays, and gamma rays. The energy transferred by this radiation produces ions, free radicals, and excited species, which take part in further reactions. In water and other polar solvents radiolysis gives rise to short-lived solvated electrons.

radiometric dating *See* radioactive dating.

radionuclide (**radioactive nuclide**) A NUCLIDE that is radioactive.

radiotherapy *See* radiology.

radio waves ELECTROMAGNETIC WAVES with wavelengths greater than a few millimeters. Radio waves are used extensively to carry information. They are also used in nuclear magnetic resonance.

radium A white radioactive luminescent metallic element of the alkaline-earth group. It has several short-lived radioisotopes and one long-lived isotope, radium-226 (half-life 1602 years). Radium is found in uranium ores, such as the oxides pitchblende and carnotite. It was formerly used in luminous paints and radiotherapy.

Symbol: Ra. Melting pt.: 700°C. Boiling pt.: 1140°C. Relative density: 5 (approx. 20°C). Proton number: 88. Relative atomic mass: 226.0254 (^{226}Ra). Electronic configuration: [Rn]7s^2.

radius of gyration Symbol: k For a body of mass m and moment of inertia I about an axis, the radius of gyration about that axis is given by $k^2 = I/m$. In other words, a point mass m rotating at a distance k from the axis would have the same moment of inertia as the body.

radon A colorless monatomic radioactive element of the rare-gas group, now known to form unstable compounds. It has 19 short-lived radioisotopes; the most stable, radon-222, is a decay product of radium-226 and itself disintegrates into an isotope of polonium with a half-life of 3.82 days. ^{222}Rn is sometimes used in radiotherapy. Radon occurs in uranium mines and is also detectable in houses built on certain types of rock.

Symbol: Rn. Melting pt.: –71°C. Boiling pt.: –61.8°C. Density: 9.73 kg m^{-3} (0°C). Proton number: 86. Electronic configuration: [Xe]4f^{14}5d^{10}6s^{2}6p^6.

Rainwater, Leo James (1917–86) American physicist who worked with Aage BOHR in the early 1950s to produce a theory of nuclear structure that unifies the SHELL MODEL with the LIQUID DROP MODEL. Rainwater shared the 1975 Nobel Prize for physics with Bohr and Ben MOTTELSON for this work.

Raman, Sir Chandrasekhara Venkata (1888–1970) Indian physicist who discovered the RAMAN EFFECT in 1928. He was awarded the 1930 Nobel Prize for physics.

Raman effect A type of scattering of light as it passes through a medium, in which the wavelength of the light is changed. This results in wavelengths being increased and decreased. Thus, light which was monochromatic before being scattered is not monochromatic after being scattered. The Raman effect occurs because of the interaction of photons with the molecules of the medium. Thus, the Raman effect is analogous to the COMPTON EFFECT (and INVERSE COMPTON EFFECT) and demonstrates the particle-like aspect of photons. It was first discovered by Sir Chandrasekhara Venkata RAMAN in 1928, having been predicted by Werner HEISENBERG and Hendrik KRAMERS in 1925.

Since the set of frequencies of light after scattering has occurred is characteristic of the material the Raman effect is used extensively to determine molecular structure and in chemical analysis. The scattering that occurs is called *Raman scattering*, with the range of new frequencies being called the *Raman spectrum*. The study of molecular structure using the Raman spectrum of the molecule is called *Raman spectroscopy*. The use of Raman spectroscopy was greatly increased by the development of the LASER. In liquids the intensity of Raman scattering is about a thousand times smaller than that of RAYLEIGH SCATTERING.

Ramsay, Norman Foster (1915–) American physicist who worked on molecular beams. In particular, he did important work on the magnetic moments of molecules and on nuclear magnetic resonance and its application to studying molecular structure. Along with his colleagues, Ramsay developed the hydrogen maser. Ramsay shared the 1989 Nobel Prize for physics with Hans DEHMELT and Wolfgang PAUL.

Ramsay, Sir William (1852–1916) Scottish chemist who discovered the inert RARE GASES argon, helium, neon, krypton, xenon, and radon in the late nineteenth and early twentieth centuries. In 1903 Ramsay, together with Frederick SODDY, showed that helium is produced by the radioactive decay of radium. He won the 1904 Nobel Prize for chemistry for his discovery of the noble gases.

range of force The distance over which a force is effective. Forces such as gravity and electricity, which are governed by an INVERSE SQUARE LAW, are *long-range forces* since they are effective over large distances, whereas strong and weak nuclear forces are *short-range forces* since they are only effective over very small distances.

rare-earths (**rare earth elements**) *See* lanthanides.

rare gases *See* noble gases.

Rarita–Schwinger equation An equation in relativistic quantum mechanics for spin-3/2 particles. It was discovered by William Rarita and Julian SCHWINGER in 1941.

rationalized units A system of units in which the equations have a logical form related to the shape of the system. SI units form a rationalized system of units. In it formulae concerned with cylindrical symmetry contain a factor of 2π; those concerned with spherical symmetry contain a factor of 4π.

ray A very narrow beam of radiation that is given an idealized representation by lines in a *ray diagram*. Such diagrams are used in geometrical optics to study the paths of rays of light in mirrors and lenses, and hence to relate the positions of objects and their images when mirrors and lenses are used.

Rayleigh, Lord (**John William Strutt**) (1842–1919) British physicist who made important contributions to several branches of physics including acoustics, optics, and fluid mechanics. In 1871 he explained the blue color of the sky in terms of the scattering of electromagnetic waves by particles in the atmosphere. One of his most important contributions to physics was the RAYLEIGH–JEANS LAW. Rayleigh performed experiments with great skill and was concerned with precise measurements. This resulted in his discovery that the density of nitrogen extracted from air is slightly greater than that of nitrogen obtained by a chemical process. This led to the discovery of argon with Sir William RAMSAY, which they announced in 1895. Both Ramsay and Rayleigh were awarded Nobel Prizes in 1904 for this discovery with Ramsay winning his for chemistry and Rayleigh winning his for physics.

Rayleigh–Jeans law The law for the distribution of energy radiated by a BLACK BODY, first stated by Lord RAYLEIGH in 1900 and modified by Sir James JEANS in 1905. It gives a good description of the distribution at long wavelengths but fails catastrophically at short wavelengths, since it predicts that the higher the frequency, the more radiation there should be. Thus, the Rayleigh–Jeans law predicts that a black body should emit very large amounts of energy in the ultraviolet part of the spectrum and beyond into higher frequencies, contrary to experimental results. This result is known as the *ultraviolet catastrophe*. It was subsequently realized by Albert EINSTEIN that the Rayleigh–Jeans law is the classical limit of PLANCK'S RADIATION LAW, thereby explaining its success at long wavelengths, and that the ultraviolet catastrophe is a serious failure of classical physics. *See* Planck radiation law.

Rayleigh scattering   A process in which there is a phase change but not a frequency change when light (and other electromagnetic radiation) is scattered by matter. Rayleigh scattering is named after Lord RAYLEIGH who analyzed it in 1871 in the course of explaining why the sky is blue.

reactor   *See* nuclear reactor.

recombination   A process in which a positive ion and an electron or negative ion combine to form a neutral atom or molecule. The neutral species that is formed in recombination is usually in an excited state, which subsequently decays with the emission of light or some other form of electromagnetic radiation.

recombination era   The time in the evolution of the EARLY UNIVERSE at which the Universe had cooled sufficiently for nuclei to combine with electrons to form neutral atoms. This occurred about 300 000 years after the big bang when the temperature of the Universe was about 6000 degrees kelvin. The name 'recombination' is a misnomer because neutral atoms are formed for the first time at the recombination era, never having existed previously.

rectifier   An electrical device that allows current to flow only in one direction to any extent. This allows a.c. conduction to be converted to d.c. conduction. Rectifiers were originally based on THERMIONIC EMISSION but have been replaced by semiconductor devices.

red giant   *See* stellar evolution.

redshift   A displacement of spectral lines towards longer wavelengths, i.e. towards the red end of the visible spectrum, which occurs for certain celestial objects such as galaxies. This can occur either because of a *Doppler redshift*, i.e. a Doppler effect associated with the movement of the galaxy away from the Earth, or a *gravitational redshift* (sometimes called an *Einstein redshift*), i.e. a redshift due to a strong gravitational field, which was predicted to occur in general relativity theory by Albert EINSTEIN.

It is customary to define a redshift as $\delta\lambda/\lambda$, where $\delta\lambda$ is the shift in wavelength of electromagnetic radiation of wavelength λ. It is thought that the gravitational redshift is responsible for the large redshifts of QUASARS.

reflection   A process in which beams of particles or waves (such as sound waves or light waves) are bounced back from a surface that is smooth (compared to the wavelengths of waves involved). The reflection of light is described by RAYS in geometrical optics with the reflection all taking place in the same plane and the angle of incidence being equal to the angle of reflection. In quantum mechanics these classical laws of light reflection emerge from the PATH INTEGRAL approach, with all other paths of light effectively cancelling each other out.

refraction   The process in which the direction of light waves changes when the light crosses the boundary between two media with different optical properties. Refraction occurs unless the wave meets the boundary at right angles. The incident ray, the refracted ray, and the normal to the boundary (i.e. an imaginary line perpendicular to the boundary between the two media) are all in the same plane. Classically, refraction is explained by the light waves moving at different speeds in the two media, as characterized by the two media having different REFRACTIVE INDICES. In quantum mechanics, refraction emerges in terms of paths of light in a way which is analogous to the quantum mechanical explanation of reflection. Refraction also occurs for other waves such as sound waves.

refractive index (refractive constant) Symbol: n The *absolute refractive index* of a medium is the ratio of the speed of light in a vacuum to the speed of light in the medium. For most materials it is sufficiently accurate to regard air as being equivalent to a vacuum. Since the value of the refractive index varies with wavelength it is necessary to specify a particular wavelength. This is usually taken to be the yellow light of sodium vapor (known as the D lines) with a wavelength of 589.3 nm. The *relative refractive index* of two adjacent media is the ratio of the speed of light in one medium to that in the adjacent medium.

regularization   The imposition of a short-distance 'cut-off' in quantum field theory, which ensures that the calculation of a quantity gives a finite answer rather than infinity. There are several ways in which this can be done. Regularization was introduced into quantum field theory by Richard FEYNMAN, Wolfgang PAULI, and others in the late 1940s. In general, it is necessary to remove the 'cut-off' before completing a calculation in quantum field theory using RENORMALIZATION. It has been speculated by Pauli and others that QUANTUM GRAVITY should give a natural physical regularization for quantum field theory. This suggestion has never been implemented successfully but there are some encouraging indications that the right type of NONCOMMUTATIVE GEOMETRY might be able to do so.

Reines, Frederick (1918–98)   American physicist who, together with Clyde COWAN, discovered the neutrino in 1956 in experiments carried out at the Savannah River nuclear reactors. Reines subsequently investigated neutrinos associated with cosmic rays and the 1987 SUPERNOVA explosion. He shared the 1995 Nobel Prize for physics with Martin PERL for his discovery of the neutrino.

Reissner–Nordstrøm solution   A solution to EINSTEIN'S FIELD EQUATION of general relativity theory that describes a nonrotating BLACK HOLE. This solution was found by Heinrich Reissner in 1916 and by Gunnar Nordstrøm in 1918.

relative atomic mass Symbol: A_r A quantity formerly called ATOMIC WEIGHT. It is defined as the mass of an atom, relative to the mass of an atom of the nuclide carbon-12, which is defined as 12 units.

relative density Symbol: d The density of a substance divided by the density of water. Usually the density of water at 4°C (the temperature at which its density is a maximum) is used. The temperature of the substance is stated or is understood to be 20°C. Relative density was formerly called *specific gravity*.

relative molecular mass Symbol: M_r A quantity formerly called *molecular weight*. It is defined as the mass of a molecule relative to the mass of an atom of the nuclide carbon-12, which is defined as 12 units. This means that the relative molecular mass of a molecule is equal to the sum of the relative atomic masses of all the atoms in the molecule.

relative permeability   *See* permeability.

relative permittivity   *See* permittivity.

relativistic mass   The mass m of a body with REST MASS m_0 which is moving with a velocity v. Special relativity theory shows that m is given by $m = m_0/\sqrt{(1-v^2/c^2)}$, where c is the speed of light in a vacuum. This equation is in very good agreement with experiment. It shows that no object with a nonzero rest mass can reach the speed of light since it would then have an infinite mass.

relativistic quantum field theory   The combination of special relativity theory and quantum field theory. All the fundamental interactions in Nature except gravity are described by relativistic quantum field theories. It is possible to obtain general results such as the SPIN–STATISTICS THEOREM and the CPT THEOREM for relativistic quantum field theory. Calculations in relativistic quantum field theories are plagued by infinities but these infinities can be removed by the procedure of RENOR-MALIZATION.

relativistic quantum mechanics The combination of special relativity theory and quantum mechanics. It is necessary to take special relativity into account using relativistic quantum mechanics if the particles are moving at speeds that are a substantial fraction of the speed of light. For example, electrons in large atoms need to be described using relativistic quantum mechanics.

In 1928 Paul DIRAC became the first person to combine quantum mechanics and special relativity theory successfully. This work led Dirac to predict that the existence of ANTIPARTICLES is a key feature of relativistic quantum mechanics.

relativistic wave equation An equation that describes the relativistic quantum mechanics of a particle. Each different spin has a different relativistic wave equation. The main relativistic wave equations are the KLEIN–GORDON EQUATION (for spin-0 particles), and the RARITA–SCHWINGER EQUATION (for spin-3/2 particles). Relativistic wave equations can be derived using the POINCARÉ GROUP. It is more satisfactory physically to regard relativistic wave equations as applying to fields than single particles.

relativity *See* special theory of relativity; general theory of relativity.

relaxation The return of a system to its state of equilibrium after a sudden change has been applied to the system. The time interval required for relaxation to take place is called the *relaxation time*. An example of relaxation is the change in the magnetic moment when a crystal is removed from a magnetic field in which its electronic spins have been aligned.

rem (*radiation equivalent man*) A former unit of dose equivalent of ionizing radiation, equal to 1/100 sievert.

renormalizable *See* renormalization.

renormalization A technique used in RELATIVISTIC QUANTUM FIELD THEORY to deal with the infinities that arise in PERTURBATION THEORY calculations beyond the first order. Renormalization was invented in the late 1940s by Richard Feynman, Julian Schwinger, and Sin-Itiro Tomonaga in quantum electrodynamics. The infinities were removed by using the observed mass and charge of the electron and a careful cancellation of infinities.

In general, quantum field theories for which finite results can be obtained to any order in perturbation theory by taking a finite number of parameters from experiment and using renormalization are called *renormalizable*. Not all quantum field theories are *renormalizable*. If a theory needs an infinite number of parameters it is said to be *nonrenormalizable* and is not regarded as a complete consistent physical theory. All the gauge theories that describe the strong, weak, and electromagnetic interactions are renormalizable whereas QUANTUM GRAVITY and the FERMI THEORY OF WEAK INTERACTIONS are nonrenormalizable.

Since renormalization involves the cancellation of infinite quantities it is difficult to make the technique mathematically rigorous. It may well be the case that the infinities in relativistic quantum field theory that necessitate the use of renormalization arise from using field theory rather than a theory involving discrete structures.

renormalization group A technique used to investigate problems involving more than one length-scale. It originated in RELATIVISTIC QUANTUM FIELD THEORY in the early 1950s when it was used to calculate how coupling constants change with energy. The renormalization group was used in this way to demonstrate the existence of the LANDAU GHOST and subsequently ASYMPTOTIC FREEDOM. In the 1970s renormalization group theory was used with great success in the theory of phase transitions. It has also been applied to many-electron systems, polymers, electron localization in disordered systems, turbulence, and chaos theory.

representation A representation of a group G is a set of operators in a vector space L that correspond to the elements of the group. A representation is said to be d-dimensional if the space L is d-dimensional. Usually, operators are defined in terms of matrices (just like the matrix mechanics formulation of quantum mechanics). The set of matrices forms a *matrix representation* of the group. An *irreducible representation* is a representation that cannot be expressed in terms of a representation of lower dimension.

In quantum mechanics one is interested in the symmetry group of a system (rotation group, point group, space group, etc.). Irreducible representations are important in quantum mechanics because the energy levels of a system are labeled by the irreducible representations of the symmetry group of the system. This enables SELECTION RULES to be derived.

resistance Symbol: R. The ratio of the potential difference across an electrical conductor to the current in it. Resistance is a measure of the opposition to the flow of charge. In metals, resistance increases with temperature as the ions move more, and hence scatter electrons more readily, thereby disrupting their flow. The quantum theory of electrons in solids enables resistance to be calculated.

resolving power A measure of the ability of an optical instrument such as a microscope to form separate images of close objects or to separate close wavelengths of radiation. The resolving power depends on the wavelength of the radiation, with the smaller the wavelength the greater the resolving power (i.e. the smaller the objects that can be distinctly observed).

resonance 1. An oscillation of a system at its natural frequency when it is given impulses at this frequency. Examples of resonance occur in mechanical systems, electrical circuits, atoms, and molecules.

2. In QUANTUM CHEMISTRY resonance refers to the representation of a molecule by two or more different molecular states or arrangements of atoms. The different structures are called *resonance structures*. For example, the bonding in the methanal molecule is intermediate between two resonance structures: the covalent structure $H_2C=O$ and the ionic structure $H_2C^+O^-$.

3. A short-lived elementary particle which decays via the strong interactions in about 10^{-23} seconds. These particles were discovered by Enrico FERMI and his colleagues in the early 1950s. There are many resonances of this type. They are excited states of hadrons.

rest energy The energy E_0 which is the energy equivalent of the REST MASS m_0 in accord with the Einstein equation $E = mc^2$. The rest energy is the minimum energy that a body can have.

rest mass The mass of a body at rest when it is measured by an observer who is also at rest in the same frame of reference. The observed mass of a body is only significantly different from its rest mass if the body is moving at a speed that is a substantial fraction of the speed of light. *See also* relativistic mass.

retardation plate A thin transparent plate of a birefringent material such as quartz, which is cut parallel to the optic axis. Light which falls on the plate normal to the optic axis is split into an ordinary ray and an extraordinary ray, with these rays passing through the plate at different speeds. It is possible to produce a specific phase difference between the rays by cutting the plate to a specific thickness. Retardation plates are used to produce circularly and elliptically polarized light.

reversible process A process that can be reversed. In a reversible process any exchange of energy, work, or matter with the environment should be reversed in the order and direction when the reverse process occurs.

rhenium A rare silvery transition metal that usually occurs naturally with molybdenum (it is extracted from flue dust in molybdenum smelters). The metal is chemically similar to manganese and is used in alloys and catalysts.

Symbol: Re. Melting pt.: 3180°C. Boiling pt.: 5630°C. Relative density: 21.02 (20°C). Proton number: 75. Relative atomic mass: 186.207. Electronic configuration: $[Xe]4f^{14}5d^56s^2$.

rhodium A rare silvery hard transition metal. It is difficult to work and highly resistant to corrosion. The metal occurs native but most is obtained from copper and nickel ores. It is used in protective finishes, alloys, and as a catalyst.

Symbol: Rh. Melting pt.: 1966°C. Boiling pt.: 3730°C. Relative density: 12.41 (20°C). Proton number: 45. Relative atomic mass: 102.90550. Electronic configuration: $[Kr]4d^85s^1$.

rho-meson Symbol: ρ. A short-lived electrically neutral spin-1 meson with a mass of about 770 MeV.

Richardson, Sir Owen Williams (1879–1959) British physicist who studied the emission of electrons from hot surfaces. He named this phenomenon THERMIONIC EMISSION and won the 1928 Nobel Prize for Physics for his pioneering work on the subject.

Richardson, Robert Coleman (1937–) American physicist who worked with David LEE and Douglas OSHEROFF on the superfluidity of helium-3 in the early 1970s. Lee, Osheroff, and Richardson shared the 1996 Nobel Prize for physics for this work.

Richardson–Dushman equation See thermionic emission.

Richter, Burton (1931–) American physicist who discovered the J/PSI PARTICLE in 1974 with his colleagues using the SPEAR accelerator which he designed. Richter shared the 1976 Nobel Prize for this discovery with Samuel TING who made the

same discovery with another team independently and almost simultaneously.

Riemann, George Friedrich Bernhard (1826–66) German mathematician who pioneered the mathematical description of curved space in a famous inaugural lecture in 1854 entitled 'Concerning the Hypotheses that Underlie Geometry'. This work initiated what came to be known as RIEMANNIAN GEOMETRY. He also did important work in other branches of mathematics and was interested in mathematical physics.

Riemannian geometry The geometry that describes curved space. This type of NON-EUCLIDEAN GEOMETRY was introduced by Bernard RIEMANN in 1854. It is a far-reaching generalization of the geometry of curved surfaces. Riemannian geometry enables curvature and distance to be defined in curved space. General relativity theory is formulated in terms of Riemannian geometry. (To be more precise, general relativity theory involves a modification of Riemannian geometry involving LORENZIAN MANIFOLDS, i.e. manifolds in which the local geometry is that of MINKOWSKI SPACETIME rather than Euclidian space.)

roentgen Symbol: R A unit of radiation, used for x-rays and γ-rays, defined in terms of the ionizing effect on air. One roentgen induces 2.58×10^{-4} coulomb of charge in one kilogram of dry air.

Rohrer, Heinrich (1933–) Swiss physicist who developed the SCANNING TUNNELLING MICROSCOPE with Gerd BINNIG, starting in 1978. Binnig and Rohrer shared the 1986 Nobel Prize for physics with Ernst RUSKA for their part in developing electron microscopes.

Röntgen, Wilhelm (1845–1923) German physicist who discovered X-RAYS in 1895. Röntgen was awarded the first ever Nobel Prize for physics in 1901.

rotational group A continuous group that is the group of rotations about an axis. In quantum mechanics the rotation group and its representations underlies the quantum theory of angular momentum and so has many applications to atoms, molecules, and nuclei.

rotational motion The motion of a body about an axis. The laws which describe rotational motion are analogous to those describing linear motion. In classical mechanics the analysis of rotational motion is important in celestial mechanics. The analysis of rotational motion is also important in the quantum mechanics of atoms, molecules, and nuclei.

R-parity *See* supersymmetry.

Rubbia, Carlo (1934–   ) Italian physicist who led the team at CERN which discovered W and Z bosons. Rubbia won the 1984 Nobel Prize for physics for this discovery. From 1989 to 1993 Rubbia was director-general of CERN.

rubidium A soft silvery highly reactive element of the alkali-metal group. Naturally occurring rubidium comprises two isotopes, one of which, ^{87}Rb, is radioactive (half-life 5×10^{10} years). It is found in small amounts in several complex silicate minerals, including lepidolite. Rubidium is used in vacuum tubes, photocells, and in making special glass.

Symbol: Rb. Melting pt.: 39.05°C. Boiling pt.: 688°C. Relative density: 1.532 (20°C). Proton number: 37. Relative atomic mass: 85.4678. Electronic configuration: [Kr]5s^1.

rubidium–strontium dating A method of radioactive dating used to estimate the age of certain geological specimens. It is based on the decay of the radioisotope rubidium-87 (which has a half-life of 5×10^{10} years) into the stable isotope strontium-87. Rubidium–strontium dating is useful for very large ages such as several thousand million years.

Rumford, Count (Benjamin Thompson) (1753–1814) American-born physicist who performed experiments which showed that heat is a form of energy and is not a substance called caloric, as was widely thought in the late eighteenth century. He also founded the Royal Institution of Great Britain in 1800, with Humphry DAVY as director.

Runge–Lenz vector A vector which is a conserved quantity in the problem of a single body orbiting a central body when there is an INVERSE SQUARE LAW of attraction between the two bodies. In celestial mechanics its physical significance is to indicate that the major axis of the ellipse of the orbit does not change with time (i.e. the orbit does not precess). The Runge–Lenz vector is associated with invariance under the four-dimensional rotation group (R4). It is named after Carl Runge, who used it in a book on vectors in 1919, and Wilhelm Lenz who used it in the old quantum theory in 1924, but was known much earlier to Pierre LAPLACE, Sir William HAMILTON, and Josiah Willard GIBBS.

A quantum mechanical version of the Runge-Lenz vector was used by Wolfgang PAULI in 1926 to solve the hydrogen atom using matrix mechanics. In 1935 Vladimir FOCK showed that the high degeneracy of the hydrogen atom in which electronic states with the same principal quantum number but different orbital quantum numbers have the same energies is related to invariance under R(4). In 1936 Valentin Bargmann showed that the R(4) invariance of the hydrogen atom is related to the conservation of the Runge–Lenz vector.

Ruska, Ernst August Friedrich (1906–1988) German physicist/electrical engineer who developed the electron microscope. He started work on developing an electron microscope in the late 1920s, soon after the wave nature of the electron had been established. By 1933 he had built an electron microscope with better resolution than an optical microscope. He shared the 1986 Nobel Prize for Physics with Gerd BINNIG and Heinrich ROHRER for his pioneering work.

Russell–Saunders coupling A type of coupling of the angular momentum of fermions in many-fermion systems such as atoms or nuclei. It is also known as *L–S coupling*. It occurs in an atomic state if the state is characterized by a resultant orbital angular momentum *L* formed as the resultant of coupling the individual orbital angular momenta *l* of fermions and a resultant spin angular momentum *S* formed as the resultant of coupling the individual spin angular momenta *s* of fermions. This is the case when the energies associated with electrostatic repulsion are much greater than the energies associated with SPIN–ORBIT COUPLING. It describes the multiplets of atoms and, to a lesser extent nuclei, of low proton number. Russell–Saunders coupling is named after Henry Russell and Frederick Albert Saunders who postulated this type of coupling in 1925 to explain the spectra of atoms with low proton number.

Atoms and nuclei with high proton number are better described by J–J COUPLING, in which the electronic states are characterized by the coupling between the individual *l* and *s* of each electron due to the energies associated with spin–orbit coupling being much greater than the energies associated with electrostatic repulsion. If the energies of electrostatic repulsion and spin–orbit coupling are similar in size then one has *intermediate coupling*.

ruthenium   A transition metal that occurs naturally with platinum. It forms alloys with platinum that are used in electrical contacts. Ruthenium is also used in jewelry alloyed with palladium.

Symbol: Ru. Melting pt.: 2310°C. Boiling pt.: 3900°C. Relative density: 12.37 (20°C). Proton number: 44. Relative atomic mass: 101.07. Electronic configuration: $[Kr]4d^7 5s^1$.

Rutherford, Ernest (Baron Rutherford of Nelson) (1871–1937)   New Zealand-born British physicist who made fundamental contributions to the study of radioactivity, the structure of the atom and nuclear physics.

Rutherford established that radioactivity occurs as three types, which he called alpha radiation, beta radiation, and gamma radiation. Together with Frederick SODDY,

he showed in the early years of the twentieth century that radioactivity involves the transmutation (change) of one chemical element into another. Further investigation of alpha radiation by Rutherford led to his realization that alpha particles are helium atoms that have lost two electrons. Along with Bertran Boltwood, Rutherford investigated RADIOACTIVE SERIES and invented RADIOACTIVE DATING, thereby enabling him to show that some rocks in the crust of the Earth are more than a billion years old.

In 1909 Rutherford suggested the experiment to Hans GEIGER and Ernest Marsden to scatter alpha particles through thin metal foils. The surprising results of this experiment led to the RUTHERFORD MODEL of the atom being postulated in 1911.

In 1919 Rutherford made one of his most significant contributions to physics when he showed that when nitrogen nuclei are bombarded by alpha particles they disintegrate. This was the first artificial transmutation of a chemical element. The last great discovery that Rutherford made was the discovery of TRITIUM with his colleagues in 1934.

Rutherford won the Nobel Prize for chemistry in 1908 for his pioneering contributions to the study of radioactivity.

rutherfordium   A radioactive element not found naturally on Earth. It can be made by bombarding atoms of ^{249}Cf with ^{12}C or by bombarding ^{248}Cm with ^{18}O.

Symbol: Rf. Melting pt.: 2100°C (est.). Boiling pt.: 5200°C (est.). Relative density: 23 (est.). Most stable isotope: ^{261}Rf (half-life 65s). Electronic configuration: $[Rn]5f^{14}6d^2 7s^2$.

Rutherford model   The model of the atom put forward by Ernest RUTHERFORD in 1911 to account for the surprising results found by Geiger and Marsden in experiments on the scattering of alpha particles by thin metal foils. It was discovered that most of the particles passed through with little or no deviation. However, a small number of particles were deflected by a large amount. Rutherford postulated that most of the mass and all of the charge was concentrated in a small NU-

CLEUS at the center of the atom, with the electrons orbiting the nucleus in a way that is analogous to the orbits of planets around the Sun.

This model has the serious difficulty that according to classical electrodynamics the electrons should very rapidly spiral into the nucleus, continually giving off electromagnetic radiation while doing so. This difficulty led to the BOHR MODEL of the atom, and hence to quantum mechanics.

Rutherford scattering The general name given to the scattering of charged particles by atomic nuclei.

Rydberg, Johannes Robert (1854–1919) Swedish physicist who brought order into the extensive nineteenth century work on atomic spectroscopy, notably by finding an empirical expression for the frequencies in a series of spectral lines which was a generalization of the expression for the BALMER SERIES. Rydberg also suggested that the PROTON NUMBER rather than ATOMIC WEIGHT was the key to the numbering of elements in the periodic table.

Rydberg constant Symbol: R A constant that occurs in the empirical expression for the wavelengths (frequencies) of lines in series of the atomic spectrum of hydrogen. The wavelength λ of a spectral line is related to the Rydberg constant R and two integers n_1 and n_2 by
$$1/\lambda = R \, (1/n_1^2 - 1/n_2^2),$$
where the value of n_2 must be greater than n_1. R has the value 1.097×10^7 m^{-1}. An expression for R was worked out by Niels BOHR when he put forward the BOHR MODEL of the atom. It was subsequently derived using both MATRIX MECHANICS and WAVE MECHANICS. It is given by $R = 2\pi^2 m e^4/ch^3$, where m is the mass of an electron, e is its charge, c is the speed of light in a vacuum and h is the Planck constant.

Sachs–Wolfe effect The process by which density fluctuations in the early Universe have left an imprint on the COSMIC MICROWAVE BACKGROUND RADIATION. It is caused by the interaction of photons with a gravitational field with density fluctuations. This causes small variations in the temperature. It is thought that the 'ripples' found by the COBE satellite were due to the Sachs–Wolfe effect. The effect was predicted by Rainer Kurt Sachs and Arthur Michael Wolfe in 1967. *See also* structure formation.

Sakharov, Andrei Dmitrievich (1921–89) Russian physicist who is best known as a dissident in the former Soviet Union but who made important contributions to cosmology, notably SAKHAROV'S CONDITIONS. After World War II he worked on the development of the Soviet hydrogen bomb. It was his concern with the environmental effects of testing nuclear weapons that led to his becoming a dissident. Sakharov was awarded the 1975 Nobel Peace Prize for his campaigns for nuclear disarmament and for promoting human rights.

Sakharov's conditions A set of three conditions in the EARLY UNIVERSE that can explain the asymmetry between matter and antimatter. These conditions are: (1) there is some process in which baryon number is not conserved; (2) CP violation occurs; (3) the early Universe is not in a state of thermal equilibrium. These conditions were put forward by Andrei SAKHAROV in 1967 and implemented in the context of GRAND UNIFIED THEORIES by a number of workers in the late 1970s.

Salam, Abdus (1926–96) Pakistani phy-sicist who made many important contributions to elementary-particle theory, especially the WEINBERG–SALAM MODEL, which unifies the weak and electromagnetic interactions. Salam put forward this model in 1968, independently of Steven WEINBERG. Salam shared the 1979 Nobel Prize for physics with Weinberg and Sheldon GLASHOW. He also played a major role in the development of SUPERSYMMETRY.

samarium A silvery element of the lanthanoid series of metals. It occurs in association with other lanthanoids. Samarium is used in the metallurgical, glass, and nuclear industries.

Symbol: Sm. Melting pt.: 1077°C. Boiling pt.: 1791°C. Relative density: 7.52 (20°C). Proton number: 62. Relative atomic mass: 150.36. Electronic configuration: $[Xe]4f^6 6s^2$.

saturation A property of strong nuclear forces that limits the number of attractive interactions that each nucleon in a nucleus can have.

s-block elements The chemical elements of groups 1 and 2 of the periodic table. The electronic configurations of these elements have a noble-gas structure plus one s electron (in the case of group 1 elements) and two s electrons (in the case of group 2 elements). This excludes elements that have incomplete inner d subshells (transition metals) or incomplete inner f subshells (lanthanides or actinides). The s-block elements are chemically reactive metals which form ions with a charge of +1 in the case of group 1 elements and +2 in the case of group 2 elements. *See also* alkali metals; alkaline-earth metals.

scalar A quantity that has a magnitude but is not associated with a direction. For example, mass, temperature, and time are scalar qualities. *See also* vector.

scalar field A field that is characterized by a magnitude at each point but not a direction. For example, the temperature at each point in a room can be regarded as a scalar field. Also, the field associated with symmetry breaking in the HIGGS MECHANISM is a scalar field. *See also* vector field.

scalar particle A particle that is associated with a scalar field. In quantum field theory scalar particles have zero spin. For example, the HIGGS BOSON is a scalar particle.

scalar potential A potential function in which the quantity that varies is a scalar. In Newtonian gravity and in electrostatics the potential functions are scalar potentials. See also VECTOR POTENTIAL.

scalar product (**dot product**) The product of two vectors that gives a scalar. If A and B are two vectors then their scalar product, denoted $A.B$ is given by $A.B = AB \cos \theta$, where A and B are the magnitudes of A and B respectively and θ is the angle between the two vectors. *See also* vector product.

scaling A term used by James BJORKEN to describe the observed pattern in DEEP INELASTIC SCATTERING experiments at SLAC in the 1960s being the same at different energies but with a different scale. These observations paved the way to the PARTON MODEL.

scandium A lightweight silvery element belonging to the first transition series. It is found in minute amounts in over 800 minerals, often associated with lanthanoids. Scandium is used in high-intensity lights and in electronic devices. The only isotope that occurs naturally (scandium-45) is not radioactive. There are also nine radioactive isotopes, all of which have relatively short half-lives.

Symbol: Sc. Melting pt.: 1541°C. Boiling pt.: 2831°C. Relative density: 2.989 (0°C). Proton number: 21. Relative atomic mass: 44.955910. Electronic configuration: [Ar]3d^{1}4s^2.

scanning tunneling microscope A microscope that probes the surfaces of materials by using quantum mechanical TUNNELING. The microscope has a probe with a tip which moves across the surface of interest. The electrons that tunnel out of the surface into the probe give an electrical signal, which changes in a way that gives information about the surface. *See also* atomic force microscope.

Schawlow, Arthur Leonard (1921–99) American physicist who made important contributions to the development of LASERS, particularly their use in spectroscopy. Schawlow shared the 1981 Nobel Prize for physics with Nicolaas BLOEMBERGEN for this work.

Schrieffer, John Robert (1921–) American physicist who put forward the successful theory of SUPERCONDUCTIVITY with John BARDEEN and Leon COOPER in 1957. They shared the 1972 Nobel Prize for physics for their theory. Schrieffer made a number of other important contributions to solid-state theory, notably regarding SOLITONS in electrically conducting polymers.

Schrödinger, Erwin (1887–1961) Austrian physicist who developed WAVE MECHANICS in a quantitative way when he put forward the SCHRÖDINGER EQUATION in 1926 and solved it for the hydrogen atom. In 1926 he also showed that wave mechanics is mathematically equivalent to MATRIX MECHANICS. In 1944 a book by Schrödinger entitled *What Is Life* was published. This book was very influential in stimulating interest in molecular biology.

Schrödinger's cat A name given to a thought experiment proposed by Erwin SCHRÖDINGER in 1935 to illustrate what he regarded as the absurdity of the COPENHAGEN INTERPRETATION of quantum mechanics. Schrödinger was concerned about the idea that in quantum mechanics a superposition of quantum states can occur and that it is necessary for there to be an observer to cause a COLLAPSE OF THE WAVE FUNCTION to a unique state.

To illustrate his concern Schrödinger suggested a system in which a particular quantum event, such as the decay of a radioactive nucleus, occurs with exactly a fifty-fifty chance. In the Copenhagen interpretation of quantum mechanics a nucleus exists in a superposition of decayed and nondecayed states unless the state of the system is measured. Schrödinger considered a sealed vessel with a live cat and a device triggered by the radioactive decay of a nucleus, which releases cyanide and kills the cat. Schrödinger concluded that the Copenhagen interpretation means that the cat is in a superposition of the states of being dead and alive until a measurement is made.

This type of argument has led many physicists and philosophers of science to look for other interpretations of quantum mechanics.

Schrödinger's equation The fundamental equation of WAVE MECHANICS. It describes the wave function ψ of a particle such as an electron. There is a *time-dependent Schrödinger equation* and a *time-independent Schrödinger equation*. The time-independent Schrödinger equation is used to calculate energy levels and wave functions of quantum-mechanical systems such as atoms, molecules, solids, and nuclei. In three dimensions the time-independent Schrödinger equation is $\nabla^2\psi + (8\pi^2 m/h^2)\,(E - V)\,\psi = 0$, where ∇^2 is the Laplace operator, m is the mass of the particle, E is its total energy and V is its potential energy. Schrödinger's equation was put forward by Erwin Schrödinger in 1926 and solved by him for the hydrogen atom. The equation is not exactly solvable for any other atomic and molecular system. In general, approximation techniques and/or the use of computers are needed to obtain accurate solutions to the Schrödinger equation for systems with more than two particles.

Schwartz, Melvin (1932–) American physicist who discovered the muon neutrino with Leon LEDERMAN and Jack STEINBERGER in 1962. Schwartz shared the 1988 Nobel Prize for physics with Lederman and Steinberger for this discovery.

Schwarz, John Henry (1941–) American physicist who pioneered superstring theory in the 1970s and 1980s.

Schwarzschild, Karl (1873–1916) German astronomer who is best known for finding what is known as the SCHWARZSCHILD SOLUTION in general relativity theory in 1916, shortly before he died of a skin disease he contracted in World War I. Even before general relativity theory he suggested that space might be non-Euclidean.

Schwarzschild radius The radius of a massive spherical body for which the ESCAPE VELOCITY is equal to c, the speed of light in a vacuum. Equivalently, it is the radius of a spherical nonrotating black hole. Its value R is given by $R = 2GM/c^2$, where G is the gravitational constant and M is the mass of the body.

Schwarzschild solution A solution to Einstein's field equation of general relativity theory, found by Karl SCHWARZSCHILD in 1916, which corresponds to a spherical nonrotating black hole.

Schwinger, Julian Seymour (1918–94) American physicist who developed QUANTUM ELECTRODYNAMICS (QED) in the late 1940s using the technique of RENORMALIZATION. Schwinger also made many other contributions to various branches of theoretical physics. Schwinger, Richard Feynman, and Sin-Itiro Tomonaga shared the 1965 Nobel Prize for physics for their pioneering work on QED.

Schwinger effect The 'sucking out' of electron–positron pairs from the vacuum by a strong electric field. This effect in quantum electrodynamics was predicted to occur by Julian SCHWINGER in 1951. It is a very small effect and has never been observed. It is thought that collisions between large nuclei may be able to give sufficiently large electric fields for the Schwinger effect to be demonstrated.

scintillation counter A type of particle detector in which a flash of light (scintillation) is emitted when an excited atom returns to its ground state, having been excited by a photon or particle. Ernest RUTHERFORD and his colleagues used a scintillation counter in their experiment which demonstrated the nuclear structure of the atom. Originally scintillations were observed and counted by eye. Now each flash of light is converted into an electrical signal, which is amplified and then recorded.

screening The reduction of the electric field of an electric charge in a polarized medium by the attraction of opposite electric charge to the initial electric charge. Screening is a feature of electrodynamics at both the classical and quantum mechanical level. By contrast, in QUANTUM CHROMODYNAMICS *antiscreening* occurs, i.e. the force due to the color charge increases with increasing distance.

Seaborg, Glenn Theodore (1912–99) American nuclear chemist who discovered and investigated several of the TRANSURANIC ELEMENTS between 1948 and 1958. Seaborg shared the 1951 Nobel Prize for chemistry with his colleague Edwin McMillan for this work.

seaborgium A radioactive metallic element not occurring naturally on Earth. It can be made by bombarding ^{249}Cf with ^{18}O nuclei. Six isotopes are known.
Symbol: Sg. Most stable isotope: ^{266}Sg (half-life 27.3 s). Electronic configuration: [Rn]$5f^{14}6d^47s^2$.

secondary emission The emission of electrons from a surface due to the impact of high-energy charged particles, particularly other (primary) electrons. It is necessary for the incident particles to impart sufficient energy to overcome the WORK FUNCTION of the solid.

second quantization A procedure in quantum field theory that formalizes the idea that particles such as electrons or photons are the quanta of fields. The technique was developed by Pascual JORDAN and his colleagues in the late 1920s.

Segrè, Emilio Gino (1905–89) Italian-born American physicist who made important experimental discoveries in nuclear and particle physics, notably the discovery of the ANTIPROTON with Owen CHAMBERLAIN in 1955 using the BEVATRON. Chamberlain and Segrè also discovered TECHNETIUM and Segrè was also involved in the discovery of ASTATINE and PLUTONIUM.

selection rule A rule that determines which transitions are possible between energy levels, in a quantum mechanical system, such as an atom, molecule, solid, or nucleus. Selection rules state which transitions take place (*allowed transitions*) and which transitions cannot take place (*forbidden transitions*). The rules are determined by the GROUP THEORY associated with the symmetry of the problem. Analysis of the spectra of quantum-mechanical systems is facilitated by selection rules.

selectron *See* supersymmetry.

selenium A metalloid element existing in several allotropic forms and belonging to group 16 of the periodic table. It occurs in minute quantities in sulfide ores and industrial sludges. The common gray metallic allotrope is very light-sensitive and is used in photocells, solar cells, some glasses, and in xerography. The red allotrope is unstable and reverts to the gray form under normal conditions.
Symbol: Se. Melting pt.: 217°C (gray). Boiling pt.: 684.9°C (gray). Relative density: 4.79 (gray). Proton number: 34. Relative atomic mass: 78.96. Electronic configuration: [Ar]$3d^{10}4s^24p^4$.

self-energy The energy associated with the interaction between a particle and the field produced by that particle. For example, the electric charge of an electron has an electric field associated with it. The strength of the electric field is inversely proportional to the square of the distance from the electron. This means that the self-energy of a point electron is infinite, both in classical and quantum field theory. Analogous difficulties occur with other fields such as the gravitational field. For all the nongravitational interactions the difficulties of the infinite self-energies can be circumvented by using RENORMALIZATION.

self-organization A system in which there is a large amount of structure and organization resulting from interactions and processes within the system. There are many examples of self-organized systems in physics, chemistry, and biology. A characteristic feature of self-organized systems is that they are associated with the flow of energy. It has been speculated that the Universe itself is a self-organized system.

semiclassical approximation An approximation technique in which the quantity of interest is expressed as an asymptotic series, with ascending powers of the Planck constant h. The first term in this series is purely classical. This technique is also known as the *Wentzel–Kramers–Brillouin* (*WKB*) approximation; it was invented independently by Gregor Wentzel, Hendrik Anton KRAMERS, and Léon BRILLOUIN in 1926. The semiclassical approximation is very useful for performing calculations involving the tunnel effect such as alpha decay and field emission.

semiconductor A solid that is an insulator at absolute zero temperature but conducts electricity, albeit less than a metal, at nonzero temperatures. Unlike metals, the conductivity of semiconductors increases with temperature. This behavior is understood in terms of energy bands of solids (*see* electronic structure of solids).

The conduction is the result of the net movement in the presence of an electric field of electrons in the conduction band and holes in the valence band. Electrons and holes are called the *charge carriers* of the semiconductor. The *majority carrier* is the type of charge carrier that predominates in a particular region or material, with the *minority carrier* being the charge carrier with the smaller concentration.

An *intrinsic semiconductor* is a pure semiconductor in which the conduction is a property of the pure material. Silicon and germanium are examples of intrinsic semiconductors. If certain impurities are added the conductivity increases dramatically. The process of adding impurities is called *doping*, with the resulting semiconductors being called *extrinsic semiconductors*.

The type of conductivity that results from doping depends on the VALENCE of the impurity atoms. Unlike intrinsic semiconductors, in which the charge carriers are equally divided between electrons and holes, the number of electrons and holes is different if the valence of the impurity atoms is different from that of the host atoms in extrinsic semiconductors. Atoms of silicon and germanium have a valence of four. Atoms such as arsenic, antimony, or phosphorus have a valence of five, which means that if such atoms are added to a lattice of atoms with a valence of four there is one extra electron per impurity atom, which is available for electrical conduction. Thus, electrons are the majority carriers in extrinsic semiconductors doped with atoms of valence five. Such semiconductors are called *n-type semiconductors*. If a semiconductor of valence four atoms is doped with atoms of valence three, such as boron, aluminum, indium, or gallium, one hole per impurity atom is created. Thus, holes are the majority carriers in extrinsic semiconductors doped with atoms of valence three. Such semiconductors are called *p-type semiconductors*.

semi-empirical calculations A technique formerly used extensively for calculating quantities of interest in atomic and molecular theory, in which quantities found from experiment such as ionization energies are used. As the power of computers has increased AB INITIO CALCULATIONS can be used, even for large molecules. This has resulted in a decline in the use of semi-empirical calculations, initially for small molecules and subsequently for large molecules.

semi-metal A metal in which the concentration of charge carriers is several orders of magnitude less than the 10^{22} cm^{-3} which is typical of ordinary metals. Arsenic and bismuth are examples of semi-metals.

shell 1. *See* electronic structure of atoms. 2. *See* shell model.

shell model A model of nuclear structure that is analogous to the shell structure of electrons in atoms. In the nuclear shell model each nucleon moves in an averaged out field due to all the other nucleons. Spin–orbit coupling is an essential feature of the nuclear shell model. The model is able to explain the stability of nuclei with MAGIC NUMBERS and to give good predictions of the spins of nuclei.

shielding 1. A barrier surrounding a region to protect that region from an electric or magnetic field.
2. A barrier surrounding a source of dangerous radiation. For example, there is a barrier in the form of a cement or lead shield around the core of a nuclear reactor to absorb neutrons and other radiation.
3. The reduction in the influence of the nuclear charge on the outer electrons of an atom due to the inner electron shells.

shiggs *See* supersymmetry.

Shockley, William Bradford (1910–89) British-born American physicist who invented the transistor with John BARDEEN and Walter BRATTAIN in the late 1940s. All three shared the 1956 Nobel Prize for physics.

shower A large number of secondary particles produced when a primary COSMIC RAY collides with nitrogen and oxygen nuclei in the atmosphere of the Earth.

Shull, Clifford Glenwood (1915–2001) American physicist who developed the technique of neutron scattering to study matter. He shared the 1994 Nobel Prize for physics with Bertram BROCKHOUSE.

Siegbahn, Kai Manne Börje (1918–   ) Swedish physicist who studied the emission of electrons from materials irradiated by x-rays (*see* photoelectron spectroscopy). Siegbahn won the 1981 Nobel Prize for physics for this pioneering work. Kai Siegbahn is the son of Karl Manne Georg Siegbahn.

Siegbahn, Karl Manne Georg (1886–1978) Swedish physicist who was a pioneer of x-ray spectroscopy. In 1925 Siegbahn and his colleagues showed that x-rays can be refracted, just as light can be. Siegbahn won the 1924 Nobel Prize for physics.

sievert Symbol: Sv. The derived SI unit of dose equivalent of ionizing radiation. It is the dose in grays multiplied by a factor that takes account of the relative effectiveness of different types of radiation of a given energy in causing biological damage. The statutory limits for exposure of the whole body, or specified parts of the body, of various categories of person are expressed in terms of this unit. *See* gray, rem, ionizing radiation.

sigma bond *See* orbital.

sigma particle Symbol: Σ The generic name for a set of three particles. These are denoted Σ^+, Σ^0, and Σ^- for the positively charged, electrically neutral, and negatively charged particles respectively. They were discovered between 1953 and 1956 and are all spin 1/2 strange baryons. The quark content of the sigma particles is that the positively charged sigma consists of two up quarks and one strange quark; the neutral sigma particle consists of one up quark, one down quark, and one strange quark; the negatively charged sigma particle consists of two down quarks and one strange quark. The masses of the sigma particles are about 10 per cent higher than that of protons; 1189.36 MeV (Σ^+), 1192.46 MeV (Σ^0), 1197.34 MeV (Σ^-). The average lifetimes of the sigma particles are 0.8×10^{-10} s (Σ^+), 5.8×10^{-20} s (Σ^0), and 1.5×10^{-10} s (Σ^-). The neutral sigma particle has a much shorter lifetime than the charged sigma particles because it can decay via the electromagnetic interactions.

signature of metric The division of space–time into space dimensions and time dimensions. It has been shown in SUPER-STRING THEORY that any combination of the number of space dimensions and time dimensions that is allowed by supersymmetry is equivalent by duality to a space–time in which there is only one time dimension.

silicon A hard brittle gray metalloid element; the second element in group 14 of the periodic table. Silicon accounts for 27.7% of the mass of the Earth's crust and occurs in a wide variety of silicates with other metals, clays, micas, and sand, which is largely SiO_2. The element is obtained on a small scale by the reduction of silicon(IV) oxide (SiO_2) by carbon or calcium carbide. For semiconductor applications very pure silicon is produced by direct reaction of silicon with an HCl/Cl_2 mixture to give silicon tetrachloride ($SiCl_4$), which can be purified by distillation. This is then decomposed on a hot wire in an atmosphere of hydrogen. For ultra-pure samples zone refining is used. Unlike carbon, silicon does not form allotropes but has only the diamond type of structure.

Symbol: Si. Melting pt.: 1410°C. Boiling pt.: 2355°C. Relative density: 2.329 (20°C). Proton number: 14. Relative atomic mass: 28.0855. Electronic configuration: [Ne]$3s^2 3p^2$.

silver A transition metal that occurs native and as the sulfide (Ag_2S) and chloride (AgCl). It is extracted as a by-product in refining copper and lead ores. It is used in coinage alloys, tableware, and jewelry. Silver compounds are used in photography.

Symbol: Ag. Melting pt.: 961.93°C. Boiling pt.: 2212°C. Relative density: 10.5 (20°C). Proton number: 47. Relative atomic mass: 107.8682. Electronic configuration: [Kr]$4d^{10} 5s^1$.

simple harmonic motion A type of periodic motion in which a body oscillates along a line about a central point, with its range being an equal distance from the central point on either side of the central point, with the body experiencing a force toward the central point that is proportional to its distance from the central point.

single-particle model A model for a many-body system in which the motion of an individual body is found by considering that body to be moving in an averaged out potential due to all the other bodies in the system.

singlet An atomic state which has a MULTIPLICITY of 1, i.e. the value of 2S + 1 is equal to one, where S is the total spin angular momentum of the electrons. Thus S = 0 for a singlet state, meaning that the spin angular momenta of the electrons cancel out.

singularity A place in space–time where quantities such as density and tidal forces become infinite. It was shown by Sir Roger PENROSE, Stephen HAWKING, and others between 1965 and 1970, that singularities inevitably occur in general relativity theory in such situations as inside a black hole and at the big bang at the beginning of the Universe. An important conjecture concerning singularities is the *cosmic censorship conjecture*, which postulates that naked singularities can never be observed. The singularities of general relativity theory are usually interpreted as a limitation of the theory. It is frequently speculated that quantum gravity should be free of singularities.

Sisyphus effect The cooling of atoms by placing them in the standing wave produced by two laser beams propagating in opposite directions. An atom in this combination of beams is like a ball rolling over a sequence of identical hills. Cooling is achieved by using the lasers and magnetic forces to ensure that the atom is always rolling uphill, and hence losing energy and becoming cooler. Temperatures of the very small fraction of a degree above absolute zero can be reached in this way. The effect is named after Sisyphus, a character in Greek mythology who was punished by the gods by having to spend his time rolling a large stone to the top of a hill, letting it roll down again, and then starting again.

SI units (*Système International* d'Unités) The internationally adopted system of units used for scientific purposes. It has six base units (the kilogram, second, kelvin, ampere, mole, and candela) and two supplementary units (the radian and steradian). Derived units are formed by multiplication and/or division of base units; a number have special names. Standard prefixes are used for multiples and submultiples of SI units. The SI system is a coherent rationalized system of units. *See table overleaf.*

skyrmion A SOLITON solution for field theory of mesons which is associated with baryons. The fermionic nature of such solitons emerges from quantization. The skyrmion is named after Anthony Skyrme who put forward this picture of the relation between mesons and baryons in the 1960s. It was revived in the context of quantum chromodynamics by Edward WITTEN in 1983. Skyrmions have been used with considerable success to describe baryons and their interactions. More generally, it has been observed that skyrmions show that fermion fields can emerge from boson fields.

SLAC The Stanford Linear Accelerator Center, an institute concerned with research in elementary particles at which there is a linear particle accelerator two miles (about three kilometers) long. This accelerator was built in the 1960s and was eventually developed to 50 GeV. *See also* SLC.

Slater, John Clark (1900–) American physicist who made important contributions to the quantum theory of atoms, molecules, and solids, particularly the SLATER DETERMINANT.

Slater determinant A determinant that is used for the ANTISYMMETRIC WAVE FUNCTION of a system with several identical fermions, such as the electrons in an atom

or molecule. This type of antisymmetric wavefunction was first used by John SLATER in 1929. It enables the use of the PERMUTATION GROUP to be avoided in many-fermion systems.

SLC (Stanford Linear Collider) A particle accelerator at SLAC that is a modification of the original linear accelerator there. In the SLC electrons and positrons are accelerated in the linear accelerator and guided in such a way that they collide head-on. Since the electrons and protons each have energies of 50 GeV the energy of particles in head-on collisions can reach 100 GeV.

slepton *See* supersymmetry.

slow neutron (**thermal neutron**) A neutron that is moving slowly and hence has a low energy. The energy of a slow neutron is roughly that of its thermal motion. Such a neutron can be captured by an atomic nucleus. Neutrons that occur naturally by radioactivity are much faster than slow neutrons. It is possible to produce slow neutrons by passing a beam of fast neutrons through a moderator. *See also* nuclear reactor.

sneutrino *See* supersymmetry.

Soddy, Frederick (1877–1956) British chemist who made important contributions to the study of radioactivity and isotopes. In his collaboration with Ernest RUTHERFORD at the beginning of the twentieth century he identified the RADIOACTIVE SERIES that begin with thorium and uranium. Rutherford and Soddy showed that the process of radioactivity changes one element into another element. In 1913 Soddy postulated the existence of isotopes of chemical elements when he showed that lead exists in several forms. He won the 1921 Nobel Prize for chemistry for his studies of radioactivity and isotopes.

sodium A soft reactive metal; the second member of the alkali metals (group 1 of the periodic table). Sodium occurs widely as NaCl in seawater and as deposits of halite

BASE AND SUPPLEMENTARY SI UNITS

Physical quantity	Name of SI unit	Symbol for SI unit
length	meter	m
mass	kilogram	kg
time	second	s
electric current	ampere	A
thermodynamic temperature	kelvin	K
luminous intensity	candela	cd
amount of substance	mole	mol
*plane angle	radian	rad
*solid angle	steradian	sr

  *supplementary units

DERIVED SI UNITS WITH SPECIAL NAMES

Physical quantity	Name of SI unit	Symbol for SI unit
frequency	hertz	Hz
energy	joule	J
force	newton	N
power	watt	W
pressure	pascal	Pa
electric charge	coulomb	C
electric potential difference	volt	V
electric resistance	ohm	Ω
electric conductance	siemens	S
electric capacitance	farad	F
magnetic flux	weber	Wb
inductance	henry	H
magnetic flux density	tesla	T
luminous flux	lumen	lm
illuminance (illumination)	lux	lx
absorbed dose	gray	Gy
activity	becquerel	Bq
dose equivalent	sievert	Sv

DECIMAL MULTIPLES AND SUBMULTIPLES USED WITH SI UNITS

Submultiple	Prefix	Symbol	Multiple	Prefix	Symbol
10^{-1}	deci-	d	10^{1}	deca-	da
10^{-2}	centi-	c	10^{2}	hecto-	h
10^{-3}	milli-	m	10^{3}	kilo-	k
10^{-6}	micro-	μ	10^{6}	mega-	M
10^{-9}	nano-	n	10^{9}	giga-	G
10^{-12}	pico-	p	10^{12}	tera-	T
10^{-15}	femto-	f	10^{15}	peta-	P
10^{-18}	atto-	a	10^{18}	exa-	E
10^{-21}	zepto-	z	10^{21}	zetta-	Z
10^{-24}	yocto-	y	10^{24}	yotta-	Y

in dried-up lakes, etc. (2.6% of the lithosphere). The element is obtained commercially by electrolysis of NaCl melts in which the melting point is reduced by the addition of calcium chloride. The metal is extremely reactive.

Symbol: Na. Melting pt.: 97.81°C. Boiling pt.: 883°C. Relative density: 0.971 (20°C). Proton number: 11. Relative atomic mass: 22.989768. Electronic configuration: [Ne]3s^1.

solar neutrino problem The problem caused by the finding that neutrino detectors on the Earth do not detect as many neutrinos coming from the Sun as models of the nuclear processes going on inside the Sun predict. This problem emerged as a result of the persistence of the theoretical astrophysicist John Bahcall and the experimentalist Ray Davis. As the experimental results accumulated in the 1970s and 1980s in a neutrino detector built down a mine in South Dakota, it became clear that only about a third of the expected number of neutrinos was being found. After careful analysis of the nuclear processes going on in the Sun it is now generally thought that some of the neutrinos emitted (which are all electron neutrinos) are converted into muon neutrinos and tau neutrinos on their way to the Earth. This is only possible if neutrinos have nonzero masses. Thus, the solar neutrino problem is consistent with other evidence that neutrinos have nonzero masses, in accord with the theoretical expectations of many GRAND UNIFIED THEORIES.

soliton A stable particle-like wave that occurs as a solution of certain nonlinear equations for propagation. It is thought that solitons have many physical applications in plasmas, fluid mechanics, solid state physics, and elementary-particle physics. INSTANTONS, MONOPOLES, and SKYRMIONS can be regarded as examples of solitons in elementary-particle physics.

Sommerfeld, Arnold Johannes Wilhelm (1868–1951) German physicist who made fundamental contributions to the old quantum theory. In a series of papers, starting in 1915 Sommerfeld generalized Bohr's theory of the atom by postulating that an energy level of an atom is characterized by an orbital quantum number and a magnetic quantum number as well as the principal quantum number. He was able to explain the 'normal' ZEEMAN EFFECT in terms of the magnetic quantum number. The orbital quantum number arose when he generalized Bohr's theory of the atom from circular orbits to elliptical orbits. By combining quantum theory with special relativity theory Sommerfeld was able to obtain agreement between theory and experiment for certain aspects of the FINE STRUCTURE of atomic spectra.

***s*-orbital** *See* orbital.

space group The set of all the symmetry operations of a crystal lattice, i.e. the rotations, reflections, translations, and all the combinations of these operations. In three dimensions there are 230 possible space groups. The theory of space groups is very useful in X-RAY CRYSTALLOGRAPHY and the quantum mechanics of electrons and lattice vibrations in crystals.

space–time The union of three space dimensions and one time dimension, which was proposed by Hermann MINKOWSKI as being the natural setting for special relativity theory and extended by Albert EINSTEIN to the curved space–time described by RIEMANNIAN GEOMETRY, which is the natural setting for general relativity theory. In general relativity theory space–time is not simply an 'arena' where physics takes place but is part of the physics, with the curvature of space–time dictating the motion of massive bodies and the presence of massive bodies dictating the curvature of space–time.

In general relativity theory space–time is viewed as a continuum but it may well be the case that at the PLANCK LENGTH, where quantum effects are likely to become very important, space–time has a discrete structure such as SPACE–TIME FOAM or a SPIN NETWORK.

space–time foam A discrete structure for SPACE–TIME at the PLANCK LENGTH, which has been proposed by John WHEELER and later by Stephen HAWKING. In this picture space–time does not have a static discrete structure but is seething with dynamic activity due to quantum mechanics. Since there is not yet a viable quantum theory of gravity this picture of space–time is speculative.

spallation *See* nucleosynthesis.

spark chamber A type of detector for charged particles that was developed in the 1960s. It consists of a chamber filled with a noble gas such as helium or neon, in which there is a stack of between 20 and 100 metal plates. When an electrically charged particle enters the chamber it leaves a trail of ionized particles in the gas. If the plates are connected alternately to the positive and negative terminals of a source of high voltage, sparks jump between the plates as a charged particle passes through the chamber.

spark counter A type of detector for counting alpha particles. It consists of a wire (or mesh) held a close distance above an earthed plate. The wire has a high positive potential, less than that required to cause a spark. When an alpha particle passes close to the wire the electric field increases and a spark occurs between the electrodes. The sparks can be counted by a suitable circuit.

sparticle *See* supersymmetry.

SPEAR (Stanford Positron–Electron Asymmetric Rings) An accelerator at Stanford. SPEAR is supplied with electrons and positrons, which move in opposite directions until they are brought together in head-on collisions. It was completed in 1972 and reaches energy of several GeV. In 1974 evidence for the existence of the J/PSI PARTICLE emerged from SPEAR.

special relativity theory A theory of motion that gives a generalization of Newtonian mechanics that is needed when dealing with objects that are moving at a substantial fraction of the speed of light. Special relativity theory applies to inertial frames of reference. The two fundamental posulates of special relativity theory are:
(1) Physical laws are identical in all frames of reference;
(2) The speed of light in a vacuum, denoted c, is a constant throughout the Universe and is independent of the speed of the observer.

Special relativity theory was put forward by Albert EINSTEIN in 1905 and expressed geometrically in terms of four-dimensional SPACE–TIME by Hermann MINKOWSKI in 1908. Special relativity theory reduces to Newtonian mechanics in the limit of low velocities. A consequence of the theory is that the mass of a body increases with its speed (*see* relativistic mass). The form of this result has the consequences that the speed of light in a vacuum is the highest possible speed and that it is impossible for a massive body to attain this speed. Einstein also found that special relativity leads to the conclusion that the mass m of a body is a measure of its energy content E, in accord with his celebrated equation $E = mc^2$. There is a vast amount of experimental evidence in support of special relativity theory.

To describe accelerated motion such as acceleration due to gravity it is necessary to extend special relativity theory to GENERAL RELATIVITY THEORY.

specific heat The amount of heat which is needed to raise a unit mass of a substance by one unit of temperature. The units for the specific heat of a substance are usually joules per kilogram per kelvin ($Jkg^{-1}K^{-1}$).

Since specific heats depend on the way that pressure and volume change during heat transfer specific heat is defined both at constant pressure and at constant volume. For solids and liquids there is not much difference between the two definitions but for gases there is a substantial difference.

spectra *See* spectrum.

spectroscope An instrument for studying spectra. The first spectroscope was constructed by Robert BUNSEN. It consisted of a hollow tube with a slit through which light entered, a lens to produce a parallel beam, a prism to disperse the light, and a telescope to observe the spectrum. For many purposes a diffraction grating is used rather than a prism.

spectroscopy The study of spectra. Spectroscopy can be divided into *nuclear spectroscopy*, *atomic spectroscopy*, and *molecular spectroscopy*. Since spectra are caused by transitions between energy levels spectroscopy allows the energy levels of systems to be determined. This, in turn, enables information about nuclear structure, atomic structure, and molecular structure to be obtained. Spectroscopy is used in chemical analysis and in determining the composition and motion of celestial bodies.

spectrum (*pl.* **spectra**) A range of electromagnetic radiation that is absorbed or emitted by a substance. In an *absorption spectrum* a continuous range of electromagnetic radiation is passed through a sample and the radiation is then analyzed to determine which wavelengths have been absorbed. In an *emission spectrum* the electromagnetic radiation that is emitted by a substance is analyzed to determine which wavelengths have been emitted.

A *continuous spectrum* has a continuous distribution of wavelengths over a wide range. It is typically produced by incandescent bodies. A *line spectrum* is a set of discontinuous lines. Line spectra are produced by transitions from excited states in atoms or ions to lower energy levels. A *band spectrum* contains bands of closely spaced lines. Band spectra are produced by molecules, with there being many quantized vibrational states of a molecule.

speed of light Symbol: *c* The speed at which electromagnetic waves, including light, travel. In a vacuum the speed of light is 2.99792458×10^8 meters per second. This is a universal constant. It is a fundamental postulate of special relativity theory that the speed of light in a vacuum is independent of the speed of the observer. Spe-

cial relativity theory shows that the speed of light in a vacuum is the highest possible speed that can be attained. Since 1983 the speed of light in a vacuum has been used to define the meter. When light travels in a material medium its speed is reduced, with it being possible to have bodies moving faster than light in a material medium (*see* Cerenkov radiation).

spin Symbol: s A quantum-mechanical property of a particle that is its intrinsic angular momentum (as opposed to its orbital angular momentum). Unlike spin in classical mechanics, such as the rotation of the Earth, spin is a quantized quantity and is limited to multiples of half the DIRAC CONSTANT ($h/2\pi$). The Dirac constant part is usually taken as the basic unit and the spin of a particle is expressed as the quantity multiplying the Dirac constant. Thus, the electron, quark, proton, and neutron all have spin 1/2. There is a *spin quantum number* associated with the spin of a particle. For example, the possible values of the spin quantum number for a spin 1/2 particle are +1/2 and −1/2. A particle has an intrinsic MAGNETIC MOMENT due to its spin, which has the consequence that it interacts with an external magnetic field. Examples of spin 1 particles are photons, gluons, and W and Z bosons.

The spin of composite systems, such as a nucleus with more than one nucleon, is determined by WIGNER'S RULE. *See also* boson; fermion; spin–statistics theorem.

spin glass An alloy in which a small amount (between 0.1 and 10 per cent) of a magnetic metal is distributed randomly in the lattice of a nonmagnetic metal. Examples of spin glasses are CuMn and AuFe. The random distribution of the magnetic atoms makes theories of the magnetic and other properties of spin glasses much more complicated than for ferromagnetic and antiferromagnetic materials.

spin network A discrete structure for space–time at the microscopic level in which it is postulated that space–time is a large network of discrete one-dimensional entities. This speculative picture was put forward in the 1960s by Sir Roger PENROSE and has been developed with much success in the 1990s in the context of QUANTUM GRAVITY.

spinor In general, a spinor can be regarded as a quantity which, rather than being returned to its original state when it is rotated by 360°, is taken to −1 times its original and is only restored to its original state when it is rotated by 720°. The quantum state of a spin 1/2 particle such as an electron is a physically important example of a spinor. In nonrelativistic quantum mechanics two-component PAULI SPINORS are used to describe the quantum states of spin-1/2 particles. In relativistic quantum mechanics DIRAC SPINORS are used to describe the quantum states of spin-1/2 particles. Spinors are also used in classical electrodynamics and general relative theory.

spin–orbit coupling The interaction between the orbital angular momentum of a particle and its spin angular momentum. Spin–orbit coupling of electrons is responsible for FINE STRUCTURE in atomic spectra. For light atoms, spin–orbit coupling is small, meaning that multiplets are well described by RUSSELL–SAUNDERS COUPLING. Spin–orbit coupling increases at atomic number increases. For heavy atoms, spin–orbit coupling is large, meaning that J–J COUPLING gives a better description of multiplets. In nuclei there is spin–orbit coupling as a result of the strong nuclear interaction. It is necessary to take this into account in the nuclear SHELL MODEL to derive the MAGIC NUMBERS.

spinor field A quantum field associated with spinors.

spin quantum number *See* electronic structure of atoms.

spin–spin interaction The interaction between the spin of one electron and the spin of another electron. This interaction gives rise to fine structure in atomic spectra. Spin–spin interaction does not increase with atomic number. This means that, for all but the lightest atoms, spin–spin interaction is much smaller than spin–orbit coupling.

spin–statistics theorem A theorem of relativistic quantum field theory that states that particles which obey BOSE–EINSTEIN STATISTICS, i.e., bosons, always have integer spins and that particles which obey FERMI–DIRAC STATISTICS, i.e., fermions, always have half-integer spins. This theorem was first proved in a general way by Wolfgang PAULI in 1940 and provides a foundation for the PAULI EXCLUSION PRINCIPLE. Since that time the spin–statistics theorem has been proved using other methods.

spin wave (**magnon**) A collective excitation associated with ferromagnetism and antiferromagnetism. A spin wave is an example of a GOLDSTONE BOSON in nonrelativistic quantum mechanics.

spontaneous compactification The way in which higher dimensions in theories such as KALUZA–KLEIN THEORY and SUPERSTRING THEORY, which involve more than four space–time dimensions, become very small. Some very plausible and attractive suggestions have been made concerning spontaneous compactification in superstring theory but these suggestions cannot be considered to be definitive at the present time.

spontaneous emission The emission of a photon when an atom returns from a higher energy level to a lower energy level. This process occurs without any external stimulus to trigger the emission. It is caused by interactions between atoms and vacuum fluctuations of the quantized electromagnetic field. This means that spontaneous emission cannot be described by the nonrelativistic Schrödinger equation but needs quantum electrodynamics to be explained properly. Spontaneous emission is responsible for the finite lifetime of an excited state of an atom before it emits a photon. *See also* quantum theory of radiation.

spontaneous symmetry breaking A situation in which the state of lowest energy of a system has a lower symmetry than the Lagrangian or Hamiltonian which specifies the system. Spontaneous symmetry breaking is a ubiquitous phenomenon in physics, particularly condensed-matter and elementary-particle physics, and is frequently associated with phase transitions. For example, when hot iron cools there is spontaneous symmetry breaking to the ferromagnetic state when it reaches the CURIE TEMPERATURE. It is thought that in the early Universe all the forces were unified and that the different forces emerged in a sequence of phase transitions associated with spontaneous symmetry breaking. *See also* broken symmetry.

sputtering   The ejection of atoms from a cathode due to the bombardment by heavy positive ions. It can be used to produce a film of metal on a surface.

squark   *See* supersymmetry.

SQUID   (superconducting quantum interference device) A device consisting of a ring of superconducting material with a radius of about 0.25 cm in which there is a very small constriction at one point where the cross-section of the ring is about one ten millionth of a square centimeter. This constriction acts as a JOSEPHSON JUNCTION. Standing waves are formed by electrons in the ring. SQUIDS are used both to investigate fundamental aspects of quantum mechanics and to measure currents and voltages.

stability of atoms   A phenomenon which can only be explained by quantum mechanics. An atom cannot be an array of static electric charges by EARNSHAW'S THEOREM. Classical electrodynamics does not permit electrons orbiting round a positively charged nucleus since the electron would very rapidly spiral towards the nucleus, giving off electromagnetic radiation continuously while doing so. However, in quantum mechanics this is not possible since it would violate the Heisenberg UNCERTAINTY PRINCIPLE. Although there is an electrostatic force that attracts the electrons to the nucleus, confining them to a

small region would increase their energy, thus allowing them to escape from this attraction. Stability in atoms is achieved by the balance between these opposing tendencies. *See also* electronic structure of atoms.

Standard Model (particle physics)   The combination of QUANTUM CHROMODYNAMICS and the WEINBERG–SALAM MODEL. In spite of its great success in describing all the nongravitational interactions it is not a complete theory of elementary particles, firstly because it does not include gravity and secondly because many quantities of interest, such as the mass of electrons, are empirical parameters rather than calculated from first principles. There are many theories of elementary particles that attempt to go beyond the Standard Model including GRAND UNIFIED THEORIES and SUPERSTRING THEORY.

star   A celestial body that shines due to nuclear fusion reactions going on inside it. The Sun is an example of a star. Stars are not distributed uniformly in the Universe but come together in galaxies, clusters of galaxies, etc. *See* structure formation.

Stark, Johannes (1874–1957)   German physicist who discovered the STARK EFFECT in 1913 and who predicted in 1902, and demonstrated in 1905, that fast-moving positively charged ions in a cathode-ray tube exhibit a DOPPLER EFFECT. Stark was awarded the 1919 Nobel Prize for physics for these discoveries.

Stark effect   The splitting of atomic spectral lines due to the presence of a strong electric field. Like the normal ZEEMAN EFFECT, the Stark effect can be explained in terms of the classical electron theory of Lorentz and the old quantum theory. In quantum mechanics the Stark effect is described using perturbation theory, with the electric field being a perturbation on the quantum states of an atom in the absence of an electric field.

state   In quantum mechanics all information about a physical system is characterized by a vector in Hilbert space (or equivalently a wave function). For that reason the terms *state*, *state vector*, and *wave function* are used interchangeably. A state in quantum mechanics is characterized by a set of quantum numbers.

stationary state A quantum mechanical state in which the expectation value of quantities such as the energy does not change with time. The state with the lowest energy is the *ground state*, with states that have higher energy being called *excited states*.

statistical mechanics The use of statistical methods applied to the large number of microscopic constituents of a system so as to predict its properties. It was initiated by James Clerk MAXWELL and Ludwig BOLTZMANN in the kinetic theory of gases. If the constituents of the system and their interactions are described by classical mechanics then this gives rise to what is called *classical statistical mechanics*. The statistical mechanics which arises when the constituents and their interactions are described by quantum mechanics is called *quantum statistical mechanics*. Statistical mechanics is divided into *equilibrium statistical mechanics*, which is concerned with systems in thermal equilibrium, and the more difficult subject of *nonequilibrium statistical mechanics*, which is concerned with systems that are not in thermal equilibrium.

steady-state theory A cosmological theory put forward in the 1940s by Sir Hermann BONDI, Thomas GOLD, and Sir Fred HOYLE in which it was postulated that the Universe always appears the same overall to all observers everywhere at all times. Since the Universe is observed to be expanding this required the continuous creation of matter. Since the steady-state theory could not account for the evidence which shows that the Universe has evolved and changed with time it fell out of favor by the late 1960s, particularly after the discovery of the cosmic microwave background radiation and other evidence in favor of the BIG-BANG THEORY, such as the

success of the hypothesis of PRIMORDIAL NUCLEOSYNTHESIS in explaining the cosmic abundances of light elements.

Stefan–Boltzmann law A result which states that the total energy radiated per unit area per unit time by a black body is proportional to the fourth power of its thermodynamic temperature. This law was discovered by Josef Stefan in 1879 and derived theoretically by his former student Ludwig BOLTZMANN in 1884 using thermodynamics and kinetic theory.

Steinberger, Jack (1921–) German-born American physicist who discovered the muon neutrino with Leon LEDERMAN and Melvin SCHWARTZ in the early 1960s. He shared the 1988 Nobel Prize for physics with Lederman and Schwartz for this discovery.

stellar evolution The evolution of stars from their birth to their final fate. It is thought that a star is born due to clouds of gas and dust in space collapsing due to gravity. Due to its gravitational contraction this cloud of matter has its internal pressure raised. This rise in pressure raises its temperature. When the temperature is sufficiently high, nuclear fusion reactions start to occur. The heat from these processes generates sufficient pressure to prevent further collapse. The main nuclear processes which fuel stars are the CARBON CYCLE and the PROTON–PROTON REACTION. The lifetime of a star and its final fate depends on its mass.

Eventually, the star must run out of hydrogen at its core. When this happens the core is mostly helium, with the core being surrounded by an envelope which is mainly hydrogen. This core collapses until the shell around the core gets sufficiently hot to start further conversion of hydrogen into helium. This causes the outer envelope of the star to expand and cool. This cooling causes the color of the star to change from white to red. For that reason a star at that stage of its evolution is called a *red giant*. A star has a relatively short part of its life as a red giant (usually between about 5 and 20 per cent of the time it spends as an ordinary hydrogen-burning star). Helium in the core is the nuclear fuel for a star at this stage of its life.

If the star has a low mass it soon runs out of helium and the core collapses to a *white dwarf*, while the outer layers drift into space. It is thought that the Sun will eventually end up as a white dwarf. For stars that have sufficient mass further nuclear reactions occur, building up elements by stellar nucleosynthesis. Iron is the heaviest element that can be formed with the production of energy (*see* nucleosynthesis). When all the nuclear fuel from a star with a large mass has been exhausted there is a dramatic collapse, causing a SUPERNOVA explosion, with the outer layers being blown away. The final fate of a large star is to become either a NEUTRON STAR or a BLACK HOLE, depending on its mass (*see* Chandrasekhar limit; Oppenheimer–Volkoff limit).

stellar nucleosynthesis *See* nucleosynthesis.

steradian Symbol: sr   The SI unit of solid angle. The surface of a sphere, for example, subtends a solid angle of 4π at its center. The solid angle of a cone is the area intercepted by the cone on the surface of a sphere of unit radius.

Stern, Otto (1888–1969) German-born American physicist who carried out important experiments with atomic and molecular beams. In the early 1920s he performed the STERN–GERLACH EXPERIMENT with Walther Gerlach. In 1931 he showed that the particles in molecular beams also have wavelike properties. In 1933 he measured the magnetic moment of the proton using molecular beams. His finding that it was substantially different from the theoretical expectation of the Dirac equation showed that the proton is not a simple pointlike charged particle. Stern won the 1943 Nobel Prize for physics for his work on the development of atomic and molecular beams.

Stern–Gerlach experiment An experiment carried out by Otto STERN and Walter Gerlach in the early 1920s in which a beam of silver atoms was passed through a nonuniform magnetic field. It was found that the beam split into two separate beams, in accord with the quantum theoretical prediction of Arnold SOMMERFELD, rather than being broadened, as would be expected from classical theory.

stimulated emission *See* quantum theory of radiation.

STM *See* scanning tunneling microscope.

Stoner, Edmund Clifton (1889–1973) English physicist who formulated the MAIN-SMITH–STONER RULER on the distribution of electrons in atoms in 1924 and who dedicated most of his career to explaining the properties of ferromagnetic materials in terms of band theory and the interactions of electrons.

Stoney, George Johnstone (1826–1911) Irish physicist who is best known for introducing the word 'electron' into science. He is said to have done so in a lecture given in 1874 but did not do so in print until 1891. He also made significant contributions to spectroscopy and the kinetic theory of gases.

stopping power A measure of the ability of a substance to reduce the kinetic energy of a particle passing through it. The *linear stopping power*, denoted S_l, is the energy loss of a particle per unit distance. It is given by $S_l = -dE/dx$, where E is the kinetic energy of the particle and x is the distance traveled in the substance. The *mass stopping power*, denoted S_m, is the linear stopping power divided by the density ρ of the substance, i.e. $S_m = S_l/\rho$. The *atomic stopping power*, denoted S_a is the energy loss per unit area normal to the motion of the particle, i.e. $S_a = S_l/n = S_l A/\rho N$, where A is the atomic weight of the atom, N is the Avogadro number and n is the number of atoms per unit volume. The *relative stopping power* is the ratio of the stopping power of a substance to that of a standard substance, which is usually taken to be air, aluminum, or oxygen.

storage ring A large evacuated toroidal ring which is used in some accelerators. It employs magnetic fields to ensure that the particles inside them, such as electrons and positrons, do not lose energy before very high-energy collisions occur.

Störmer, Horst L. (1949–) German-born American physicist who discovered the fractional QUANTUM HALL EFFECT with Daniel TSUI. Störmer and Tsui shared the 1998 Nobel Prize for physics for this discovery with Robert LAUGHLIN, who provided a theoretical explanation of it.

STP Standard temperature and pressure. It is defined as 0°C and one atmosphere (101 325 pascals). It is used to provide standard conditions when measuring quantities such as the density of a gas that vary with both temperature and pressure.

strange attractor A path in phase space that is not closed. Strange attractors are characteristic of chaotic behavior. See attractor, phase space.

strange matter Matter that consists of up, down, and strange quarks. Normal matter does not contain strange quarks since protons and neutrons are made up from up and down quarks. It has been suggested that strange matter might have been formed in the EARLY UNIVERSE. It has also been suggested that some neutron stars might have strange matter at their cores or even mostly consist of strange matter. A neutron star made of normal matter cannot spin faster than once every few milliseconds without falling apart. However, because a neutron star made of strange matter would be more compact than a neutron star made of normal matter it could spin twice as fast. Thus, if a pulsar is found which spins faster than about half a millisecond this would be good evidence for strange matter.

strangeness Symbol: s A quantum number given to STRANGE PARTICLES.

strange quark See quark.

strange particles Particles that contain at least one STRANGE QUARK. The KAON and the LAMBDA PARTICLE are examples of strange particles. The strange particles were so named because when they were first found in cosmic-ray showers in the late 1940s and early 1950s it was observed that they were always produced in pairs and that their lifetimes were much longer than expected. It was postulated that if one of the members of a pair is assigned a strangeness of +1 then the other particle is assigned a strangeness of –1, thus ensuring that strangeness is conserved in the production of the pair. Particles with one strange quark are assigned a strangeness s = –1 and particles with one anti-strange quark are assigned a strangeness s = +1. Strangeness is conserved in the fast strong interaction process which creates the pair of strange particles (and in processes involving electromagnetic interactions) but is not conserved in the slow weak interaction processes of the decay of strange particles. The concept of strangeness and strange particles was put forward by Murray GELL-MANN and independently by Kazuhiko NISHIJIMA in 1953.

Strassmann, Fritz (1902–80) German chemist who collaborated with Otto HAHN on the experiment which first clearly demonstrated the phenomenon of nuclear fission.

string A one-dimensional object that appears in elementary-particle theory. Strings sometimes appear as solutions of the equations of quantum field theories. For example, cosmic strings are solutions for certain grand unified theories. Strings are the fundamental entities in *string theories*, i.e. theories in which pointlike elementary particles are replaced by strings of a finite length (*open strings*) or loops (*closed strings*). SUPERSTRING THEORY is the combination of string theory and SUPERSYMMETRY.

stripping reaction   *See* nuclear reaction.

strong CP problem   The problem that when INSTANTON effects are taken into account QUANTUM CHROMODYNAMICS (QCD) is CP violating. This is a problem because CP is observed to be a conserved symmetry in strong interaction processes and QCD is thought to be the theory describing the strong interactions. Solutions to the strong CP problem have been proposed both in the framework of QCD and of more general theories. At present it has not been established as to which, if any, of these solutions is correct.

strong interactions   *See* fundamental interactions.

strong nuclear interactions   *See* fundamental interactions.

strontium   A soft low-melting reactive metal; the fourth member of group 2 of the periodic table and a typical alkaline-earth element. Strontium has a low abundance in the Earth's crust, occurring as strontianite ($SrCO_3$) and celestine ($SrSO_4$). The isotope strontium-90 was produced in nuclear tests in the 1950s, and occurs in the environment. It is absorbed by the body and can be incorporated into bone.

Symbol: Sr. Melting pt.: 769°C. Boiling pt.: 1384°C. Relative density: 2.54 (20°C). Proton number: 38. Relative atomic mass: 87.62. Electronic configuration: $[Kr]5s^2$.

structure formation   The formation of large-scale structure such as galaxies, clusters of galaxies, etc. in the Universe. Most theories of structure formation are based on the JEANS INSTABILITY in the context of general relativity theory, with the large-scale structure emerging from the inhomogeneity of matter in the Universe. It is thought that quantum fluctuations in the EARLY UNIVERSE were responsible for the necessary inhomogeneities, with INFLATION expanding these fluctuations. Evidence that this theory is correct comes from the small variations in the cosmic microwave background radiation found by COBE. At present none of the detailed quantitative theories of structure formation can be regarded as definitive. One of the difficulties is that the nature of DARK MATTER is not known.

subcritical   Describing an arrangement of fissile material that does not permit a sustained chain reaction because too many neutrons are absorbed without causing fission or otherwise lost. *Compare* supercritical. *See also* multiplication factor.

subshell   *See* electronic structure of atoms.

sulfur   A low-melting nonmetallic solid, yellow colored in its common forms; the second member of group 16 of the periodic table. Sulfur occurs in the elemental form in Sicily and some southern states of the USA, and in large quantities in combined forms such as sulfide ores (FeS_2) and sulfate rocks ($CaSO_4$). It forms about 0.5% of the Earth's crust. The element exhibits allotropy and its structure in all phases is quite complex. The common crystalline modification, *rhombic* sulfur, is in equilibrium with a *triclinic* modification above 96°C. Both have structures based on S_8-rings but the crystals are quite different. If molten sulfur is poured into water a dark red 'plastic' form is obtained in a semi-elastic form. The structure appears to be a helical chain of S atoms. Selenium and tellurium both have a gray 'metallike' modification but sulfur does *not* have this form. The British spelling is 'sulphur'.

Symbol: S. Melting pt.: 112.8°C. Boiling pt.: 444.6°C. Relative density: 2.07. Proton number: 16. Relative atomic mass: 32.066. Electronic configuration: $[Ne]3s^23p^4$.

Sunyaev–Zel'dovich effect   A distortion in the COSMIC MICROWAVE BACKGROUND radiation caused by photons in the background radiation being scattered to slightly higher energies by electrons in the hot plasma in clusters of galaxies along the line of sight. This effect was predicted by Rashid Sunyaev and Yakov ZEL'DOVICH in 1972. It predicts that the temperatue of the background radiation is changed by about one part in ten thousand. The Sunyaev–Zel'dovich effect has been measured for some clusters of galaxies.

superconductivity The loss of electrical resistance that occurs in certain substances below a certain critical temperature which is characteristic of that substance. Superconductivity is also characterized by the MEISSNER EFFECT, i.e. the expulsion of magnetic flux. Superconductivity occurs in 26 metallic elements and many alloys and compounds. It was discovered by Kamerlingh ONNES in 1911.

Most superconductors have transition temperatures that are a few degrees above the absolute zero temperature. In the mid 1980s a number of ceramic materials which are superconductors up to much higher temperatures (liquid nitrogen temperatures) were found. This phenomenon is known as *high-temperature superconductivity*.

There are certain practical applications of superconductivity, such as the production of very strong magnetic fields, and the discovery of high-temperature superconductivity gives the possibility of far more applications, such as high-speed computers and power transmission.

Superconductivity is explained (with the exception of high-temperature superconductivity) by the BARDEEN–COOPER–SCHRIEFFER THEORY (BCS theory), in which the interaction between an electron and the lattice give rise to an attraction to another electron. Thus, electrons are paired, with these pairs being called *Cooper pairs*. In the BCS theory the superconducting state can be regarded as a Bose–Einstein condensate of the Cooper pairs. At present, there is not a consensus as to what the correct theory of high-temperature superconductivity is.

supercooling The cooling of a liquid to below its freezing temperature without freezing it. Such a supercooled state is metastable, with any disturbance causing rapid freezing. There is also an analogous cooling of a vapor which is used in a CLOUD CHAMBER.

supercritical Describing an arrangement of fissile material that sustains a branching chain reaction; i.e. more neutrons are being produced than are wasted and escape. *Compare* subcritical. *See also* multiplication factor.

superfluidity The property of liquid helium that at very low temperatures it flows without friction. Helium-4 becomes superfluid at about 2.2 K. The Bose–Einstein condensation of helium-4 atoms is a key ingredient in understanding superfluidity.

Very feeble interatomic forces between He-3 atoms enables pairs of such atoms to be formed, with these pairs undergoing a Bose–Einstein condensation to a superfluid state. This occurs a few thousandths of a degree above the absolute zero temperature. The properties of superfluid helium-3 and superfluid helium-4 are very different.

supergravity A SUPERSYMMETRIC FIELD THEORY that incorporates gravity. It is most natural to formulate supergravity theory as a KALUZA–KLEIN THEORY in 11 space–time dimensions. It is widely thought, but has never been proved, that supergravity has infinities that cannot be removed by the procedure of RENORMALIZATION. For that reason, and others, it is now thought that supergravity is not capable of being a successful UNIFIED FIELD THEORY. However, supergravity theory is related to SUPERSTRING THEORY by DUALITY. *See also* superstring theory.

superheated liquid A liquid that has been heated to a temperature which is above the normal boiling temperature at the applied pressure.

supermembrane theory A theory that combines SUPERSYMMETRY with the basic entities being two-dimensional extended objects (called *membranes*). There are serious difficulties with supermembrane theory as a starting point for a unified theory of forces and particles but it may emerge from M-THEORY.

supermodel A model that describes elementary particles that incorporates SUPER-

SYMMETRY. Supersymmetric GUTs and the supersymmetric Standard Model are examples of supermodels.

supernova (*pl.* **supernovae**) An explosive brightening of a star in which the energy radiated by it increases by a very large amount. For a short time a supernova shines more brightly than a whole galaxy. Supernova explosions are rare events. It is thought that only a few supernovae occur in a galaxy like the Milky Way each century. Nevertheless, supernovae are very important because they both manufacture elements heavier than iron and disperse all heavy elements through space when they explode. A supernova explosion occurs when a star has used up all its available nuclear fuel and suddenly collapses to a NEUTRON STAR.

Super Proton Synchrotron An accelerator that was constructed at CERN in the 1970s so as to accelerate protons to high energies (about 400 GeV). It was modified in the 1980s to a proton–antiproton collider, with energies of 900 GeV being obtained.

supersaturation The state of a vapor in which the pressure exceeds that at which condensation normally occurs (at the temperature being considered). The supersaturated state is a metastable state and readily disturbed (by electrically charged particles for example).

superstring theory A unified theory of the fundamental interactions that combines STRING THEORY with SUPERSYMMETRY. In superstring theory the basic entities are one-dimensional objects called *superstrings* which are about 10^{-35} m long. This very short distance is equivalent to energies of about 10^{19} GeV.

A purely *bosonic string*, i.e. a string that is associated with bosons and does not contain fermions, is only consistent as a quantum theory in 26-dimensional space–time. A superstring contains both bosons and fermions and is only consistent as a quantum theory in 10-dimensional space–time. It is postulated that four large space–time dimensions emerge from superstring theory by SPONTANEOUS COMPACTIFICATION.

A great merit of superstring theory is that it leads to massless spin-2 particles. This means that superstring theory automatically contains a quantum theory of the gravitational interactions. However, an unsatisfactory feature of the way that superstring theory is currently formulated is that it envisages a string as moving through space–time, whereas in a true theory of quantum gravity space–time should emerge as a derived concept. Some work has been done which attempts to remedy this deficiency by combining spin networks in quantum gravity and superstring theory. There is some evidence that superstring theory is free of the infinities that plague attempts to construct a quantum field theory of gravity and which cannot be dealt with by renormalization. However, a complete proof of this has not yet been established.

Superstrings can either be open strings or closed strings. By the mid 1980s it was found that there were five superstring theories which are consistent. Some of these theories are associated with the gauge group SO(32), while the others are associated with the gauge group E(8) × E(8), (*see* Lie group). In the mid 1990s it was found that the five types of superstring theory and also 11-dimensional supergravity theory are related to each other by a set of dualities. This has led to the postulate that there is a theory in 11-dimensional space–time called *M-theory* which contains all the five superstring theories and supergravity theory as limiting cases. M-theory is not well understood at the time of writing. It appears to involve SUPERMEMBRANES and more generally *branes*, i.e. extended objects, with a one-brane being a string, a two-brane being a membrane, etc.

supersymmetric field theory A quantum field theory that incorporates supersymmetry. There are cancellations in such theories which result in the difficulties of infinities being less severe than in nonsupersymmetric field theories. Some supersymmetric field theories are completely finite. Supersymmetric gauge theories have more mathematical structure associated with them than nonsupersymmetric gauge theories. This allows more nonperturbative information to be obtained from supersymmetric gauge theories.

supersymmetric GUTs (supersymmetric grand unified theories) Grand unified theories that incorporate supersymmetry. There is tenuous experimental evidence in support of supersymmetric GUTs from the evolution of coupling constants as a function of energy. In addition, supersymmetric GUTs predict a much longer lifetime for the decay of the proton, thus providing an explanation of why proton decay has not been observed.

supersymmetric Standard Model The extension of the Standard Model of particle physics to incorporate supersymmetry. The supersymmetric Standard Model makes a number of predictions about particle physics. In particular, it predicts that there should be several Higgs bosons. Although there is no experimental evidence for the supersymmetric Standard Model at the time of writing it is hoped that future accelerators will enable its predictions to be tested.

supersymmetry (SUSY) A symmetry that combines fermions and bosons into one mathematical structure. Supersymmetry has been a key ingredient in attempts to find a unified theory of all the fundamental interactions and elementary particles. In supersymmetry all the known particles have *partners*, sometimes called *superpartners*. The partners of existing fermions are bosons which are named by adding 's' to the beginning of the name of the fermion. For example, one has *selectron, squark*, and *slepton*. The partners of existing bosons are fermions which are named by having 'ino' at the end of the name. For example, one has *gluino, photino*, and *zino*. Some of the particles predicted by supersymmetry are good candidates for much of the DARK MATTER in the Universe.

For supersymmetry to be relevant to the elementary particles that are presently observed it must be a broken symmetry (*see*

supersymmetry breaking). At present, the experimental evidence for supersymmetry is tenuous, with none of the superpartners having been discovered. It is hoped that evidence for supersymmetry will come from future accelerators. *See also* superstring theory.

supersymmetry breaking The spontaneous symmetry breaking of supersymmetry. Since the observed fermions and bosons are not the supersymmetric partners of each other supersymmetry must be a spontaneously broken symmetry. Supersymmetry cannot be a GLOBAL SYMMETRY since its realization as a broken symmetry would give rise to a *Goldstone fermion*, i.e. the fermionic analog of a GOLDSTONE BOSON, and no such particle has been observed. Thus, if supersymmetry is a symmetry of Nature there must be an analog of the Higgs mechanism that gives the supersymmetric particles masses beyond those which can be reached by particle accelerators at the time of writing. Many detailed mechanisms have been proposed for supersymmetry breaking, particularly in SUPERSTRING THEORY, but none of these mechanisms are compelling. It may well be the case that supersymmetry breaking will not be properly understood until accelerators discover supersymmetric particles.

supertransuranics (superheavy elements) A set of elements with atomic numbers of about 114 and mass numbers of about 298 that is predicted by nuclear shell theory to exist as an 'island of stability' due to 114 protons and 184 neutrons both being stable. It has been claimed that element 114 has been synthesized but, at the time of writing, this claim cannot be regarded as well-established since superheavy elements are very difficult to synthesize.

susceptibility *See* electric susceptibility; magnetic susceptibility.

SUSY *See* supersymmetry.

symmetric wave function A wave function for a system with more than one identical particle in which interchanging the coordinates of any two of these particles does not change the wave function. Systems with more than one boson are described by symmetric wave functions. Because of this, unlike fermions, there is no restriction on the number of bosons in a given quantum state. *See also* antisymmetric wave function.

symmetry The set of *invariances* of a system, i.e. the set of operations, known as *symmetry operations*, on a system that leave the system unchanged. Examples of symmetry operations include rotations and reflections for molecules and translations for crystals. Symmetry is studied systematically by a branch of mathematics called GROUP THEORY. Symmetry and BROKEN SYMMETRY are two of the most important concepts throughout physics and chemistry.

synchrocyclotron An accelerator that modifies the CYCLOTRON by synchronizing the field which is causing the acceleration with the relativistic increase in the mass of the particle as its speed increases.

synchrotron A particle accelerator that accelerates particles in a toroidal ring, with the fields which produce the acceleration being synchronized with the relativistic increase in the mass of the particle as its speed increases.

synchrotron radiation The electromagnetic radiation that is emitted by charged particles when they are moving at speeds which are a substantial fraction of the speed of light. The greater the speed of the charged particles is, the shorter the wavelength of the radiation is. Synchrotron radiation gets its name from the fact that it is produced by charged particles accelerated in a synchrotron but also occurs naturally when rapidly moving electrons spiral along magnetic lines of force. For example, synchrotron radiation is produced around pulsars in this way.

T

tachyon A hypothetical type of particle that moves faster than the speed of light in a vacuum. No such particle has ever been observed and it is generally thought that tachyons cannot exist.

Tamm, Igor Yevgenyevich (1895–1971) Russian physicist who, together with Ilya FRANK, explained CERENKOV RADIATION soon after its discovery. Cerenkov, Frank, and Tamm shared the 1958 Nobel Prize for physics for this work. Tamm made a number of other important contributions to physics including the theory of surface states in solids.

tandem generator An arrangement of two linear accelerators end to end. The ends are earthed and there is a common anode at the center at a potential difference of up to 20 million volts. Negative ions are accelerated from earth potential at one end to the center, where the surplus electrons are stripped off to convert them into positive ions, which then continue to accelerate to the far end. They thus get the acceleration due to the high potential twice over. Protons can be given energies up to 40 MeV.

tantalum A silvery transition element. It is strong, highly resistant to corrosion, and is easily worked. Tantalum is used in turbine blades and cutting tools and in surgical and dental work. It has a stable isotope (tantalum-181) and a long-lived radioactive isotope (tantalum-180, with a half-life of $>10^7$ years). It also has several short-lived isotopes.
   Symbol: Ta. Melting pt.: 2996°C. Boiling pt.: 5425 ± 100°C. Relative density: 16.654 (20°C). Proton number: 73. Relative atomic mass: 180.9479. Electronic configuration: [Xe]4f^{14}5d^{3}6s^2.

tau Symbol: τ A spin-1/2 electrically charged lepton which is in the third FAMILY (GENERATION) of elementary particles. It has a mass of 1.784 GeV, with a lifetime of 3×10^{-13} s. The tau lepton was discovered in 1975 at SLAC.

tau neutrino Symbol: ν_τ The neutrino associated with the tau particle. There is evidence that the tau neutrino has a nonzero mass, but at present its mass has not been determined.

Taylor, Joseph Hooton (1941–) American astrophysicist who discovered the BINARY PULSAR with Russell HULSE in 1974. Hulse and Taylor shared the 1993 Nobel Prize for physics

Taylor, Richard Edward (1929–) Canadian physicist who, together with Jerome FRIEDMAN and Henry KENDALL, performed scattering experiments at SLAC in the 1960s that established that nucleons have pointlike constituents inside them (which were subsequently identified as quarks). Friedman, Kendall, and Taylor shared the 1990 Nobel Prize for physics for this work.

technetium A radioactive transition metal that does not occur naturally on Earth. It is produced artificially by bombarding molybdenum with neutrons and also during the fission of uranium.
   Symbol: Tc. Melting pt.: 2172°C. Boiling pt.: 4877°C. Relative density: 11.5 (est.). Proton number: 43. Relative atomic mass: 98.9063 (^{99}Tc). Most stable isotope: ^{98}Tc (half-life 4.2×10^6 years). Electronic configuration: [Kr]4d^{5}5s^2.

technicolor A theory that attempts to

go beyond the STANDARD MODEL of particle physics by postulating that the Higgs boson is not a fundamental scalar particle but is a bound state of strongly-interacting spin-1/2 fermions called *techni-fermions* in a way that is analogous to the way that mesons are made up of quarks. This theory, at least in its simplest form, is not in accord with observations and does not have any evidence in its favor.

Teller, Edward (1908–) Hungarian-born American physicist who is best known for promoting the development of the hydrogen bomb in the United States. He also made many important contributions to molecular, solid-state, and nuclear physics.

tellurium A brittle silvery metalloid element belonging to group 16 of the periodic table. It is found native and in combination with metals. Tellurium is used mainly as an additive to improve the qualities of stainless steel and various metals.
Symbol: Te. Melting pt.: 449.5°C. Boiling pt.: 989.8°C. Relative density: 6.24 (20°C). Proton number: 52. Relative atomic mass: 127.6. Electronic configuration: $[Kr]4d^{10}5s^25p^4$.

TEM *See* transmission electron microscope.

temperature A measure of the heat there is in a body, with heat flowing from bodies of higher temperature to those at lower temperature. If there is not a heat flow between two bodies then they are said to be in THERMAL EQUILIBRIUM, with the two bodies having the same temperature. Temperature used to be defined in terms of certain physical properties, such as the freezing point and boiling point of water, but is now defined in terms of thermodynamics, with there being an absolute zero temperature.

tensor A mathematical quantity that can be regarded as a generalization of a vector in which a quantity has different values in different directions. The physical and

mathematical significance of a tensor in one coordinate system is not changed by a transformation to a different coordinate system. In general relativity theory each observer is associated with a coordinate system. It is therefore natural to express general relativity theory in terms of tensors since this theory is concerned with effects that are independent of the observers and their associated coordinate systems.

tensor field A field in which each point is characterized by the value of a TENSOR. The gravitational field in general relativity theory is a tensor field.

tensor particle A spin-2 particle associated with a TENSOR FIELD. A GRAVITON is an example of a tensor particle.

terbium A soft ductile malleable silvery rare element of the lanthanoid series of metals. It occurs in association with other lanthanoids. One of its few uses is as a dopant in solid-state devices.
Symbol: Tb. Melting pt.: 1356°C. Boiling pt.: 3123°C. Relative density: 8.229 (20°C). Proton number: 65. Relative atomic mass: 158.92534. Electronic configuration: $[Xe]4f^96s^2$.

term a name given to an atomic energy level.

tesla Symbol: T The SI unit of magnetic flux density, equal to a flux density of one weber of magnetic flux per square meter. 1 T = 1 Wb m^{-2}.

Thales (*c.* 625 BC–550 BC) Greek philosopher, mathematician, and astronomer who is frequently credited as being the first person to explain natural phenomena scientifically rather than the behavior of the gods. He postulated that all matter is ultimately made of water.

thallium A soft malleable grayish metallic element belonging to group 13 of the periodic table. It is found in lead and cadmium ores, and in pyrites (FeS_2). Thallium is highly toxic and was used previously as a rodent and insect poison.

Various compounds are now used in photocells, infrared detectors, and low-melting glasses.

Symbol: Tl. Melting pt.: 303.5°C. Boiling pt.: 1457°C. Relative density: 11.85 (20°C). Proton number: 81. Relative atomic mass: 204.3833. Electronic configuration: $[Xe]4f^{14}5d^{10}6s^26p^1$.

theory of everything A unified theory of all the fundamental interactions, elementary particles, and cosmology. It is hoped that SUPERSTRING THEORY will be developed into such a theory. However, many scientists regard the concept of a 'theory of everything' as contrary to the spirit of science and think that it could only be claimed that such a theory gives a unified description of all the phenomena known at the time. In other words, it could not guarantee the impossibility of some future discovery being made which could not be described by the theory.

thermal equilibrium *See* temperature.

thermal field theory Quantum field theory at nonzero temperatures. Thermal field theory is particularly useful in analyzing the sequence of phase transitions that is thought to have occurred in the EARLY UNIVERSE.

thermal neutron *See* slow neutron.

thermal pressure The pressure of a gas as a result of the motion of its particles, with the higher the temperature the greater the energies of the particles.

thermal reactor *See* nuclear reactor.

thermionic emission The emission of electrons from a heated metal or metal-oxide surface. The emitted current density J is related to the thermodynamic temperature T of the emitter by the *Richardson–Dushman equation*, i.e. $J = AT^2\exp(-W/kT)$, where A is a constant, W is the WORK FUNCTION of the emitter and k is the Boltzmann constant. This equation was derived in the early years of the twentieth century by Sir Owen RICHARDSON using classical statistical mechanics and in a modified form by Saul Dushman in the 1920s using quantum statistical mechanics.

thermionics The branch of electronics concerned with THERMIONIC EMISSION.

thermodynamics The branch of science that deals with heat and its relationships with other forms of energy. It has applications in chemistry, engineering, and throughout physics. It is governed by four laws. The *first law of thermodynamics* is the law of conservation of energy. It states that the total energy of a closed system is constant, although it can be converted from one form to another (such as from kinetic energy into heat). The *second law of thermodynamics* can be stated in a number of equivalent ways. One statement is that heat can never flow spontaneously from a system at a lower temperature to a system at a higher temperature, i.e. work needs to be done to make this happen, with a *refrigerator* being an example of this. An equivalent statement is that ENTROPY always increases in a spontaneous process. The *third law of thermodynamics* states that the entropy of all perfect crystals tends to zero as the absolute temperature tends to zero. Another law, that is assumed by the other laws, is known as the *zeroth law of thermodynamics*. It states that if two systems are each in thermal equilibrium with a third system then the two systems are in thermal equilibrium with each other.

thermoluminescent dating A method for dating the firing of samples of pottery. As a result of absorbed alpha radiation, electrons become trapped at energy levels higher than normal. If the temperature is raised, they revert to the ground state with emission of photons (light). The amount of light emitted can be measured. This depends on the number of trapped electrons, which in turn depends on the lapse of time since the pottery was fired. It also depends on the amount of irradiation by alpha particles during that period and the type of material.

thermonuclear reaction *See* nuclear fusion; thermonuclear reactor.

thermonuclear reactor A reactor in which NUCLEAR FUSION takes place in a controlled way. The two major problems in constructing such a reactor are: (1) reaching the very high temperatures required and (2) containing the fusion plasma. Two methods are used to deal with these problems. In *magnetic containment* the fusion plasma is contained in a *Tokamak*, i.e. a reactor with a toroidal shape in which magnetic fields guide the electrically charged particles of the plasma round the toroid while preventing them from touching the walls of the container. In *pellet fusion* the aim is to compress and heat a very small pellet of the nuclear fuel using lasers or electron beams in the hope that fusion occurs before the pellet flies apart. In spite of a great deal of effort, commercially viable thermonuclear reactors have not yet been constructed.

theta vacuum The vacuum state for non-Abelian gauge fields that are not coupled to either fermion fields or Higgs fields. It is analogous to a Bloch function in a crystal (*see* Bloch's theorem), with there being tunneling between an infinite number of degenerate states. In the presence of a massless fermion field the tunneling is completely suppressed. If the fermion has a small mass then the tunneling is not completely suppressed but is much smaller than for pure gauge fields. The theta vacuum is important in gauge theories such as quantum chromodynamics.

Thomas factor A factor of 1/2 which emerges in the calculation of fine structure in atomic spectra due to spin when a subtle feature of special relativity theory is taken into account. The Thomas factor was pointed out and shown to be necessary to calculate the fine structure splitting by Llewellyn Hilleth Thomas in 1926. The factor appears automatically in the DIRAC EQUATION.

Thomas–Fermi theory An approximation technique for calculating the electronic structure of atoms, in which the electrons of an atom are regarded as a gas of fermions in the external potential of the atomic nucleus. This method gives reasonably good results and becomes exact in the limit of infinite atomic number. It is not as accurate as HARTREE–FOCK THEORY but is considerably simpler.

Thomson, Sir George Paget (1892–1975) British physicist who demonstrated the wave nature of electrons in the 1920s (independently of Clinton DAVISSON and Lester GERMER). Thomson and Davisson shared the 1937 Nobel Prize for physics for this discovery. George Thomson was the son of J. J. Thomson.

Thomson, Sir Joseph John (J. J.) (1856–1940) British physicst who discovered the electron in 1897 using cathode rays. Thus, Thomson initiated the study of elementary particles and the structure of the atom. His later work included attempts to construct models of the atom and studies of positively charged ions (then known as *canal rays*). Thomson won the 1906 Nobel Prize for physics for his research into the conduction of electricity through gases.

Thomson model A model of atomic structure put forward by J. J. THOMSON soon after his discovery of the electron. In this model, frequently called the *plum pudding model*, the electrons are embedded in a sphere of positive electricity, with the electrons being like the raisins in a plum pudding. This model of atoms was shown to be incorrect by the scattering experiments of Ernest RUTHERFORD and his colleagues, which demonstrated the existence of the atomic nucleus (*see* Rutherford model).

Thomson scattering The scattering of electromagnetic radiation by electrons analyzed by J. J. THOMSON in 1906 to try to determine how many electrons there are in an atom. Thomson scattering can be considered to be the nonrelativistic analog of the COMPTON EFFECT.

Thomson, William *See* Kelvin, Lord.

't Hooft, Gerardus (1946–) Dutch physicist who showed in 1971 that it is possible to renormalize non-Abelian gauge theories, including the WEINBERG–SALAM MODEL. This was a key discovery, which led to the very rapid development of gauge theories of elementary particles in the 1970s. After his initial breakthrough he developed renormalization theory of gauge theories with Martinus VELTMAN, investigated MAGNETIC MONOPOLES, and proposed a very plausible picture of QUARK CONFINEMENT based on a superconductor-like picture of the vacuum state in QUANTUM CHROMODYNAMICS. Since the 1980s his main interest has been QUANTUM GRAVITY. In the mid 1990s he put forward the HOLOGRAPHIC HYPOTHESIS. The 1999 Nobel Prize for physics was shared by 't Hooft and Veltman for their pioneering work on the renormalization of gauge theories.

thorium A toxic radioactive element of the actinoid series that is a soft ductile silvery metal. It has several long-lived radioisotopes found in a variety of minerals including monazite. Thorium is used in magnesium alloys, incandescent gas mantles, and nuclear fuel elements.

Symbol: Th. Melting pt.: 1780°C. Boiling pt.: 4790°C (approx.). Relative density: 11.72 (20°C). Proton number: 90. Relative atomic mass: 232.0381. Electronic configuration: [Rn]$6d^27s^2$.

thorium series *See* radioactive series.

Thorne, Kip S. (1940–) American physicist who was one of the main figures in the revival of interest in general relativity theory in the 1960s and 1970s. In particular, he investigated the astrophysics of black holes and methods for detecting gravitational radiation.

thought experiment (gedanken experiment) An experiment that is not actually carried out but is analyzed theoretically. Such thought experiments were used extensively by Niels Bohr, Albert Einstein, and others in discussions on the foundations of quantum mechanics. Perhaps the most famous thought experiment was that of SCHRÖDINGER'S CAT. Some experiments which started out as thought experiments subsequently became real experiments when this became technologically possible. Most notably, the EPR EXPERIMENT was realized in modified form, much later as the Aspect experiment.

thulium A soft malleable ductile silvery element of the lanthanoid series of metals. It occurs in association with other lanthanoids.

Symbol: Tm. Melting pt.: 1545°C. Boiling pt.: 1947°C. Relative density: 9.321 (20°C). Proton number: 69. Relative atomic mass: 168.93421. Electronic configuration: [Xe]$4f^{13}6s^2$.

time A measure of the interval that has elapsed between two events. Scientific standards of time used to be based on astronomical measurements but are now based on the frequency of the electromagnetic radiation in an ATOMIC CLOCK.

The nature of time is dramatically altered both in special and general relativity theory with it being combined with space to form space–time. It is not clear what concept of time will emerge from the combination of general relativity theory and quantum mechanics. *See also* arrow of time.

time dependence of fundamental constants The possible variation in the value of fundamental constants, such as the fine structure constant, with time. For many years it appeared from evidence such as the OKLO REACTOR that the fundamental constants do not change with time. However, at present, there appears to be some evidence that the value of the fine structure constant has changed very slightly since the big bang. So far there is no convincing theoretical explanation for any time dependence of fundamental constants.

time dilation The slowing down of clocks. This occurs in special relativity theory when a clock is moving at a high speed relative to an observer and in general relativity theory when a clock is in a strong gravitational field.

time reversal Symbol: T The replacement of time t by time $-t$. Invariance under this operation is known as *T invariance*. *T violation* occurs in weak interactions of kaon decays (as does CP violation). *See also* CPT theorem.

time travel The ability to travel in time. This is theoretically allowed in general relativity theory by possibilities such as WORMHOLES. There are serious paradoxes involved in the possibility of traveling backward in time. Stephen HAWKING has postulated the *chronology protection conjecture* to forbid the possibility of time travel but this conjecture has not been convincingly proved.

tin A white lustrous metal of low melting point; the fourth member of group 14 of the periodic table. Tin has three crystalline modifications or allotropes, α-tin or 'gray tin' (diamond structure), β-tin or 'white tin', and γ-tin; the latter two are metallic with close-packed structures. Tin also has several isotopes. It is used in a large number of alloys including Babbit metal, bell metal, Britannia metal, bronze, gun metal, and pewter as well as several special solders.

Symbol: Sn. Melting pt.: 232°C. Boiling pt.: 2270°C. Relative density: 7.31 (20°C). Proton number: 50. Relative atomic mass: 118.710. Electronic configuration: $[Kr]4d^{10}5s^25p^2$.

Ting, Samuel Chao Chung (1936–) American physicist who discovered the J/PSI PARTICLE with his colleagues in 1974 (independently of Burton RICHTER and his colleagues). Ting and Richter shared the 1976 Nobel Prize for physics for this discovery.

titanium A silvery transition metal that occurs in various ores as titanium(IV) oxide and also in combination with iron and oxygen. It is used in the aerospace industry as it is strong, resistant to corrosion, and has a low density.

Symbol: Ti. Melting pt.: 1660°C. Boiling pt.: 3287°C. Relative density: 4.54 (20°C). Proton number: 22. Relative atomic mass: 47.867. Electronic configuration: $[Ar]3d^24s^2$.

Tokamak *See* thermonuclear reactor.

Tomonaga, Sin-Itiro (1906–79) Japanese physicist who developed the RENORMALIZATION of quantum electrodynamics in the 1940s, independently of Richard FEYNMAN and Julian SCHWINGER. Feynman, Schwinger, and Tomonaga shared the 1965 Nobel Prize for physics for this work.

topological defect An object such as a COSMIC STRING or a DOMAIN WALL.

topological invariant *See* topology.

topological quantization condition The quantization of a quantity because of TOPOLOGY. The quantum numbers that occur as the result of topology are not associated with group theory or Noether's theorem. Magnetic monopoles have topological quantization conditions associated with them.

topology A branch of geometry concerned with the properties of geometrical objects that are unchanged by continuous deformations such as twisting or stretching. If a geometrical quantity remains unchanged by continuous deformations it is called a *topological invariant*. Topology is used extensively in theories of fundamental interactions and elementary particles.

top quark *See* quark.

Townes, Charles Hard (1915–) American physicist who constructed the first MASER in 1953. Townes shared the 1964 Nobel Prize for physics with Nicolai BASOV and Aleksandr PROKHOROV for this discovery. Townes later became interested in astronomy and was the first person to discover polyatomic molecules in space.

transactinide elements Elements that have an atomic number that is greater than

103, i.e. elements that are beyond LAWREN-CIUM in the periodic table. These elements are all radioactive and have very short half-lives.

transistor An electronic device which is capable of both amplification and rectification. It incorporates junctions between *n*-type and *p*-type SEMICONDUCTORS. The transistor has almost replaced the THERMIONIC TUBE in electronic devices.

transition metals A set of elements in the period table in which *d* or *f* electron shells are filled going from left to right along this part of the periodic table. *See also* actinides; lanthanides.

transition probability The probability of a transition occurring from one quantum state to another quantum state. Transition probabilities are described by the QUANTUM THEORY OF RADIATION.

transition temperature *See* critical temperature. *See also* phase transition.

transmission electron microscope *See* ELECTRON MICROSCOPE.

transmutation The conversion of one chemical element into another either by the natural processes of radioactivity or NU-CLEOSYNTHESIS or artificially produced nuclear processes. For example, plutonium is produced by bombarding uranium with neutrons. Such artificially produced nuclear processes enable nuclear physicists to realize the alchemists' old dream of converting one element into another.

transport coefficients Quantities that give a measure for flow in a system. Electrical and thermal conductivity are examples of transport coefficients. Since it is very difficult to calculate such coefficients exactly from first principles for interacting systems it is necessary to use approximation techniques. A measure of resistance to flow in a system is given by an inverse transport coefficient.

transuranic elements Elements that have an atomic number greater than 92; i.e. elements that are beyond URANIUM in the periodic table. These elements are radioactive and have short half-lives.

transverse wave *See* wave.

trap A device for confining isolated atoms or electrons in a small region. This is frequently done using a combination of electric and magnetic fields. Such traps enable single atoms and electrons to be studied in great detail.

triple alpha process An important process in NUCLEOSYNTHESIS in which three helium nuclei combine to form one nucleus of carbon. It is a key stage in the production of heavy elements in stars because when two helium nuclei combine the resulting beryllium nucleus is very unstable, lasting less than 10^{-19} seconds. For carbon-12 to be made it is necessary for a third alpha particle to arrive and combine with the beryllium nucleus during its brief existence. This triple alpha process was first proposed by Edwin Salpeter in the early 1950s. At first it appeared that a third alpha particle colliding with a beryllium nucleus would make it even less stable. However, Fred HOYLE postulated that there is a 'resonance' energy level in carbon-12, which allows the energy of the third particle to be absorbed. This was subsequently found to be the case, thereby giving credibility to the concept of the triple alpha process. This process is the main source of energy in red giants.

triple bond A CHEMICAL BOND in which three pairs of electrons are shared.

triplet state An atomic state in which the spin angular momenta of two electrons have combined so as to give a total nonzero spin. In accordance with HUND'S RULES a triplet state almost always has a lower energy than a singlet state.

tritium Symbol: T An isotope of hydrogen that contains one proton and two neutrons. It is radioactive, with a half-life of

12.3 years, and undergoes beta decay to helium-3. Tritium is used as a tracer in chemical reactions.

truth quark A name formerly given to the TOP QUARK.

Tsui, Daniel C. (1939–) Chinese-born American physicist who discovered the fractional QUANTUM HALL EFFECT with Horst STÖRMER. Tsui, Störmer, and Robert LAUGHLIN shared the 1998 Nobel Prize for physics for the discovery and explanation of the effect.

tungsten A transition metal occurring naturally in wolframite ($(Fe,Mn)WO_4$) and scheelite ($CaWO_4$). It was formerly called *wolfram*. It is used as the filaments in electric lamps and in various alloys.
    Symbol: W. Melting pt.: $3410 \pm 20°C$. Boiling pt.: $5650°C$. Relative density: 19.3 ($20°C$). Proton number: 74. Relative atomic mass: 183.84. Electronic configuration: $[Xe]4f^{14}5d^46s^2$.

tunnel diode A semiconductor diode which makes use of the TUNNEL EFFECT. It consists of a highly doped *p–n* semiconductor junction in which there is a large current associated with electrons tunneling from the valence band to the conduction band. The tunnel diode is used in many electronic devices.

tunnel effect (tunneling) The passage of a particle through a barrier which it does not have enough energy to surmount classically. This is a purely quantum-mechanical phenomenon. There are many examples of tunneling in quantum mechanics, including alpha decay.

T violation *See* time reversal (T).

twistor A mathematical entity involving complex numbers which has been devised by Sir Roger PENROSE in the hope of combining quantum mechanics and general relativity theory. Although it has not succeeded in its original aim twistor theory has been used to find INSTANTON and MAGNETIC MONOPOLE solutions in gauge theories and has intriguing connections with SUPERSYMMETRY and SUPERSTRING THEORY.

U V

Uhlenbeck, George Eugene (1900–88) Dutch-born American physicist who, together with Samuel GOUDSMIT, put forward the idea of electron spin in 1925. He subsequently made important contributions to statistical mechanics.

ultraviolet catastrophe *See* Rayleigh–Jeans law.

ultraviolet radiation ELECTROMAGNETIC RADIATION in which the wavelengths are shorter than those of visible light but longer than those of x-rays.

ultraviolet spectroscopy The spectroscopy associated with ULTRAVIOLET RADIATION. Ultraviolet spectroscopy is used extensively in chemical analysis and the determination of molecular structure.

unbroken symmetry *See* symmetry.

uncertainty principle (Heisenberg's uncertainty principle) According to quantum mechanics the laws of physics can control only the *probability* of certain events or values, and are not deterministic. Heisenberg's principle expresses approximately this inherent uncertainty in physical laws. It can be expressed by the relations:
$$\Delta x \Delta p_x \geq h/4\pi \sim h/2\pi$$
$$\Delta t \Delta E \geq h/4\pi \sim h/2\pi$$
Here h is the Planck constant, Δx is the inherent uncertainty in the x-coordinate of the position of a particle, Δp_x is the uncertainty in the x-coordinate of the momentum (similar rules apply to the y and z components), ΔE is the uncertainty in the energy, and Δt the uncertainty in the time. The uncertainties refer to the inherent nature of the laws and not to limitations of

experimental method. They are expressed as root-mean-square deviations.

For example, consider the second relation. An excited state of an atom may typically last 10^{-9} s. Any measurement of the excitation energy must in principle be made within this time, so the uncertainty Δt cannot be greater than about 10^{-9} s. Hence the uncertainty ΔE cannot be less than about $(h/2\pi) \div 10^{-9} \sim 10^{-25}$ joule. Since the excitation energy will be about 10^{-18} joule its value can only be determinate to within about one in 10^7, however refined the measurements are.

uniaxial crystal A double-refracting crystal that only has one optic axis.

unified field theory A field theory that would give a unified description of all the elementary particles and fundamental interactions. Originally, the expression referred to the unification of general relativity theory and classical electrodynamics. Such a theory has not yet been found but progress has been made in unifying the weak and electromagnetic interactions (*see* Weinberg–Salam model).

One of the aims of Albert EINSTEIN in his many attempts to find a unified field theory was to derive quantum mechanics from unified field theory. It is now thought that this cannot be done. The failure of many promising attempts to find a unified field theory with a relativistic quantum field theory such as SUPERGRAVITY theory has led most physicists to think that it is necessary to have a unified theory with extended objects such as SUPERSTRING THEORY.

unitary symmetry A generalization of isospin. It uses a group SU(3) and predicts that elementary particles form multiplets

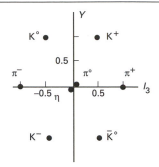

Meson octet ($J = 0$, $P = -1$, $B = 0$)

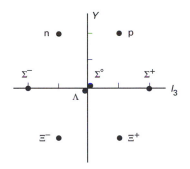

Baryon octet ($J = \frac{1}{2}$, $P = +1$, $B = 1$)

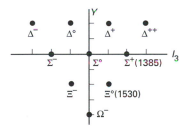

Baryon decuplet ($J = \frac{2}{3}$, $P = +1$, $B = 1$)

of 1, 8, 10, or 27 particles. They can be illustrated by plotting hypercharge (Y) against isospin (I_3). *See also* eightfold way.

unit cell The set of particles (atoms, molecules, or ions) in a crystal that is repeated in a crystal lattice.

universal constants *See* fundamental constants.

universal gas equation An equation relating the pressure p, the volume V, and the thermodynamic (absolute) temperature T of an ideal gas. It states that $pV = nRT$, where n is the amount of gas in the specimen and R is the gas constant. This equation is only obeyed approximately by real gases, with it being obeyed best at high temperatures and low pressures.

Universe All the space and matter that exists. *See also* cosmology; early Universe; multiverse.

Unruh effect Any body which is accelerating must experience itself to be embedded in a hot gas of particles, with the temperature being proportional to the acceleration. This result was stated by William Unruh in 1976. The temperature produced in this way, which is ultimately caused by quantum fluctuations of the vacuum, is too small to be detected so far.

up quark *See* quark.

upsilon An electrically neutral spin-1 meson that consists of a bottom quark–anti-bottom quark pair. Its mass is 9.46 GeV and its lifetime is 10^{-20} s. It was discovered at Fermilab in 1977.

uranium A toxic radioactive silvery element of the actinoid series of metals. Its three naturally occurring radioisotopes, ^{238}U (99.283% in abundance), ^{235}U (0.711%), and ^{234}U (0.005%), are found in numerous minerals including the uranium oxides pitchblende, uraninite, and carnotite. The readily fissionable ^{235}U is a major nuclear fuel and nuclear explosive, while ^{238}U is a source of fissionable ^{239}Pu.
  Symbol: U. Melting pt.: 1132.5°C. Boiling pt.: 3745°C. Relative density: 18.95 (20°C). Proton number: 92. Relative atomic mass: 238.0289. Electronic configuration: [Rn]5f³6d¹7s².

uranium–lead dating A radioactive dating technique used to estimate that age of certain rocks. This technique is based on the decay of uranium -238 to lead-206 (half-life 4.5×10^9 years) or the decay of uranium-235 to lead-207 (half-life 7.1 ×

10^8 years). This technique can be used for rocks with ages between 10^7 and 10^9 years.

uranium series *See* radioactive series.

Urey, Harold Clayton (1893–1981) American physical chemist who discovered deuterium in 1932 with his colleagues. He was awarded the 1934 Nobel Prize for chemistry for this discovery. Urey also isolated heavy isotopes of other elements such as carbon, nitrogen, oxygen, and sulfur and investigated the effects of isotopes on the rates of chemical reactions.

vacuum A space containing very few atoms and molecules. There are no atoms or molecules in a *perfect vacuum*, but this is an unobtainable ideal.

vacuum fluctuation A fluctuation in the value of a quantity that is associated with the vacuum state having ZERO-POINT ENERGY. The existence of vacuum fluctuations is a specifically quantum-mechanical effect. It is thought that quantum fluctuations in the EARLY UNIVERSE were responsible for the large-scale structure in the Universe.

vacuum polarization A change in the structure of the vacuum state due to the presence of a charged particle. For example, a real electron is surrounded by a 'mist' of virtual electron–positron pairs, with the virtual positron being attracted to the real electron and the virtual electron being repelled from the real electron. This results in a shielding of the real electron, thus altering its measured charge. Vacuum polarization is associated with measurable observable phenomena.

vacuum state The state with the lowest energy (ground state) in a relativistic quantum field theory. The vacuum state in relativistic quantum field theory can have a complicated structure, including BROKEN SYMMETRY. *See also* vacuum fluctuation; vacuum polarization; virtual particle; virtual process.

valence (valency) The combining power of an atom or radical. It is equal to the number of hydrogen atoms that the atom could combine with or displace in a chemical reaction. The concept of valence was put forward in the nineteenth century but was not understood until quantum mechanics explained the ELECTRONIC STRUCTURE OF ATOMS.

valence band *See* electronic structure of solids.

valence bond theory A theory of chemical bonding that emphasizes bonding between atoms involving pairs of electrons. It is nearer to traditional ideas concerning bonding and structure than MOLECULAR ORBITAL theory. Valence bond theory was developed extensively by Linus PAULING in the 1930s.

valence electron An electron in an incompletely filled shell (usually an outer shell) that can take part in chemical bonding.

valency *See* valence.

vanadium A silvery transition element occurring in complex ores in small quantities. It is used in alloy steels.
   Symbol: V. Melting pt.: 1890°C. Boiling pt.: 3380°C. Relative density: 6.1 (20°C). Proton number: 23. Relative atomic mass: 50.94. Electronic configuration: $[Ar]3d^34s^2$.

Van de Graaff generator An electrostatic generator capable of producing high p.d.s (up to millions of volts). It consists of a large smooth metal sphere on top of a hollow insulating cylinder. An endless insulating belt runs between pulleys at each end of the cylinder. Charge is sprayed from metal points connected to a high-voltage source on to the bottom of the belt. It is then carried up to the top of the belt where it is collected by other metal points and accumulated on the outside of the sphere, which becomes highly charged.

van der Meer, Simon (1925–) Dutch physicist and engineer who developed a technique known as *stochastic cooling* that helped to focus beams of protons and antiprotons. This enabled the colliding power of the beams to be maximized. This work at CERN was a crucial contribution to the discovery of the W and Z bosons. Van der Meer shared the 1984 Nobel Prize for physics with Carl RUBBIA for this work.

van der Waals, Johannes Diderik (1837–1923)   Dutch physicist who is best known for deriving an equation of state that describes real gases (VAN DER WAALS' EQUATION). Van der Waals won the 1910 Nobel Prize for physics for this work.

van der Waals' equation   An equation of state for real gases. For one mole of gas the equation is
$$(p + a/V_m^2)(V_m - b) = RT,$$
where p is the pressure, V_m the molar volume, and T the thermodynamic temperature. a and b are constants for a given substance and R is the gas constant. The equation gives a better description of the behavior of real gases than the perfect gas equation ($pV_m = RT$).

The equation contains two corrections: b is a correction for the non-negligible size of the molecules; a/V_m^2 corrects for the fact that there are attractive forces between the molecules, thus slightly reducing the pressure from ideal.

van der Waals forces   Attractive intermolecular forces. These forces are much weaker than chemical bonds and are short range. They were explained by Fritz LONDON in 1930 in terms of quantum mechanics applied to the dipoles of molecules. These dipoles can be: (1) two molecules with permanent dipole moments; (2) dipole–induced dipole interactions, with the dipole of one molecule polarizing a neighboring molecule; (3) *dispersion forces* caused by instantaneous dipoles in molecules or atoms.

Van't Hoff, Jacobus Henricus (1852–1911)   Dutch chemist who initiated the subject of stereochemistry and related it to optical activity. He also made important contributions to thermodynamics, the

theory of solutions, and the kinetics of chemical reactions. He won the first Nobel Prize for chemistry in 1901.

Van Vleck, John Hasbrouck (1899–1980)   American physicst who made major contributions to understanding the electrical, magnetic, and optical properties of matter in terms of quantum mechanics. He shared the 1977 Nobel Prize for physics with Philip ANDERSON and Sir Nevill MOTT.

variational principle   A mathematical principle that says that some quantity is either a maximum or minimum. Examples of variational principles are FERMAT'S PRINCIPLE OF LEAST TIME and the PRINCIPLE OF LEAST ACTION. Variational principles are very useful in all branches of physics. Considerable insight into how the variational principles of classical physics emerge from quantum mechanics is given by the PATH INTEGRAL formulation of quantum mechanics.

vector   A quantity that has both a magnitude and a direction. For example, force and velocity are both vectors. It is possible to write a three dimensional vector V in terms of components V_1, V_2, and V_3 along the x, y, and z axes as $V_1\mathbf{i} + V_2\mathbf{j} + V_3\mathbf{k}$, where $\mathbf{i}$, $\mathbf{j}$, and $\mathbf{k}$ are *unit vectors* (i.e. vectors of unit length) along the x, y, and z axes.

vector boson (vector particle)   A spin-1 particle that is associated with a VECTOR FIELD. All the nongravitational interactions are mediated by vector bosons: the electromagnetic interactions by photons, the strong interactions by gluons and the weak interactions by W and Z bosons.

vector field   A field that is characterized by a vector quantity at each point. An example of a vector field is the electric field, with the direction of the field being indicated by lines of force.

vector particle   *See* vector boson.

vector potential   A potential function that is a vector quantity at each point in space. For example, the potential associated with a magnetic field is a vector potential.

vector product (cross product) If two vectors U and V have components U_1, U_2, U_3, and V_1, V_2, V_3 respectively then the vector product $V \times V$ is given by $U \times V = (U_2V_3 - U_3V_2)\mathbf{i} + (U_3V_1 - U_1V_3)\mathbf{j} + (U_1V_2 - U_2V_1)\mathbf{k}$, where $\mathbf{i}$, $\mathbf{j}$, $\mathbf{k}$ are unit vectors. The vector product is itself a vector which is perpendicular to both U and V and has the length $UV \sin \theta$, where U and V are the lengths of U and V respectively and θ is the angle between them. *See also* scalar product.

vector space A set of vectors from which: (1) an operation of addition is defined such that if U and V are vectors then the sum $U + V$ is also a vector, (2) an operation of scalar multiplication is defined so that if V is a vector and a is a scalar then the product aV is also a vector. A HILBERT SPACE is a particular type of vector space.

Veltman, Martinus Justinus Godefridus (1931–) Dutch physicist who made key contributions to the development of the RENORMALIZATION of gauge theories. Veltman and 't Hooft shared the 1999 Nobel Prize for physics for this work.

vibrational motion Regular to-and-fro motion. The vibrational motion of molecules is quantized. The analysis of molecular vibrations using infrared spectroscopy gives information about molecular structure.

virial coefficients *See* virial equation.

virial equation An equation of state that attempts to describe real gases rather than ideal gases. It can be written in the form: $pV/RT = 1 + B/V + C/V^2 + \ldots$, where p and V are the pressure and volume of the gas respectively, R is the universal gas constant, T is the absolute (thermodynamic) temperature and the constants B, C, ... are called *virial coefficients*. This equation of

state was put forward by Heike Kamerlingh ONNES in 1901.

virtual particle A particle that emerges from the vacuum state because of quantum fluctuations. Because certain quantities, such as electric charge, have to be conserved it is necessary for many types of virtual particles to be produced in *virtual pairs*. For example, quantum fluctuations in the vacuum state of the electromagnetic field produce virtual pairs of electrons and positrons, which exist briefly before recombining. Processes involving virtual particles are called *virtual processes*. For example, in quantum electrodynamics an electrically charged particle such as an electron is thought of as being surrounded by a 'mist' of *virtual photons*, with emission and reabsorption going on all the time. Virtual processes have real physical significance. It is necessary to take them into acount in theoretical calculations of quantities such as the magnetic moment of the electron to obtain agreement with the very accurate experimental determinations of these quantities.

virtual process *See* virtual particle.

visible light *See* electromagnetic spectrum.

volt Symbol: V The SI unit of electrical potential, potential difference, and e.m.f., defined as the potential difference between two points in a circuit between which a constant current of one ampere flows when the power dissipated is one watt. $1 V = 1 J C^{-1}$.

Volta, Count Alessandro Giuseppe Antonio Anastasio (1745–1827) Italian physicist who built the first electric battery. The unit of electric potential, i.e. the *volt* is named after him.

von Laue, Max Theodor Felix (1879–1960) German physicist who is best known for his prediction of X-RAY DIFFRACTION by crystals. This prediction was confirmed experimentally by his students Walter Friedrich and Paul Knipping in

1912. This showed conclusively that x-rays are a type of electromagnetic radiation and that crystals consist of regular arrays of atoms. This discovery was soon used by William BRAGG and his son Lawrence BRAGG in the determination of the structure of crystals (*see* x-ray crystallography). Von Laue won the 1914 Nobel Prize for physics for his discovery.

von Neumann, John (1903–57)   Hungarian-born American mathematician who put forward the HILBERT SPACE formulation of quantum mechanics in the late 1920s. Von Neumann was one of the greatest mathematicians of the twentieth century. He made major contributions to set theory, abstract algebra, and the theory of operators.

Walton, Ernest Thomas Sinton
(1903–95) Irish physicist who developed
what came to be known as the COCK-
CROFT–WALTON MACHINE with John COCK-
CROFT in 1932. Cockcroft and Walton
shared the 1951 Nobel Prize for physics for
'their pioneer work on the transmutation
of atomic nuclei by artifically accelerated
atomic particles'.

wave Waves may be *progressive* or *stationary*. A progressive (or *traveling* wave)
is an oscillatory disturbance propagated
through a medium or, in the case of electromagnetic waves, through space. In each
type of wave there are two kinds of disturbance. Electromagnetic waves involve electric and magnetic fields oscillating in
directions at right angles to each other and
to the direction of propagation, equal
amounts of energy being transferred by the
electric and magnetic fields. Waves on the
surface of a liquid involve motion of particles in a vertical plane, there being
horizontal (longitudinal) and vertical
(transverse) oscillations. In the case of regular waves such as ripples and ocean rollers
the water particles move very nearly in circles so the longitudinal and transverse amplitudes are equal. Sound waves involve
fluctuations of pressure and particle velocity. Waves in a stretched string involve oscillations of particles at right angles to the
string and fluctuations of the tension.

A simple harmonic plane progressive
wave can be expressed by the equation
$$y = a\sin 2\pi(ft - x/\lambda)$$
where y is the displacement at time t at a
distance from the origin x. The maximum
value of the displacement, a, is called the
amplitude. f is the frequency and λ is the
wavelength. The relationship can also be
expressed in the forms

$$y = a\sin 2\pi(vt - x/\lambda)$$
or
$$y = a\sin 2\pi(t/T - x/\lambda)$$
where v is the wave speed and T is the period. Such waves can equally be expressed
in terms of cosines. If the minus sign is replaced by a plus sign, these equations represent waves traveling in the opposite
direction.

For electromagnetic waves two similar
equations are required, the displacement
being the electric field in one, and the magnetic field in the other. The two distur-
bances are in step. For waves on liquids
two equations are needed, one representing
the vertical displacement and the other the
horizontal displacement. These are a quarter of a wavelength out of step. The wave
profile has narrow peaks and broad
troughs. In practice, waves never conform
exactly to the ideal simple harmonic form
since this implies infinite duration and infinite space, without any attenuation.

A stationary (or *standing*) wave is an
oscillation equivalent to the resultant of
two equal progressive waves traveling in
opposite directions. Combining the equations
$$y = a\sin 2\pi(ft - x/\lambda)$$
and
$$y = a\sin 2\pi(ft + x/\lambda)$$
we obtain a resultant displacement
$$Y = 2a\sin 2\pi ft \cos 2\pi x/\lambda$$
This represents a simple harmonic disturbance whose amplitude varies with x according to a cosine law. Any given type of
wave requires two similar equations, one
for each kind of disturbance such that
where one has its maximum amplitude the
other has its minimum.

wave equation A partial differential
equation that describes the propagation of

a wave. In three dimensions it has the form: $\nabla^2\phi = (1/c^2)\partial^2\phi/\partial t^2$, where ϕ is the displacement and c is the speed of propagation of the wave.

wave function Symbol: Ψ A function that describes the quantum state of a system in WAVE MECHANICS. Thus, a wave function is an EIGENFUNCTION of the SCHRÖDINGER EQUATION. The physical interpretation of a wave function is given by BORN'S INTERPRETATION.

wave function of the Universe A wave function that satisfies the form of the SCHRÖDINGER EQUATION for quantum gravity (known as the *Wheeler–De Witt equation*) applied to cosmology. Each different solution of this equation represents a possible Universe. The concept of wave function of the Universe raises serious doubts about the COPENHAGEN INTERPRETATION of quantum mechanics since the Universe, by definition, cannot have any observer outside it.

wavelength *See* wave.

wave mechanics A formulation of quantum mechanics put forward by Erwin SCHRÖDINGER in 1926, following the suggestion of Louis DE BROGLIE that particles such as electrons might also have wavelike properties. The basic equation of wave mechanics is the SCHRÖDINGER EQUATION. In 1926 it was shown by Schrödinger and others that wave mechanics and matrix mechanics are equivalent.

wave number The reciprocal of the WAVELENGTH; i.e. it is the number of wave cycles per unit length.

wave packet A superposition of waves in which there is one dominant WAVE NUMBER k but several other wave numbers not far from k. It is convenient to analyze scattering processes in quantum mechanics in terms of wave packets.

wave–particle duality The key feature of quantum mechanics that objects have both wave and particle properties. Thus,

objects such as electrons, which behave like particles classically, exhibit wave properties such as ELECTRON DIFFRACTION in quantum mechanics. Similarly, electromagnetic radiation is well described in terms of waves in classical theory but exhibits the properties of particles in phenomena such as the COMPTON EFFECT, the PHOTOELECTRIC EFFECT, and the RAMAN EFFECT.

W boson An electrically charged spin-1 boson that mediates weak interaction processes that involve a change in electric charge. It has a mass of 80.4 GeV. The electric charge of the W boson can either be +1 (W^+) or –1 (W^-). The W bosons were discovered at CERN in 1983, with their properties being in accord with the theoretical expectations of the WEINBERG–SALAM MODEL. W bosons decay into either lepton–anti-lepton or quark–anti-quark pairs within 10^{-23} second.

weak interactions *See* fundamental interactions.

weber Symbol: Wb The SI unit of magnetic flux, equal to the magnetic flux that, linking a circuit of one turn, produces an e.m.f. of one volt when reduced to zero at a uniform rate in one second. 1 Wb = 1 V s.

Weinberg, Steven (1933–) American physicist who has made many major contributions to quantum field theory, particularly the model he put forward in 1967 to unify the weak and electromagnetic interactions. Weinberg shared the 1979 Nobel Prize for physics with Abdus SALAM and Sheldon GLASHOW for the development of this model.

Weinberg–Salam model A model put forward by Steven WEINBERG in 1967, and independently by Abdus SALAM in 1968, which unifies the weak and electromagnetic interactions. This model is a non-Abelian gauge theory with the gauge group SU(2) × U(1) in which the Higgs mechanism operates. This mechanism gives the W and Z bosons, which mediate the weak interactions nonzero masses. The predictions of the Weinberg–Salam model are in very good agreement with all experiments, particularly as regards the masses of the W and Z bosons. However, the HIGGS BOSON has not yet been found.

Wentzel–Kramers–Brillouin (WKB) approximation *See* semiclassical approximation.

Werner, Alfred (1866–1919) French chemist who made important contributions to the theory of molecular structure and valence in inorganic chemistry. Werner won the 1913 Nobel Prize for chemistry.

Weyl, Hermann (1885–1955) German-born American mathematician who made important contributions to several branches of mathematics and theoretical physics.

Wheeler, John Archibald (1911–) American physicist. In 1939 he wrote a classic paper with Niels BOHR that explained nuclear fission in terms of the LIQUID DROP MODEL. In the 1940s he worked with Richard FEYNMAN on formulating classical electrodynamics in terms of action at a distance.

white dwarf A type of compact star that is supported against further gravitational collapse by the degeneracy pressure of electrons. White dwarfs are the final state of stars with fairly low mass (similar to that of the Sun). A white dwarf consists of nuclei (mostly helium) and a degenerate electron gas. The density of a typical white dwarf is about 10^9 kg m^{-3}. The maximum mass a white dwarf can have is given by the CHANDRASEKHAR LIMIT.

white hole The hypothetical time reverse of a BLACK HOLE, i.e. a region of space–time from which matter explodes. Although the Einstein field equation of general relativity theory is symmetric under time reversal it is generally thought that white holes do not exist. They have sometimes been postulated to explain various violent phenomena in the Universe but there is no observational evidence for them.

Wieman, Carl E. (1951–) American physicist who was one of the co-discoverers of Bose–Einstein condensates. Wieman shared the 2001 Nobel Prize for physics with Eric CORNELL and Wolfgang KETTERLE for this discovery.

Wien's law A relationship in blackbody radiation which states that the product of the wavelength λ_m at which the maximum radiation of energy occurs and the absolute (thermodynamic) temperature is a constant, i.e. $\lambda_m T$ = constant. Thus the higher the temperature the smaller the wavelength (and hence the higher the frequency) at which the maximum radiation of energy occurs. Wien's law was derived by Wilhelm Wien in 1893 using a combination of thermodynamics and classical electrodynamics.

Wigner, Eugene Paul (1902–95) Hungarian-born American physicist who made many important contributions to physics, particularly the application of symmetry and group theory to quantum mechanics. Wigner was awarded a share of the 1963 Nobel Prize for physics for his work on quantum mechanics.

Wigner energy The energy that is in a crystal as a result of radiation. For example, in a nuclear reactor some of the energy lost by neutrons is stored by the graphite of the moderator. This alters the moderator. Wigner energy is named after Eugene WIGNER.

Wigner nuclides A pair of nuclei with the same odd nucleon number in which the proton number and neutron number differ by one. For example, the pair hydrogen-3 (one proton, two neutrons) and helium-3 (two protons, one neutron) are Wigner nuclides.

Wigner's rule If a system (such as a nucleus) contains N fermions then the whole system acts as a fermion if N is odd and as a boson if N is even. This rule was discovered by Eugene WIGNER in 1929 and independently by Paul Ehrenfest and Robert OPPENHEIMER in 1931.

Wigner–Witmer rules A set of rules that determine which molecular electronic states can arise from the electronic states of the separate atoms that make up the molecule. Because of the correlation between atomic states and molecular states these rules are known as *correlation rules*. They were found by Eugene WIGNER and E. E. Witmer in 1928 using group theory.

Wilson, Charles Thomson Rees (1869–1959) British physicist who invented the CLOUD CHAMBER in 1911. He was awarded a share of the 1927 Nobel Prize for physics with Arthur COMPTON.

Wilson, Kenneth Geddes (1936–) American physicist who developed RENORMALIZATION GROUP theory in the 1970s, particularly its application to the theory of phase transitions. Wilson was awarded the 1982 Nobel Prize for physics for his contributions to the theory of phase transitions.

Wilson, Robert Woodrow (1936–) American radio astronomer who, together with Arno PENZIAS discovered the COSMIC MICROWAVE BACKGROUND in the mid 1960s. Wilson won a share of the 1978 Nobel Prize for physics with Penzias for this discovery.

WIMP (weakly interacting massive particle) Such particles are thought to be necessary in cosmology to explain the amount of DARK MATTER, which appears to exist in the Universe. The name is misleading since the particles do not have to take part in weak interactions. WIMPs are nonbaryonic particles. It is thought that SUPERSYMMETRY provides good candidates for WIMPs. At the time of writing WIMPs have not yet been detected.

wino *See* supersymmetry.

wire chamber An improvement of the SPARK CHAMBER which eventually led to the DRIFT CHAMBER. In a wire chamber there are sheets of parallel wires rather than the parallel metal plates of a spark chamber. The distance between the wires is about a millimeter.

Witten, Edward (1951–) Prolific American mathematical physicist who has made important contributions to many topics in fundamental physics. In 1990 Witten's work in stimulating mathematics by results obtained from quantum field theory/superstring theory was recognized by the award of the Fields Medal (the equivalent of a Nobel Prize in mathematics).

WKB approximation *See* Wentzel–Kramers–Brillouin approximation.

Woodward–Hoffmann rules A set of rules from which conclusions about the course of chemical reactions can be drawn. These rules were put forward in the 1960s by Robert Burns Woodward and Roald Hoffmann. They were derived from consideration of the way that orbitals of the reactants change continuously into orbitals of the products during the course of the reaction. This is known as *conservation of orbital symmetry. See also* frontier orbital theory.

work function The energy (usually expressed in electron volts) required to remove an electron from a metal surface. The work function is a key quantity in the theory of the PHOTOELECTRIC EFFECT and of thermionic emission.

world line A curve in a space–time diagram that represents the history of an object. If a body moves uniformly then its world line is straight. If there is acceleration its world line is curved.

wormhole A solution of Einstein's field equation of general relativity theory that gives a 'short cut' connecting two different regions of space–time. This has led to speculative discussions concerning time travel via wormholes. Most wormholes are predicted to have very short life-times. However, Kip THORNE and his colleagues discovered in the late 1980s that some solutions of Einstein field equations are wormhole solutions with long lifetimes. Wormholes have also been conjectured to exist in quantum gravity. There is no observation evidence for the existence of wormholes and it would be exceedingly difficult, if not impossible, to construct one.

Wu, Chien-Shiung (1912–97) Chinese-born American physicist who performed the experiments in 1956 that demonstrated that Tsung Dao LEE and Chen Ning YANG were correct in their suggestion that parity is violated in the weak interactions. In 1963 Wu and her colleagues found experimental evidence in support of the V–A THEORY of weak interactions.

XYZ

X boson A type of massive spin 1 gauge boson that is postulated to exist in GRAND UNIFIED THEORIES (GUTs). The masses of the X bosons are predicted to be about 10^{15} GeV. Although this is far too high to be observed directly their existence could be inferred if the proton decayed since the X boson is thought to mediate the decay of quarks into leptons (and hence bring about proton decay). However, proton decay has not been observed, thus ruling out GUTs without supersymmetry. Since there is (tenuous) evidence for SUPERSYMMETRIC GUTS it may well be the case that X bosons do exist but with properties that are somewhat different from those predicted by non-supersymmetric GUTs.

xenon A colorless odorless monatomic element of the rare-gas group. It occurs in trace amounts in air. Xenon is used in electron tubes and strobe lighting.

Symbol: Xe. Melting pt.: $-111.9°C$. Boiling pt.: $-107.1°C$. Density: 5.8971 ($0°C$) kg m^{-3}. Proton number: 54. Relative atomic mass: 131.29. Electronic configuration: $[Kr]4d^{10}5s^25p^6$.

xi particles (**cascade particles**) Symbol: Ξ A pair of baryons, each of which contains two strange quarks. The *xi-minus particle* consists of one down quark and two strange quarks. It is a spin-1/2 particle with electric charge −1, a mass of 1.321 GeV and a lifetime of 1.6×10^{-10} s. It was discovered in 1952. The *xi-zero* consists of one up quark and two strange quarks. It is an electrically neutron spin-1/2 particle, with a mass of 1.315 GeV and a lifetime of 3×10^{-10} s. It was discovered in 1959.

x-radiation A form of electromagnetic radiation with short wavelength. The wavelength is commonly in the range 10^{-10} m to 10^{-11} m but much shorter or longer waves can be produced.

X-radiation is generated whenever high-energy electrons strike matter. In an x-ray tube electrons from a hot filament are accelerated through a large potential difference (typically 10^5 V) and focused upon a *target* or *anticathode* usually made of a high-melting-point metal. Radiation is emitted from the region where the electron beam strikes the target. The radiation is also caused by high-voltage cathode-ray tubes (such as are used in some television receivers and computer terminals) and some other electrical equipment. X-rays are also generated when beta radiation is absorbed in matter.

X-radiation is absorbed in matter mostly by the *photoelectric effect*, the probability of which is proportional to the fourth power of the proton number Z. Thus bone, which contains calcium (Z = 20) and phosphorus (Z = 15), absorbs far more x-radiation than does soft tissue, which contains few atoms with greater proton number than oxygen (Z = 8). Hence a body exposed to x-radiation casts a shadow on a photographic emulsion or fluorescent screen, which can be used in diagnosis. The radiation can be detected using ionization chambers, proportional counters, Geiger-Müller tubes, photography, fluorescence, and other methods.

The spectrum of x-radiation shows lines superimposed upon a continuum. The continuous spectrum is caused by the abrupt slowing-down of electrons as they pass through matter, and is called *bremsstrahlung* (German for 'braking radiation'). The short-wavelength limit λ_0 is produced by the whole kinetic energy of an electron going to generate a single quantum of radiation:

$$\lambda_0 = hc/qV$$

where h is the Planck constant, c is the speed of light, q is the electron charge, and V is the p.d. across the x-ray tube. The line spectrum is caused by atoms that have had inner electrons ejected by electron impact. As electrons from higher energy quantum states enter the empty inner states, radiation is emitted with wavelengths characteristic of the element. Hence these lines can be used in x-ray spectroscopy.

x-ray crystallography   The determination of crystal structure using X-RAY DIFFRACTION. A beam of x-rays is directed at a crystal and a photographic plate records the pattern of the diffracted x-rays. The crystal structure can be determined from this diffraction pattern.

x-ray diffraction   The diffraction of x-rays by a crystal, due to the distances between atoms in a crystal having about the same size as the wavelengths of x-rays. X-ray diffraction is used to determine the structure of crystals in X-RAY CRYSTALLOGRAPHY. *See also* Bragg's law.

x-ray spectroscopy   The study of spectra in the x-ray region. These spectra correspond to transitions involving inner electron states of atoms. *See also* Moseley's law.

x-ray tube   A device that produces x-rays. It consists of a vacuum tube in which electrons are accelerated to high speeds by a strong electrostatic field and directed to a metal target. A beam of electrons is produced by a white-hot filament. The x-rays are produced when the electrons hit the metal target.

Yang, Chen Ning (1922–   )   Chinese-born American physicist who has made many important contributions to theoretical physics, particularly the theory of elementary particles. Together with Tsung Dao LEE, he predicted that parity is not conserved in the weak interactions, that there is more than one type of neutrino and

that neutral currents and W bosons exist. Lee and Yang also contributed to the theory of phase transitions and the many-body problem in quantum mechanics. Together with Robert Mills he initiated non-Abelian gauge theories in 1954 (*see* Yang–Mills theory). Lee and Yang shared the 1957 Nobel Prize for physics for their swiftly confirmed prediction of parity violation in weak interactions, which they made the previous year.

Yang–Mills theory   A name sometimes given to non-Abelian gauge theories after Chen Ning YANG and the American physicist Robert Mills who initiated this type of theory in 1954. Sometimes, the term 'Yang–Mills theory' is taken to refer to the gauge group SU(2) specifically.

Young, Thomas (1773–1829)   British scholar and physician who helped to establish the wave theory of light. In the early years of the nineteenth century he demonstrated the wave nature of light by passing light through a pinhole and then through two more pinholes, thereby setting up an interference pattern. *See* double-slit experiment.

ytterbium   A soft malleable silvery element having two allotropes and belonging to the lanthanoid series of metals. It occurs in association with other lanthanoids. Ytterbium has been used to improve the mechanical properties of steel.

Symbol: Yb. Melting pt.: 824°C. Boiling pt.: 1193°C. Relative density: 6.965 (20°C). Proton number: 70. Relative atomic mass: 173.04. Electronic configuration: [Xe]4f^{14}6s^2.

yttrium   A silvery metallic element belonging to the second transition series. It is found in almost every lanthanoid mineral, particularly monazite. Yttrium is used in various alloys, in yttrium–aluminum garnets used in the electronics industry and as gemstones, as a catalyst, and in superconductors. A mixture of yttrium and europium oxides is widely used as the red phosphor on television screens.

Symbol: Y. Melting pt.: 1522°C. Boiling pt.: 3338°C. Relative density: 4.469 (20°C). Proton number: 39. Relative atomic mass: 88.90585. Electronic configuration: $[Kr]4d^15s^2$.

Yukawa, Hideki (1907–81) Japanese physicist who in 1935 put forward the idea that the strong interactions of protons and neutrons are mediated by pi-mesons (pions). This theory was vindicated by the discovery of the pion in 1947, with the properties that Yukawa had predicted. As a result Yukawa was awarded the 1949 Nobel Prize for physics.

Z boson An electrically neutral spin-1 boson that mediates weak interaction processes in which the electric charge is not changed. It has a mass of about 91.2 GeV. Because the Z boson is electrically neutral it is frequently denoted Z^0. The particle was discovered at CERN in 1983, with its properties being in accord with the expectations of the WEINBERG–SALAM MODEL. It has a lifetime of about 10^{-23} second.

Zeeman, Pieter (1865–1943) Dutch physicist who discovered and investigated the ZEEMAN EFFECT in 1896–97. He shared the 1902 Nobel Prize for physics for this discovery with Hendrik Antoon LORENTZ who provided a theoretical explanation of it.

Zeeman effect The splitting of atomic spectral lines in a magnetic field. This phenomenon was discovered and investigated by Pieter ZEEMAN in 1896–97. Some of this splitting was explained by Hendrik Antoon LORENTZ in terms of his theory of the electron. The part of the splitting which could be explained in this way is called the *normal Zeeman effect*. However, part of the splitting could not be explained by Lorentz's theory and was therefore called the *anomalous Zeeman effect*.

The normal Zeeman effect could also be explained by the BOHR–SOMMERFELD MODEL of electrons in atoms. The anomalous Zeeman effect is explained and quantitatively described in terms of spin in quantum mechanics. An important development in unraveling the anomalous Zeeman effect came from the semi-empirical theoretical analysis of it by the German-born American physicist Alfred Landé in several papers in the early 1920s. This led to a quantity, now known as the *Landé splitting factor*, appearing in atomic spectroscopy. The problem of the anomalous Zeeman effect helped to pave the way for the concept of electron spin, and hence to the correct value of the magnetic moment of the electron.

Zeilinger's principle A principle put forward by the Austrian physicist Anton Zeilinger in 1999 as a foundational principle of quantum mechanics. Zeilinger's principle can be stated as: *an elementary system carries one bit of information*. Zeilinger and his colleagues have explained key features of quantum mechanics in terms of this idea.

Zel'dovich, Yakov Borisovitch (1914–87) Russian physicist who made many important contributions to cosmology and astrophysics. He was one of the first people to suggest that supermassive black holes power quasars. He predicted the existence of the cosmic microwave background radiation soon before its discovery. He predicted what became known as the SUNYAEV–ZEL'DOVICH EFFECT with Rashid Sunyaev in 1972. He also performed calculations on the amount of helium produced in the EARLY UNIVERSE and investigated the problem of the formation of large-scale structure in the Universe. Zel'dovich also worked on the Soviet atom and hydrogen bombs.

Zener breakdown The emergence of electrical conductivity when an insulator is put in an electric field that is so strong that electrons make transitions directly from the valence band to the conduction band. This phenomenon is used in a device called the *Zener diode*. Zener breakdown and the Zener diode are named after the American physicist Clarence Zener, who investigated this type of breakdown in 1934.

zero-point energy The energy which a

substance has at the absolute zero temperature (0 K). This is a nonzero quantity since it would then be possible to know both the position and the momentum of the particle exactly, contrary to Heisenberg's UNCERTAINTY PRINCIPLE.

Zewail, Ahmed H. (1946–) Egyptian-born American chemist who developed FEMTOCHEMISTRY, thereby enabling the dynamics of chemical reactions to be studied experimentally. Zewail was awarded the 1999 Nobel Prize for chemistry for this work.

zinc A bluish-white transition metal occurring naturally as the sulfide (zinc blende) and carbonate (smithsonite). Zinc is used to galvanize iron, in alloys (e.g. brass), and in dry batteries.
Symbol: Zn. Melting pt.: 419.58°C. Boiling pt.: 907°C. Relative density: 7.133 (20°C). Proton number: 30. Relative atomic mass: 65.39. Electronic configuration: $[Ar]3d^{10}4s^2$.

zirconium A hard lustrous silvery transition element that occurs in a gemstone, zircon ($ZrSiO_4$). It is used in some strong alloy steels.
Symbol: Zr. Melting pt.: 1850°C. Boiling pt.: 4380°C. Relative density: 6.506 (20°C). Proton number: 40. Relative atomic mass: 91.224. Electronic configuration: $[Kr]4d^25s^2$.

Zino *See* supersymmetry.

Zweig, George (1937–) Russian-born American physicist who put forward the idea of fractionally charged constituents of baryons, independently of Murray GELL-MANN, in 1964. Zweig called his particles *aces* but the name *quarks* put forward by Gell-Mann quickly became firmly established.

Zwicky, Fritz (1898–1974) Swiss-American astronomer who made a number of important contributions to astronomy, such as the suggestion with Walter Baade in 1934 that a supernova is the transition of an ordinary star into a neutron star. Zwicky was one of the first people to consider the possibility of dark matter and that galaxies could act as gravitational lenses.

APPENDIXES

Appendix I

Fundamental Constants

speed of light	c	$2.997\ 924\ 58 \times 10^8$ m s^{-1}
permeability of free space	μ_o	$4\pi \times 10^{-7}$
		$= 1.256\ 637\ 0614 \times 10^{-6}$ H m^{-1}
permittivity of free space	$\varepsilon_0 = \mu_0^{-1}c^{-2}$	$8.854\ 187\ 817 \times 10^{-12}$ F m^{-1}
charge of electron or proton	e	$\pm 1.602\ 177\ 33 \times 10^{-19}$ C
rest mass of electron	m_e	$9.109\ 39 \times 10^{-31}$ kg
rest mass of proton	m_p	$1.672\ 62 \times 10^{-27}$ kg
rest mass of neutron	m_n	$1.674\ 92 \times 10^{-27}$ kg
electron charge-to-mass ratio	e/m	$1.758\ 820 \times 10^{11}$ C kg^{-1}
electron radius	r_e	$2.817\ 939 \times 10^{-15}$ m
Planck constant	h	$6.626\ 075 \times 10^{-34}$ J s
Boltzmann constant	k	$1.380\ 658 \times 10^{-23}$ J K^{-1}
Faraday constant	F	$9.648\ 531 \times 10^4$ C mol^{-1}

Elementary Particles

Leptons

electron	e–	−1	0.511	½
neutrino (electron)	ν_e	0	0	½
neutrino (muon)	ν_μ	0	0	½
muon	μ^-	−1	105.66	½

Baryons

proton	p	+1	938.26	½
neutron	n	0	939.55	½
xi particle	Ξ^0	0	1314.9	½
	Ξ^-	−1	1321.3	½
sigma particle	Σ^+	+1	1189.5	½
	Σ^0	0	1192.5	½
	Σ^-	−1	1197.4	½
lambda particle	Λ	0	1115.5	½
omega particle	Ω^-	−1	1672.5	¾

Mesons

kaon	κ^-	-1	493.8	0
	κ^+	+1	493.8	0
pion	Π^+	+1	139.6	0
	Π^0	0	135	0
	Π^-	−1	139.6	0
phi particle	Φ	0	1020	1
psi particle	Ψ	0	3095	1
eta particle	η^0	0	548.8	0

Note: It is common in particle physics to measure the mass of a particle in units of energy/c^2, where c is the speed of light. The above values are in units of MeV/c^2, where 1 MeV/c^2 = 1.78×10^{-30} kilogram.

Appendix II

The Chemical Elements
(* indicates the nucleon number of the most stable isotope)

Element	Symbol	p.n.	r.a.m	Element	Symbol	p.n.	r.a.m
actinium	Ac	89	227*	europium	Eu	63	151.965
aluminum	Al	13	26.982	fermium	Fm	100	257*
americium	Am	95	243*	fluorine	F	9	18.9984
antimony	Sb	51	112.76	francium	Fr	87	223*
argon	Ar	18	39.948	gadolinium	Gd	64	157.25
arsenic	As	33	74.92	gallium	Ga	31	69.723
astatine	At	85	210	germanium	Ge	32	72.61
barium	Ba	56	137.327	gold	Au	79	196.967
berkelium	Bk	97	247*	hafnium	Hf	72	178.49
beryllium	Be	4	9.012	hassium	Hs	108	265*
bismuth	Bi	83	208.98	helium	He	2	4.0026
bohrium	Bh	107	262*	holmium	Ho	67	164.93
boron	B	5	10.811	hydrogen	H	1	1.008
bromine	Br	35	79.904	indium	In	49	114.82
cadmium	Cd	48	112.411	iodine	I	53	126.904
calcium	Ca	20	40.078	iridium	Ir	77	192.217
californium	Cf	98	251*	iron	Fe	26	55.845
carbon	C	6	12.011	krypton	Kr	36	83.80
cerium	Ce	58	140.115	lanthanum	La	57	138.91
cesium	Cs	55	132.905	lawrencium	Lr	103	262*
chlorine	Cl	17	35.453	lead	Pb	82	207.19
chromium	Cr	24	51.996	lithium	Li	3	6.941
cobalt	Co	27	58.933	lutetium	Lu	71	174.967
copper	Cu	29	63.546	magnesium	Mg	12	24.305
curium	Cm	96	247*	manganese	Mn	25	54.938
dubnium	Db	105	262*	meitnerium	Mt	109	266*
dysprosium	Dy	66	162.50	mendelevium	Md	101	258*
einsteinium	Es	99	252*	mercury	Hg	80	200.59
erbium	Er	68	167.26	molybdenum	Mo	42	95.94

Appendix II

Element	Symbol	p.n.	r.a.m	Element	Symbol	p.n.	r.a.m
neodymium	Nd	60	144.24	scandium	Sc	21	44.956
neon	Ne	10	20.179	seaborgium	Sg	106	263*
neptunium	Np	93	237.048	selenium	Se	34	78.96
nickel	Ni	28	58.69	silicon	Si	14	28.086
niobium	Nb	41	92.91	silver	Ag	47	107.868
nitrogen	N	7	14.0067	sodium	Na	11	22.9898
nobelium	No	102	259*	strontium	Sr	38	87.62
osmium	Os	76	190.23	sulfur	S	16	32.066
oxygen	O	8	15.9994	tantalum	Ta	73	180.948
palladium	Pd	46	106.42	technetium	Tc	43	99*
phosphorus	P	15	30.9738	tellurium	Te	52	127.60
platinum	Pt	78	195.08	terbium	Tb	65	158.925
plutonium	Pu	94	244*	thallium	Tl	81	204.38
polonium	Po	84	209*	thorium	Th	90	232.038
potassium	K	19	39.098	thulium	Tm	69	168.934
praseodymium	Pr	59	140.91	tin	Sn	50	118.71
promethium	Pm	61	145*	titanium	Ti	22	47.867
protactinium	Pa	91	231.036	tungsten	W	74	183.84
radium	Ra	88	226.025	uranium	U	92	238.03
radon	Rn	86	222*	vanadium	V	23	50.94
rhenium	Re	75	186.21	xenon	Xe	54	131.29
rhodium	Rh	45	102.91	ytterbium	Yb	70	173.04
rubidium	Rb	37	85.47	yttrium	Y	39	88.906
ruthenium	Ru	44	101.07	zinc	Zn	30	65.39
rutherfordium	Rf	104	261*	zirconium	Zr	40	91.22
samarium	Sm	62	150.36				

Appendix III

Periodic Table of the Elements

Periodic Table of the Elements - giving group, atomic number, and chemical symbol

Period	1	2	3	4	5	6	7	8	9	10	11	12	13	14	15	16	17	18
1	1 H																	2 He
2	3 Li	4 Be											5 B	6 C	7 N	8 O	9 F	10 Ne
3	11 Na	12 Mg											13 Al	14 Si	15 P	16 S	17 Cl	18 Ar
4	19 K	20 Ca	21 Sc	22 Ti	23 V	24 Cr	25 Mn	26 Fe	27 Co	28 Ni	29 Cu	30 Zn	31 Ga	32 Ge	33 As	34 Se	35 Br	36 Kr
5	37 Rb	38 Sr	39 Y	40 Zr	41 Nb	42 Mo	43 Tc	44 Ru	45 Rh	46 Pd	47 Ag	48 Cd	49 In	50 Sn	51 Sb	52 Te	53 I	54 Xe
6	55 Cs	56 Ba	57-71 La-Lu	72 Hf	73 Ta	74 W	75 Re	76 Os	77 Ir	78 Pt	79 Au	80 Hg	81 Tl	82 Pb	83 Bi	84 Po	85 At	86 Rn
7	87 Fr	88 Ra	89-103 Ac-Lr	104 Rf	105 Db	106 Sg	107 Bh	108 Hs	109 Mt									

	57 La	58 Ce	59 Pr	60 Nd	61 Pm	62 Sm	63 Eu	64 Gd	65 Tb	66 Dy	67 Ho	68 Er	69 Tm	70 Yb	71 Lu
Lanthanides															
Actinides	89 Ac	90 Th	91 Pa	92 U	93 Np	94 Pu	95 Am	96 Cm	97 Bk	98 Cf	99 Es	100 Fm	101 Md	102 No	103 Lr

The above is the modern recommended form of the table using 18 groups. Older group designations are shown below.

Modern form	1	2	3	4	5	6	7	8	9	10	11	12	13	14	15	16	17	18
European convention	IA	IIA	IIIA	IVA	VA	VIA	VIIA	VIII (or VIIA)			IB	IIB	IIIB	IVB	VB	VIB	VIIB	0 (or VIIIB)
N. American convention	IA	IIA	IIIB	IVB	VB	VIB	VIIB	VIII (or VIIB)			IB	IIB	IIIA	IVA	VA	VIA	VIIA	VIIIA (or 0)

247

Appendix IV

Webpages

The following organizations all have information on atomic, nuclear, and particle physics

Brookhaven National Laboratory (BNL)	www.bnl.gov
CERN (European Laboratory for Particle Physics)	www.cern.ch
Cornell University	www.lns.cornell.edu
European Institute for Transuranium Elements	itu.jrc.cec.eu.int
Fermi National Accelerator Laboratory (Fermilab)	www.fnal.gov
JET (Joint European Torus)	www.jet.efd.org
JINR (Joint Institute for Nuclear Research in Dubna, Russia)	www.jinr.ru
Kurchatov Institute (Russia)	www.kiae.ru
Lawrence Berkeley National Laboratory (LBNL)	www.lbl.gov
Los Alamos National Laboratory	www.lanl.gov
Stanford Linear Accelerator Center (SLAC)	www.slac.stanford.edu
UCLA Particle Beam Physics Laboratory	pbpl.physics.ucla.edu

Bibliography

Barnett, Michael R., Mühry, Henry, & Quinn, Helen R. *The Charm of Strange Quarks*. New York: Springer-Verlag, 2000.

Barrow, John D., *The World within the World*. Oxford, U.K.: Oxford University Press, 1988.

Born, Max. *Atomic Physics*. 8th ed. Glasgow, U.K.: Blackie, 1969.

Chown, Marcus. *The Magic Furnace: The Search for the Origins of Atoms*. London: Jonathan Cape, 1999.

Cushing, James T. *Philosophical Concepts in Physics*. Cambridge, U.K.: Cambridge University Press, 1998.

Davies, Paul. *The Forces of Nature*. Cambridge, U.K.: Cambridge University Press, 1979.

Einstein, Albert. *Relativity: The Special and the General Theory*. 15th ed. New York: Bonanza Books, 1952.

Einstein, Albert, & Infeld, Leopold. *The Evolution of Physics: The Growth of Ideas from Early Concepts to Relativity and Quanta*. New York: Simon and Schuster, 1938.

Emsley, John. *Nature's Building Blocks: An A–Z Guide to the Elements*. Oxford, U.K.: Oxford University Press, 2001.

Feynman, Richard P. *QED: The Strange Theory of Light and Matter*. Princeton, N.J.: Princeton University Press, 1985.

Greene, Brian. *The Elegant Universe: Superstrings, Hidden Dimensions and the Quest for the Ultimate Theory*. London: Jonathan Cape, 1999.

Gribbin, John. *Companion to the Cosmos*. London: Weidenfeld & Nicolson, 1996.

Gribbin, John. *Q is for Quantum*. London: Weidenfeld & Nicolson, 1998.

Harrison, Edward R. *Cosmology: The Science of the Universe*. 2nd ed. Cambridge, U.K.: Cambridge University Press, 2000.

Hawking, Stephen. *A Brief History of Time*. New York: Bantam, 1988.

Mackintosh, Ray, Al-Khalili, Jim, Jonson, Björn, & Peña, Teresa. *Nucleus: A Trip into the Heart of Matter*. Bath, U.K.: Canopus Publishing, 2001.

Milburn, Gerard J. *Schrödinger's Machines: The Quantum Technology Reshaping Everyday Life*. New York: W. H. Freeman, 1997.

Pais, Abraham. *Inward Bound*. Oxford, U.K.: Oxford University Press, 1986.

Penrose, Sir Roger. *The Road to Reality: The Mathematics and Physics of the Universe*. London: Vintage, 2002.

Pullman, Bernard. *The Atom in the History of Human Thought*. Oxford, U.K.: Oxford University Press, 1998.

Schwinger, Julian. *Einstein's Legacy: The Unity of Space and Time*. New York: Scientific American Library, 1986.

Smolin, Lee. *Three Roads to Quantum Gravity*. London: Weidenfeld & Nicolson, 2000.

Stewart, Ian. *Does God Play Dice? The New Mathematics of Chaos*. 2nd ed. London: Penguin, 1997.

Taylor, John C. *Hidden Unity in Nature's Laws*. Cambridge, U.K.: Cambridge University Press, 2001.

't Hooft, Gerard. *In Search of the Ultimate Building Blocks*. Cambridge, U.K.: Cambridge University Press, 1997.

Weinberg, Steven. *The First Three Minutes: A Modern View of the Origin of the Universe*. London: André Deutsch, 1977.

Weinberg, Steven. *The Discovery of Subatomic Particles*. New York: Scientific American Library, 1983.

Weyl, Hermann. *Symmetry*. Princeton, N.J.: Princeton University Press, 1952.